Lecture Notes in Mathematics

C.I.M.E. Foundation Subseries

Volume 2281

Editors-in-Chief

Jean-Michel Morel, CMLA, ENS, Cachan, France

Bernard Teissier, IMJ-PRG, Paris, France

Series Editors

Karin Baur, University of Leeds, Leeds, UK

Michel Brion, UGA, Grenoble, France

Camillo De Lellis, IAS, Princeton, NJ, USA

Alessio Figalli, ETH Zurich, Zurich, Switzerland

Annette Huber, Albert Ludwig University, Freiburg, Germany

Davar Khoshnevisan, The University of Utah, Salt Lake City, UT, USA

Ioannis Kontoyiannis, University of Cambridge, Cambridge, UK

Angela Kunoth, University of Cologne, Cologne, Germany

Ariane Mézard, IMJ-PRG, Paris, France

Mark Podolskij, University of Luxembourg, Esch-sur-Alzette, Luxembourg

Sylvia Serfaty, NYU Courant, New York, NY, USA

Gabriele Vezzosi, UniFI, Florence, Italy

Anna Wienhard, Ruprecht Karl University, Heidelberg, Germany

More information about this series at http://www.springer.com/series/304

Fondazione C.I.M.E., Firenze

FONDAZIONE CIME
ROBERTO CONTI
CENTRO INTERNAZIONALE MATEMATICO ESTIVO
INTERNATIONAL MATHEMATICAL SUMMER CENTER

C.I.M.E. stands for *Centro Internazionale Matematico Estivo*, that is, International Mathematical Summer Centre. Conceived in the early fifties, it was born in 1954 in Florence, Italy, and welcomed by the world mathematical community: it continues successfully, year for year, to this day.

Many mathematicians from all over the world have been involved in a way or another in C.I.M.E.'s activities over the years. The main purpose and mode of functioning of the Centre may be summarised as follows: every year, during the summer, sessions on different themes from pure and applied mathematics are offered by application to mathematicians from all countries. A Session is generally based on three or four main courses given by specialists of international renown, plus a certain number of seminars, and is held in an attractive rural location in Italy.

The aim of a C.I.M.E. session is to bring to the attention of younger researchers the origins, development, and perspectives of some very active branch of mathematical research. The topics of the courses are generally of international resonance. The full immersion atmosphere of the courses and the daily exchange among participants are thus an initiation to international collaboration in mathematical research.

C.I.M.E. Director (2002 – 2014)
Pietro Zecca
Dipartimento di Energetica "S. Stecco"
Università di Firenze
Via S. Marta, 3
50139 Florence
Italy
e-mail: zecca@unifi.it

C.I.M.E. Director (2015 –)
Elvira Mascolo
Dipartimento di Matematica "U. Dini"
Università di Firenze
viale G.B. Morgagni 67/A
50134 Florence
Italy
e-mail: mascolo@math.unifi.it

C.I.M.E. Secretary
Paolo Salani
Dipartimento di Matematica "U. Dini"
Università di Firenze
viale G.B. Morgagni 67/A
50134 Florence
Italy
e-mail: salani@math.unifi.it

CIME activity is carried out with the collaboration and financial support of INdAM (Istituto Nazionale di Alta Matematica)

For more information see CIME's homepage: **http://www.cime.unifi.it**

Yves Achdou • Pierre Cardaliaguet •
François Delarue • Alessio Porretta •
Filippo Santambrogio

Mean Field Games

Cetraro, Italy 2019

Pierre Cardaliaguet • Alessio Porretta

Editors

 Springer

FONDAZIONE
CIME
ROBERTO CONTI
CENTRO INTERNAZIONALE MATEMATICO ESTIVO
INTERNATIONAL MATHEMATICAL SUMMER CENTER

Authors

Yves Achdou
Laboratoire Jacques-Louis Lions (LJLL)
Université Paris-Diderot
Paris, France

Pierre Cardaliaguet
Ceremade, UMR CNRS 7534
Université Paris-Dauphine
Paris, France

François Delarue
Lab J.A. Dieudonné
Université Côte d'Azur, CNRS
Nice, France

Alessio Porretta
Dipartimento di Matematica
Università di Roma Tor Vergata
Roma, Italy

Filippo Santambrogio
Institut Camille Jordan
Université Claude Bernard
Villeurbanne, France

Editors

Pierre Cardaliaguet
Ceremade, UMR CNRS 7534
Université Paris-Dauphine
Paris, France

Alessio Porretta
Dipartimento di Matematica
University of Rome Tor Vergata
Roma, Italy

ISSN 0075-8434
Lecture Notes in Mathematics
C.I.M.E. Foundation Subseries
ISBN 978-3-030-59836-5
https://doi.org/10.1007/978-3-030-59837-2

ISSN 1617-9692 (electronic)

ISBN 978-3-030-59837-2 (eBook)

Mathematics Subject Classification: Primary: 49-XX; Secondary: 91-XX

© The Editor(s) (if applicable) and The Author(s), under exclusive license to Springer Nature Switzerland AG 2020
This work is subject to copyright. All rights are solely and exclusively licensed by the Publisher, whether the whole or part of the material is concerned, specifically the rights of translation, reprinting, reuse of illustrations, recitation, broadcasting, reproduction on microfilms or in any other physical way, and transmission or information storage and retrieval, electronic adaptation, computer software, or by similar or dissimilar methodology now known or hereafter developed.
The use of general descriptive names, registered names, trademarks, service marks, etc. in this publication does not imply, even in the absence of a specific statement, that such names are exempt from the relevant protective laws and regulations and therefore free for general use.
The publisher, the authors, and the editors are safe to assume that the advice and information in this book are believed to be true and accurate at the date of publication. Neither the publisher nor the authors or the editors give a warranty, expressed or implied, with respect to the material contained herein or for any errors or omissions that may have been made. The publisher remains neutral with regard to jurisdictional claims in published maps and institutional affiliations.

This Springer imprint is published by the registered company Springer Nature Switzerland AG.
The registered company address is: Gewerbestrasse 11, 6330 Cham, Switzerland

Preface

The volume is dedicated to the theory of Mean Field Games. This theory aims at describing differential games with a large number of interacting agents. The number of applications of the theory is huge, ranging from macroeconomics to crowd motions and from finance to power grid models. In all these models, each agent controls his/her own dynamical state, which evolves in time according to a deterministic or stochastic differential equation. The individual goal is to minimize some cost depending not only on his/her own control but also on the behavior of the whole population of agents, which is described through the distribution law of the dynamical states. In this setting, the central concept is the notion of Nash equilibria, which describes how agents play in an optimal way by taking into account the others' strategies.

The theory of Mean Field Games has been introduced and largely developed by J.-M. Lasry and P.L. Lions through a series of papers around 2005 and during the famous lectures of Lions at the Collège de France. At about the same time, M. Huang, P. Caines, and R. Malhamé discussed similar models under the terminology of "Nash certainty equivalence principle."

The first motivation of Lasry and Lions' early works was to study the limit of Nash equilibria in N-players differential games, as N goes to infinity, under suitable conditions of symmetry and coupling. The mean field approach developed to this purpose led to the construction of a macroscopic model, which is now well suited to describe, in many different contexts, the equilibria between individual strategies and collective behavior in large population dynamics. Thus, the theory has known an impressive growth so far, from a theoretical point of view as well as from the point of view of applications. This is not surprising because, in terms of mathematics, the theory is very rich and involves several fields: the analysis of partial differential equations (PDEs), stochastic analysis, calculus of variations, mean field theory,...

In June 2019, a CIME School on Mean Field Games was organized in Cetraro, Italy. The goal was to cover some of the most important aspects of the theory and most recent developments. This volume collects the notes of the CIME courses and contains 4 contributions: the first one (by P. Cardaliaguet and A. Porretta) is a general introduction to the theory, mostly focused on the PDEs' ingredients; the second contribution (by F. Santambrogio) is dedicated to some variational aspect of the theory; the third part (by F. Delarue) focuses on the master equation with common noise; and the last contribution (by Y. Achdou and M. Laurière) is devoted to the issue of numerics, theory and simulations, and applications to concrete models. We now explain more in detail the contents of the volume.

- The first chapter, by P. Cardaliaguet and A. Porrretta, is a general presentation of the theory of Mean Field Games through its two representative PDEs. Both equations describe, though in a different way, the Nash equilibria of differential games with infinitely many players. The first one is the MFG system, which couples a forward Fokker–Planck equation with a backward Hamilton–Jacobi equation and for which a detailed analysis is presented. The second one is the master equation, a kind of transport equation on the space of measures, for which mostly the heuristic ideas are presented.
- In the second chapter, F. Santambrogio describes a class of MFGs having a variational structure: in this case the MFG equilibria can be obtained by minimizing an energy functional. The chapter is mostly focused on the Lagrangian approach to first-order MFG systems with local couplings. The main goal is to prove that minimizers of a suitably expressed global energy are equilibria in the sense that a.e. trajectory solves a control problem with a running cost depending on the density of all the agents. This requires a fine regularity analysis of the minimizers and involves tools from Calculus of Variations and Optimal Transportation theory.
- The third chapter, by François Delarue, is dedicated to the case that, in the MFG model, the agents are affected by a common noise. The goal of this chapter is to address in a rigorous way the solvability of the master equation for mean field games on a finite state space with a common noise. The results in their own but also the structure of the underpinning common noise are new in the literature on Mean Field Games.
- The last part of the volume, by Y. Achdou and M. Laurière, is devoted to the numerical approximation of the solution to MFG problems. This topic is all the more important that there are very few explicit or semi-explicit solutions to MFGs and numerical simulations are often the only way to obtain quantitative information for this class of models. The chapter focuses on the MFG system and presents several aspects of a finite difference method used to approximate this system, including convergence, variational aspects, and algorithms for solving the resulting systems of nonlinear equations. It also discusses in detail two concrete applications: a model of crowd motion and a model with heterogeneous agents in macroeconomics.

Paris, France   Pierre Cardaliaguet
Roma, Italy   Alessio Porretta

Contents

Chapter 1
An Introduction to Mean Field Game Theory

Pierre Cardaliaguet and Alessio Porretta

Abstract These notes are an introduction to Mean Field Game (MFG) theory, which models differential games involving infinitely many interacting players. We focus here on the Partial Differential Equations (PDEs) approach to MFGs. The two main parts of the text correspond to the two emblematic equations in MFG theory: the first part is dedicated to the MFG system, while the second part is devoted to the master equation.

The MFG system describes Nash equilibrium configurations in the mean field approach to differential games with infinitely many players. It consists in the coupling between a backward Hamilton-Jacobi equation (for the value function of a single player) and a forward Fokker-Planck equation (for the distribution law of the individual states). We discuss the existence and the uniqueness of the solution to the MFG system in several frameworks, depending on the presence or not of a diffusion term and on the nature of the interactions between the players (local or nonlocal coupling). We also explain how these different frameworks are related to each other. As an application, we show how to use the MFG system to find approximate Nash equilibria in games with a finite number of players and we discuss the asymptotic behavior of the MFG system.

Pierre Cardaliaguet was partially supported by the ANR (Agence Nationale de la Recherche) project ANR-12-BS01-0008-01, by the CNRS through the PRC grant 1611 and by the Air Force Office for Scientific Research grant FA9550-18-1-0494.
Alessio Porretta was partially supported by the Tor Vergata project "Mission: Sustainability" (2017) E81I18000080005 (DOmultiage—Dynamic Optimization in Multi-Agents phenomena).

P. Cardaliaguet
CEREMADE, UMR CNRS 7534, Université Paris-Dauphine, Paris Cedex, France
e-mail: cardaliaguet@ceremade.dauphine.fr

A. Porretta (✉)
Dipartimento di Matematica, Università di Roma Tor Vergata, Roma, Italy
e-mail: porretta@mat.uniroma2.it

© The Editor(s) (if applicable) and The Author(s), under exclusive license to Springer Nature Switzerland AG 2020
P. Cardaliaguet, A. Porretta (eds.), *Mean Field Games*,
Lecture Notes in Mathematics 2281, https://doi.org/10.1007/978-3-030-59837-2_1

1

The master equation is a PDE in infinite space dimension: more precisely it is a kind of transport equation in the space of measures. The interest of this equation is that it allows to handle more complex MFG problems as, for instance, MFG problems involving a randomness affecting all the players. To analyse this equation, we first discuss the notion of derivative of maps defined on the space of measures; then we present the master equation in several frameworks (classical form, case of finite state space and case with common noise); finally we explain how to use the master equation to prove the convergence of Nash equilibria of games with finitely many players as the number of players tends to infinity.

As the works on MFGs are largely inspired by P.L. Lions' courses held at the Collège de France in the years 2007–2012, we complete the text with an appendix describing the organization of these courses.

1.1 Introduction

Mean field game (MFG) theory is devoted to the analysis of optimal control problems with a large number of small controllers in interaction. As an example, they can model crowd motions, in which the evolution of a pedestrian depends on the crowd which is around. Similar models are also used in economics: there, macroeconomic quantities are derived from the microeconomic behavior of the agents who interact through aggregate quantities, such as the prices or the interest rates. In the Mean Field Game formalism, the controllers are assumed to be "rational" (in the sense that they optimize their behavior by taking into account the behavior of the other controllers), therefore the central concept of solution is the notion of Nash equilibrium, in which no controller has interest to deviate unilaterally from the planned control. In general, playing a Nash equilibrium requires for a player to anticipate the other players's responses to his/her action. For large population dynamic games, it is unrealistic for a player to collect detailed information about the state and the strategies of the other players. Fortunately this impossible task is useless: mean field game theory explains that one just needs to implement strategies based on the distribution of the other players. Such a strong simplification is well documented in the (static) game community since the seminal works of Aumann [20]. However, for differential games, this idea has been considered only very recently: the starting point is a series of papers by Lasry and Lions [143–145, 150], who introduced the terminology in around 2005. The term *mean field* comes for an analogy with the mean field models in mathematical physics, which analyse the behavior of many identical particles in interaction (see for instance [111, 176, 177]). Here the particles are replaced by agents or players, whence the name of mean field games. Related ideas have been developed independently, at about the same time, by Caines, Huang and Malhamé [132–135], under the name of Nash certainty equivalence principle. In the economic literature, similar models (often in discrete time) were introduced in the 1990s as "heterogeneous agent models" (see, for instance, the pioneering works of Aiyagari [13] and Krussell and Smith [138]).

Since these seminal works, the study of mean field games has known a quick growth. There are by now several textbooks on this topic: the most impressive one is the beautiful monograph by Carmona and Delarue [68], which exhaustively covers the probability approach of the subject. One can also quote the Paris-Princeton Lectures by Gueant, Lasry and Lions [128] where the authors introduce the theory with sample of applications, the monograph by Bensoussan, Frehse and Yam [30], devoted to both mean field games and mean field control with a special emphasis on the linear-quadratic problems, and the monograph by Gomes, Pimentel and Voskanyan [122], on the regularity of the MFG system. Finally, [65] by the first author with Delarue, Lasry and Lions studies the master equation (with common noise) and the convergence of Nash equilibria as the number of players tends to infinity.

This text is a basic introduction to mean field games, with a special emphasis on the PDE aspects. The central ideas were largely developed in Pierre-Louis Lions' series of lectures at the Collège de France [149] during the period 2007–2012. As these courses contain much more material than what is developed here, we added in the appendix some notes on the organization of these courses in order to help the interested reader.

The main mathematical object of the text is the so-called mean field game system, which takes the form

$$\begin{cases} (i) & -\partial_t u - \nu \Delta u + H(x, Du, m) = 0 & \text{in } (0, T) \times \mathbb{R}^d \\ (ii) & \partial_t m - \nu \Delta m - \text{div}\left(H_p(x, Du, m)m\right) = 0 & \text{in } (0, T) \times \mathbb{R}^d \\ (iii) & m(0) = m_0 , \ u(x, T) = G(x, m(T)) & \text{in } \mathbb{R}^d \end{cases} \quad (1.1)$$

In the above system, the unknowns u and m are scalar and depend on time $t \in [0, T]$ and space $x \in \mathbb{R}^d$. The two equations are of (possibly degenerate) parabolic type (i.e., $\nu \geq 0$); the first equation is backward in time while the second one is forward in time. There are two other crucial structure conditions for this system: the first one is the convexity of $H = H(x, p, m)$ with respect to the second variable. This condition means that the first equation (a Hamilton-Jacobi equation) is associated with an optimal control problem and is interpreted as the value function associated with a typical small player. The second structure condition is that m_0 (and therefore $m(t, \cdot)$) is (the density of) a probability measure on $\mathbb{R}^d$. The Hamiltonian $H = H(x, p, m)$, which couples the two equations, depends on space, on the variable $p \in \mathbb{R}^d$ and on the probability measure m.

Let us briefly explain the interpretation of this system as a Nash equilibrium problem in a game with infinitely many small players. An agent (=a player) controls through his/her control α the stochastic differential equation (SDE)

$$dX_s = b(X_s, \alpha_s, m(s))ds + \sqrt{2\nu}dB_s \quad (1.2)$$

where (B_t) is a standard Brownian motion. He/She aims at minimizing the quantity

$$\mathbb{E}\left[\int_0^T L(X_s, \alpha_s, m(s))ds + G(X_T, m(T))\right],$$

where the running cost $L = L(x, \alpha, m)$ and the terminal cost $G = G(x, m)$ depend on the position x of the player, the control α and the distribution m of the other players. Note that in this cost the evolution of the measure $m(s)$ enters as a parameter. To solve this problem one introduces the value function:

$$u(t, x) = \inf_\alpha \mathbb{E}\left[\int_t^T L(X_s, m(s), \alpha_s)ds + G(X_T, m(T))\right],$$

where the infimum is taken over admissible controls α and where X solves the SDE (1.2) with initial condition $X_t = x$. The value function u then satisfies the PDE (1.1-(i)) where

$$H(x, p, m) = \sup_\alpha [-b(x, \alpha, m) \cdot p - L(x, m, \alpha)].$$

Given the value function u, it is known that the agent plays in the optimal way by using the feedback control $\alpha^* = \alpha^*(t, x)$ such that the drift is of the form $b(x, \alpha^*(t, x), m(t)) = -H_p(x, Du(t, x), m(t))$. Now, if all agents argue in this way and if their associated noises are independent, the law of large numbers implies that their distribution evolves with a velocity which is due, on the one hand, to the diffusion, and, on the other hand, on the drift term $-H_p(x, Du(t, x), m(t))$. This leads to the Kolmogorov-Fokker-Planck equation (1.1-(ii)). The fact that system (1.1) describes a Nash equilibrium can be seen as follows. As the single player is "small" (compared to the collection of the other agents), his/her deviation does not change the population dynamics. Hence the behavior of the other agents, and therefore their time dependent distribution $m(t)$, can be taken as given in the individual optimization. This corresponds to the concept of Nash equilibrium where all players play an optimal strategy while freezing the others' choices.

The main part of these notes (Sect. 1.3) is devoted to the analysis of the mean field game system (1.1): we discuss the existence and uniqueness of the solution in various settings and the interpretation of the system. This analysis takes some time since the PDE system behaves in a quite different way according to whether the system is parabolic or not (i.e., ν is positive or zero) and according to the regularity of H with respect to the measure. These various regimes correspond to different models: for instance, in many applications in finance, the diffusion is nondegenerate (i.e., $\nu > 0$), while ν often vanishes in macroeconomic models. In most applications in economy the dependence of the Hamiltonian $H = H(x, p, m)$ with respect to the probability measure m is through some integral form of m (moments, variance), but in models of crowd motion it is very often through the value at position x of the density $m(x)$ of m. We will discuss these different features of the mean field game system, with, hopefully, a few novelties in the treatment of the equations. To keep these notes as simple as possible, the analysis is done for systems with periodic in space coefficients: the analysis for other boundary problems follows the same lines, with additional technicalities. We will also mention other relevant aspects of the MFG systems: their application to differential games with finitely many players, the long time ergodic behavior, the vanishing viscosity limits, ...

The second focus of these notes is a (short and mostly formal) introduction to the "master equation" (Sect. 1.4). Indeed, it turns out that, in many applications, the MFG system (1.1) is not enough to describe the MFG equilibria. On the one hand, the MFG system does not explain how the agents take their decision in function of their current position and of the current distribution of the players (in "feedback form"). Secondly, it does not explain why one can expect the system to appear as the limit of games with finitely many players. Lastly, the PDE system does not allow to take into account problems with common noise, in which the dynamic of the agents is subject to a common source of randomness. All these issues can be overcome by the introduction of the master equation. This equation (introduced by Lions in his courses at Collège de France) takes the form of a partial differential equation in the space of measures which reads as follows (in the simplest setting):

$$
\begin{cases}
-\partial_t U(t, x, m) - \nu \Delta_x U(t, x, m) + H(x, D_x U(t, x, m), m) - \nu \int_{\mathbb{R}^d} \operatorname{div}_y D_m U(t, x, m, y) \, m(dy) \\
\quad + \int_{\mathbb{R}^d} D_m U(t, x, m, y) \cdot D_p H(y, D_x U(t, y, m), m) m(dy) = 0 \\
\qquad\qquad \text{in } (0, T) \times \mathbb{R}^d \times \mathcal{P}_2 \\
U(T, x, m) = G(x, m) \qquad \text{in } \mathbb{R}^d \times \mathcal{P}_2
\end{cases}
$$

where $\mathcal{P}_2$ is the space of probability measures on $\mathbb{R}^d$ (with finite second order moment). Here the unknown is the scalar quantity $U = U(t, x, m)$ depending on time and space and on the measure m (representing the distribution of the other players). This equation involves the derivative $D_m U$ of the unknown with respect to the measure variable (see Sect. 1.4.2). We will briefly explain how to prove the existence and the uniqueness of a solution to the master equation and its link with the MFG system. We will also discuss how to extend the equation to problems with a common noise (a noise which affects all the players). Finally, we will show how to use this master equation to prove that Nash equilibria in games with finitely many players converge to MFG equilibria.

These notes are organized as follows: in a preliminary part (Sect. 1.2), we introduce fundamental tools for the understanding and the analysis of MFG problems: a brief recap of the dynamic programming approach in optimal control theory, the description of the space of probability measures and some basic aspects of mean field theory. Then we concentrate on the MFG system (1.1) (Sect. 1.3). Finally, the analysis on the space of measures and the master equation are discussed in the last part (Sect. 1.4). We complete the text by an appendix on the organization of P.L. Lions' courses on MFGs at the Collège de France (Appendix).

1.2 Preliminaries

In this section we recall some basic notion on optimal control and dynamic programming, on the space of probability measures and on mean field limits. As mean field games consist in a combination of these three topics, it is important to collect some preliminary knowledge of them.

1.2.1 *Optimal Control*

We briefly describe, in a very formal way, the optimal control problems we will meet in these notes. We refer to the monographs by Fleming and Rischel [106], Fleming and Soner [107], Yong and Zhou [180] for a rigorous treatment of the subject.

Let us consider a stochastic control problem where the state (X_s) of the system is governed by the stochastic differential equation (SDE) with values in $\mathbb{R}^d$:

$$X_s^\alpha = x + \int_t^s b(r, X_r^\alpha, \alpha_r)dr + \int_t^s \sigma(r, X_r^\alpha, \alpha_r)dB_r. \tag{1.3}$$

In the above equation, $B = (B_s)_{s \geq 0}$ is a N-dimensional Brownian motion (starting at 0) adapted to a fixed filtration $(\mathcal{F}_t)_{t \geq 0}$, $b : [0, T] \times \mathbb{R}^d \times A \to \mathbb{R}^d$ and $\sigma : [0, T] \times \mathbb{R}^d \times A \to \mathbb{R}^{d \times N}$ satisfy some regularity conditions given below and the process $\alpha = (\alpha_s)$ is progressively measurable with values in some set A. We denote by $\mathcal{A}$ the set of such processes. The elements of $\mathcal{A}$ are called the control processes.

A generic agent controls the process X through the control α in order to reach some goal: here we consider optimal control problems, in which the controller aims at minimizing some cost J. We will mostly focus on the finite horizon problem, where J takes the form:

$$J(t, x, \alpha) = \mathbb{E}\left[\int_t^T L(s, X_s^\alpha, \alpha_s)ds + g(X_T)\right].$$

Here $T > 0$ is the finite horizon of the problem, $L : [0, T] \times \mathbb{R}^d \times A \to \mathbb{R}$ and $g : \mathbb{R}^d \to \mathbb{R}$ are given continuous maps (again we are more precise in the next section on the assumptions on L and g). The controller minimizes J by using controls in $\mathcal{A}$. We introduce the value function as the map $u : [0, T] \times \mathbb{R}^d \to \mathbb{R}$ defined by

$$u(t, x) = \inf_{\alpha \in \mathcal{A}} J(t, x, \alpha).$$

1.2.1.1 Dynamic Programming and the Verification Theorem

The main interest of the value function is that it indicates how the controller should choose his/her control in order to play in an optimal way. We explain the key ideas in a very informal way. A rigorous treatment of the question is described in the references mentioned above.

Let us start with the **dynamic programming principle**, which states the following identity: for any $t_1 \leq t_2$,

$$u(t_1, x) = \inf_{\alpha \in \mathcal{A}} \mathbb{E}\left[\int_{t_1}^{t_2} L(s, X_s^\alpha, \alpha_s)ds + u(t_2, X_{t_2}^\alpha) \right]. \tag{1.4}$$

The interpretation is that, to play optimally at time t_1, the controller does not need to predict in one shot the whole future strategy provided he/she knows what would be the best reward at some future time t_2, in which case it is enough to focus on the optimization between t_1 and t_2. So far, the optimization process can be built step by step like in semigroup theory. This relation has a fundamental consequence: to play in an optimal way the agent only needs to know the current state and play accordingly (and not the whole filtration at time t).

Fix now $t \in [0, T)$. Choosing $t_1 = t$, $t_2 = t + h$ (for $h > 0$ small) and assuming that u is smooth enough, we obtain by Itô's formula and (1.4) that

$$u(t, x) = \inf_{\alpha \in \mathcal{A}} \mathbb{E}\Big[\int_t^{t+h} L(s, X_s^\alpha, \alpha_s)ds + u(t, x) + \int_t^{t+h} (\partial_t u(s, X_s^\alpha) + Du(s, X_s^\alpha) \cdot b(s, X_s^\alpha, \alpha_s)$$
$$+ \frac{1}{2}\text{Tr}(\sigma\sigma^*(s, X_s^\alpha, \alpha_s)D^2 u(s, X_s^\alpha)))ds \Big].$$

Simplifying by $u(t, x)$, dividing by h and letting $h \to 0^+$ gives (informally) the Hamilton-Jacobi equation

$$0 = \inf_{a \in A}\left[L(t, x, a) + \partial_t u(t, x) + Du(t, x) \cdot b(t, x, a) + \frac{1}{2}\text{Tr}(\sigma\sigma^*(t, x, a)D^2 u(t, x)) \right].$$

Let us introduce the Hamiltonian H of our problem: for $p \in \mathbb{R}^d$ and $M \in \mathbb{R}^{d \times d}$,

$$H(t, x, p, M) := \sup_{a \in A}\left[-L(t, x, a) - p \cdot b(t, x, a) - \frac{1}{2}\text{Tr}(\sigma\sigma^*(t, x, a)M) \right].$$

Then the **Hamilton-Jacobi equation** can be rewritten as a terminal value problem:

$$\begin{cases} -\partial_t u(t, x) + H(t, x, Du(t, x), D^2 u(t, x)) = 0 & \text{in } (0, T) \times \mathbb{R}^d, \\ u(T, x) = g(x) & \text{in } \mathbb{R}^d. \end{cases}$$

The first equation is backward in time (the map H being nonincreasing with respect to $D^2 u$). The terminal condition comes just from the definition of u for $t = T$.

Let us now introduce $\alpha^*(t, x) \in A$ as a maximum point in the definition of H when $p = Du(t, x)$ and $M = D^2u(t, x)$. Namely

$$H(t, x, Du(t, x), D^2u(t, x)) = -L(t, x, \alpha^*(t, x)) - Du(t, x) \cdot b(t, x, \alpha^*(t, x))$$

$$- \frac{1}{2}\text{Tr}(\sigma\sigma^*(t, x, \alpha^*(t, x))D^2u(t, x)). \quad (1.5)$$

We assume that α^* is sufficiently smooth to justify the computation below. We are going to show that α^* is the **optimal feedback**, namely the optimal strategy to play at time t in the state x. Indeed, one has the following "**Verification Theorem**": Let $(X_s^{\alpha^*})$ be the solution of the stochastic differential equation

$$X_s^{\alpha^*} = x + \int_t^s b(r, X_r^{\alpha^*}, \alpha^*(r, X_r^{\alpha^*}))dr + \int_t^s \sigma(r, X_r^{\alpha^*}, \alpha^*(r, X_r^{\alpha^*}))dB_r$$

and set $\alpha_s^* = \alpha^*(s, X_s^{\alpha^*})$. Then

$$u(t, x) = J(t, x, \alpha^*).$$

Note that, with a slight abuse of notation, here $\alpha^* = (\alpha_s^*)$ is a control, namely it belongs to $\mathcal{A}$. Strictly speaking, (α_t^*) is the optimal control, $\alpha^*(t, x)$ being the optimal feedback.

Heuristic Argument By Itô's formula, we have

$$g(X_T^{\alpha^*}) = u(T, X_T^{\alpha^*}) = u(t, x) + \int_t^T (\partial_t u(s, X_s^{\alpha^*}) + Du(s, X_s^{\alpha^*}) \cdot b(s, X_s^{\alpha^*}, \alpha_s^*)$$

$$+ \frac{1}{2}\text{Tr}(\sigma\sigma^*(s, X_s^{\alpha^*}, \alpha_s^*)D^2u(s, X_s^{\alpha^*})))ds + \int_t^T \sigma^*(s, X_s^{\alpha^*}, \alpha_s^*)Du(s, X_s^{\alpha^*}) \cdot dB_s.$$

Taking expectation, using first the optimality of α^* in (1.5) and then the Hamilton-Jacobi equation satisfied by u, we obtain

$$\mathbb{E}\left[g(X_T^{\alpha^*})\right] = u(t, x) + \mathbb{E}\left[\int_t^T (\partial_t u(s, X_s^{\alpha^*}) - H(s, X_s^{\alpha^*}, Du(s, X_s^{\alpha^*}), D^2u(s, X_s^{\alpha^*})) - L(s, X_s^{\alpha^*}, \alpha_s^*))ds\right]$$

$$= u(t, x) - \mathbb{E}\left[\int_t^T L(s, X_s^{\alpha^*}, \alpha_s^*)ds\right].$$

Rearranging terms, we find

$$u(t, x) = \mathbb{E}\left[\int_t^T L(s, X_s^{\alpha^*}, \alpha_s^*)ds + g(X_T^{\alpha^*})\right],$$

which shows the optimality of α^*. $\square$

The above arguments, although largely heuristic, can be partially justified. Surprisingly, the dynamic programming principle is the hardest step to prove, and only holds under strong restrictions on the probability space. In general, the value function is smooth only under very strong assumptions on the system. However, under milder conditions, it is at least continuous and then it satisfies the Hamilton-Jacobi equation in the viscosity sense. Besides, the Hamilton-Jacobi has a unique (viscosity) solution so that it characterizes the value function. If the diffusion is strongly non degenerate (e.g. if $N = d$ and σ is invertible with a smooth and bounded inverse) and if the Hamiltonian is smooth, then the value function is smooth as well. In this setting the above heuristic argument can be justified and the verification Theorem can be proved to hold.

We finally recall that, whenever α^* is uniquely defined from (1.5), then the Hamiltonian H is differentiable at (Du, D^2u) and

$$
\begin{cases}
H_p(t, x, Du(t, x), D^2u(t, x)) = -b(t, x, \alpha^*(t, x)), \\
H_M(t, x, Du(t, x), D^2u(t, x)) = -\frac{1}{2}\mathrm{Tr}(\sigma\sigma^*(t, x, a^*(t, x))D^2(\cdot))
\end{cases}
\tag{1.6}
$$

This is a consequence of the so-called Envelope Theorem:

Lemma 1.1 *Let A be a compact metric space, $\mathcal{O}$ be an open subset of $\mathbb{R}^d$ and $f : A \times \mathcal{O} \to \mathbb{R}$ be continuous and such that $D_x f$ is continuous on $A \times \mathcal{O}$. Then the marginal map*

$$
V(x) = \inf_{a \in A} f(a, x)
$$

is differentiable at each point $x \in \mathcal{O}$ such that the infimum in $V(x)$ is a unique point $a_x \in A$, and we have

$$
DV(x) = D_x f(a_x, x).
$$

Proof Let $x \in \mathcal{O}$ be such that the infimum in $V(x)$ is a unique point $a_x \in A$. Then an easy compactness argument shows that, if a_y is a minimum point of $V(y)$ for $y \in \mathcal{O}$ and $y \to x$, then $a_y \to a_x$.

Fix $y \in \mathcal{O}$. Note first that, as $a_x \in A$,

$$
V(y) \leq f(a_x, y) = f(a_x, x) + D_x f(a_x, z_y) \cdot (y - x) = V(x) + D_x f(a_x, z_y) \cdot (y - x),
$$

for some $z_y \in [x, y]$.

Conversely,

$$
V(x) \leq f(a_y, x) = f(a_y, y) + D_x f(a_y, z'_y) \cdot (x - y) = V(y) + D_x f(a_y, z'_y) \cdot (x - y),
$$

for some $z'_y \in [x, y]$.

By continuity of $D_x f$ and convergence of a_y, we infer that

$$\lim_{y \to x} \frac{|V(y) - V(x) - D_x f(a_x, x) \cdot (y - x)|}{|y - x|}$$

$$\leq \liminf_{y \to x} \left| D_x f(a_x, z_y) - D_x f(a_x, x) \right| + \left| D_x f(a_y, z'_y) - D_x f(a_x, x) \right| = 0.$$

$$\square$$

1.2.1.2 Estimates on the SDE

In the previous introduction, we were very fuzzy about the assumptions and the results. A complete rigorous treatment of the problem is beyond the aim of these notes. However, we need at least to clarify a bit the setting of our problem. For this, we assume the maps b and σ to be continuous on $[0, T] \times \mathbb{R}^d \times A$ and Lipschitz continuous in x independently of t and a: there is a constant $K > 0$ such that, for any $x, y \in \mathbb{R}^d, t \in [0, +\infty), a \in A$,

$$|b(t, x, a) - b(t, y, a)| + |\sigma(t, x, a) - \sigma(t, y, a)| \leq K|x - y|.$$

Under these assumptions, for any bounded control $\alpha \in \mathcal{A}$, there exists a unique solution to (1.3). By a solution we mean a progressively measurable process X such that, for any $T > 0$,

$$\mathbb{E}\left[\int_t^T |X_s^\alpha|^2 ds \right] < +\infty$$

and (1.3) holds $\mathbb{P}$-a.s. More precisely, we have:

Lemma 1.2 *Let α be a bounded control in $\mathcal{A}$. Then there exists a unique solution X^α to (1.3) and this solution satisfies, for any $T > 0$ and $p \in [2, +\infty)$,*

$$\mathbb{E}\left[\sup_{t \in [t,T]} |X_t^\alpha|^p \right] \leq C(1 + |x|^p) + \|b(\cdot, 0, \alpha.)\|_\infty^p + \|\sigma(\cdot, 0, \alpha.)\|_\infty^p),$$

where $C = C(T, p, d, K)$.

Remark 1.1 In view of the above result, the cost J is well-defined provided, for instance, that the maps $L : [0, T] \times \mathbb{R}^d \times A \to \mathbb{R}$ and $g : \mathbb{R}^d \to \mathbb{R}$ are continuous with at most a polynomial growth.

Proof The existence can be proved by a fixed point argument, exactly as in the more complicated setting of the McKean-Vlasov equation (see the proof of Theorem 1.2 below). Let us show the bound. We set $M := \|b(\cdot, 0, \alpha.)\|_\infty + \|\sigma(\cdot, 0, \alpha.)\|_\infty$. We have, by Hölder's inequality

$$|X_s^\alpha|^p \leq C(p, T, d) \left(|x|^p + \int_t^s |b(r, X_r^\alpha, \alpha_r)|^p dr + \left| \int_t^s \sigma(r, X_r^\alpha, \alpha_r) dB_r \right|^p \right)$$

where the constant $C(p, T, d)$ depends only on p, T and d.

Thus

$$\mathbb{E} \left[\sup_{t \leq r \leq s} |X_r^\alpha|^p \right] \leq C(p, T, d) \left(|x|^p + \int_t^s \mathbb{E} \left[|b(r, X_r^\alpha, \alpha_r)|^p \right] dr + \mathbb{E} \left[\sup_{t \leq r \leq s} \left| \int_t^r \sigma(u, X_u^\alpha, \alpha_u) dB_u \right|^p \right] \right).$$

Note that

$$|b(s, X_s^\alpha, \alpha_s)| \leq |b(s, 0, \alpha_s)| + L|X_s^\alpha| \leq M + L|X_s^\alpha|$$

and, in the same way,

$$|\sigma(s, X_s^\alpha, \alpha_s)| \leq M + L|X_s^\alpha|. \tag{1.7}$$

So we have

$$\int_t^s \mathbb{E} \left[|b(r, X_r^\alpha, \alpha_r)|^p \right] dr \leq 2^{p-1} (M^p (s - t) + L^p \int_t^s \mathbb{E} \left[|X_r^\alpha|^p \right] dr).$$

By the Burkholder-Davis-Gundy inequality (see Theorem IV.4.1 in [172]), we have

$$\mathbb{E} \left[\sup_{t \leq r \leq s} \left| \int_t^r \sigma(u, X_u^\alpha, \alpha_u) dB_u \right|^p \right] \leq C_p \mathbb{E} \left[\left(\int_t^s \mathrm{Tr}(\sigma \sigma^*(r, X_r^\alpha, \alpha_r)) dr \right)^{p/2} \right],$$

where the constant C_p depends on p only. Combining Hölder's inequality (since $p/2 \geq 1$) with (1.7), we then obtain

$$\mathbb{E} \left[\sup_{t \leq r \leq s} \left| \int_t^r \sigma(u, X_u^\alpha, \alpha_u) dB_u \right|^p \right] \leq C_p (s - t)^{p/2 - 1} 2^{p-1} \left(M^p (s - t) + L^p \int_t^s \mathbb{E} \left[|X_r^\alpha|^p \right] dr \right).$$

Putting together the different estimates we get therefore, for $s \in [t, T]$,

$$\mathbb{E} \left[\sup_{t \leq r \leq s} |X_r^\alpha|^p \right] \leq C(p, T, d) \left(1 + |x|^p + M^p + \int_t^s \mathbb{E} \left[|X_r^\alpha|^p \right] dr \right)$$

$$\leq C(p, T, d) \left(1 + |x|^p + M^p + \int_t^s \mathbb{E} \left[\sup_{t \leq u \leq r} |X_u^\alpha|^p \right] dr \right),$$

We can then conclude by Gronwall's Lemma.

$$\square$$

1.2.2   The Space of Probability Measures

In this section we describe the space of probability measures and a notion of distance on this space. Classical references on the distances over the space of probability measures are the monographs by Ambrosio, Gigli and Savaré [18], by Rachev and Rüschendorf [171], Santambrogio [174], and Villani [178, 179].

1.2.2.1   The Monge-Kantorovich Distance

Let (X, d) be a Polish space (= complete metric space). We have mostly in mind $X = \mathbb{R}^d$ endowed with the usual distance. We denote by $\mathcal{P}(X)$ the set of Borel probability measures on X. Let us recall that a sequence (m_n) of $\mathcal{P}(X)$ narrowly converges to a measure $m \in \mathcal{P}(X)$ if, for any test function $\phi \in C_b^0(X)$ (= the set of continuous and bounded maps on X), we have

$$\lim_n \int_X \phi(x) m_n(dx) = \int_X \phi(x) m(dx).$$

Let us recall that the topology associated with the narrow convergence corresponds to the weak-* topology of the dual of $C_b^0(X)$: for this reason we will also call it weak-* convergence. According to *Prokhorov compactness criterium*, a subset $\mathcal{K}$ of $\mathcal{P}(X)$ is (sequentially) relatively compact for the narrow convergence if and only if it is tight: for any $\varepsilon > 0$ there exists a compact subset K of X such that

$$\sup_{\mu \in \mathcal{K}} m(X \backslash K) \leq \varepsilon.$$

In particular, for any $\mu \in \mathcal{P}(X)$ and any $\varepsilon > 0$, there is some X_ε compact subset of X with $\mu(X \backslash X_\varepsilon) \leq \varepsilon$ (Ulam's Lemma).

We fix from now on a point $x_0 \in X$ and we denote by $\mathcal{P}_1(X)$ the set of measures $m \in \mathcal{P}(X)$ such that

$$\int_X d(x, x_0) m(dx) < +\infty.$$

By the triangle inequality, it is easy to check that the set $\mathcal{P}_1(X)$ does not depend on the choice of x_0. We endow $\mathcal{P}_1(X)$ with the Monge-Kantorovich distance:

$$\mathbf{d}_1(m_1, m_2) = \sup_\phi \int_X \phi(x)(m_1 - m_2)(dx) \qquad \forall m_1, m_2 \in \mathcal{P}_1(X),$$

where the supremum is taken over the set of maps $\phi : X \to \mathbb{R}$ such that ϕ is 1-Lipschitz continuous. Note that such a map ϕ is integrable against any $m \in \mathcal{P}_1(X)$ because it has at most a linear growth.

We note for later use that, if $\phi : X \to \mathbb{R}$ is $Lip(\phi)$-Lipschitz continuous, then

$$\int_X \phi(x)(m_1 - m_2)(dx) \le Lip(\phi)\mathbf{d}_1(m_1, m_2).$$

Moreover, if X_1 and X_2 are random variables on some probability space $(\Omega, \mathcal{F}, \mathcal{P})$ such that the law of X_i is m_i, then

$$\mathbf{d}_1(m_1, m_2) \le \mathbb{E}[|X_1 - X_2|], \tag{1.8}$$

because, for any 1-Lipschitz map $\phi : X \to \mathbb{R}$,

$$\int_X \phi(x)(m_1 - m_2)(dx) = \mathbb{E}[\phi(X_1) - \phi(X_2)] \le \mathbb{E}[|X_1 - X_2|].$$

Taking the supremum in ϕ gives the result. Actually one can show that, if the probability space $(\Omega, \mathcal{F}, \mathcal{P})$ is "rich enough" (namely it is a "standard probability space"), then

$$\mathbf{d}_1(m_1, m_2) = \inf_{X_1, X_2} \mathbb{E}[|X_1 - X_2|],$$

where the infimum is taken over random variables X_1 and X_2 such that the law of X_i is m_i.

Lemma 1.3 $\mathbf{d}_1$ *is a distance over* $\mathcal{P}_1(X)$.

Proof The reader can easily check the triangle inequality. We now note that $\mathbf{d}_1(m_1, m_2) = \mathbf{d}_1(m_2, m_1) \ge 0$ since one can always replace ϕ by $-\phi$ in the definition. Let us show that $\mathbf{d}_1(m_1, m_2) = 0$ implies that $m_1 = m_2$. Indeed, if $\mathbf{d}_1(m_1, m_2) = 0$, then, for any 1-Lipschitz continuous map ϕ, one has $\int_X \phi(x)(m_1 - m_2)(dx) \le 0$. Replacing ϕ by $-\phi$, one has therefore $\int_X \phi(x)(m_1 - m_2)(dx) = 0$. It remains to show that this equality holds for any continuous, bounded map $\phi : X \to \mathbb{R}$. Let $\phi \in C_b^0(X)$. We show in Lemma 1.4 below that there exists a sequence of maps (ϕ_k) such that ϕ_k is k-Lipschitz continuous, with $\|\phi_k\|_\infty \le \|\phi\|_\infty$, and the sequence (ϕ_k) converges locally uniformly to ϕ. By Lipschitz continuity of ϕ_k, we have $\int_X \phi_k d(m_1 - m_2) = 0$. Since we can apply Lebesgue convergence theorem (because the ϕ_k are uniformly bounded and m_1 and m_2 are probability measures), we obtain that $\int_X \phi(x)(m_1 - m_2)(dx) = 0$. This proves that $m_1 = m_2$. $\square$

Lemma 1.4 *Let* $\phi \in C_b^0(X)$ *and let us define the sequence of maps* (ϕ_k) *by*

$$\phi_k(x) = \inf_{y \in X} \phi(y) + kd(y, x) \qquad \forall x \in X$$

Then $\phi_k \le \phi$, ϕ_k *is* k-*Lipschitz continuous with* $\|\phi_k\|_\infty \le \|\phi\|_\infty$, *and the sequence* (ϕ_k) *converges locally uniformly to* ϕ.

Proof We have

$$\phi_k(x) = \inf_{y \in X} \phi(y) + kd(y, x) \le \phi(x) + kd(x, x) = \phi(x),$$

so that $\phi_k \le \phi$. Let us now check that ϕ_k is k-Lipschitz continuous. Indeed, let $x_1, x_2 \in X, \varepsilon > 0$ and y_1 be ε-optimal in the definition of $\phi_k(x_1)$. Then

$$\phi_k(x_2) \le \phi(y_1) + kd(y_1, x_2) \le \phi(y_1) + kd(y_1, x_1) + kd(x_1, x_2) \le \phi_k(x_1) + \varepsilon + kd(x_1, x_2).$$

As ε is arbitrary, this shows that ϕ_k is k-Lipschitz continuous. Note that $\phi_k(x) \ge -\|\phi\|_\infty$. As $\phi_k \le \phi$, this shows that $\|\phi_k\|_\infty \le \|\phi\|_\infty$.

Finally, let $x_k \to x$ and y_k be $(1/k)$-optimal in the definition of $\phi_k(x_k)$. Our aim is to show that $(\phi_k(x_k))$ converges to $\phi(x)$, which will show the local uniform convergence of (ϕ_k) to ϕ. Let us first remark that, by the definition of y_k, we have

$$kd(y_k, x_k) \le \phi_k(x_k) - \phi(y_k) + 1/k \le 2\|\phi\|_\infty + 1.$$

Therefore

$$d(y_k, x) \le d(y_k, x_k) + d(x_k, x) \to 0 \qquad \text{as } k \to +\infty.$$

This shows that $(\phi(y_k))$ converges to $\phi(x)$ and thus

$$\liminf_k \phi_k(x_k) \ge \liminf_k \phi(y_k) + kd(y_k, x_k) - 1/k \ge \liminf_k \phi(y_k) - 1/k = \phi(x).$$

On the other hand, since $\phi_k \le \phi$, we immediately have $\limsup_k \phi_k(x_k) \le \phi(x)$, from which we conclude the convergence of $(\phi_k(x_k))$ to $\phi(x)$. $\qquad\square$

Proposition 1.1 *Let (m_n) be a sequence in $\mathcal{P}_1(X)$ and $m \in \mathcal{P}_1(X)$. There is an equivalence between:*

i) $\mathbf{d}_1(m_n, m) \to 0$,

ii) (m_n) *narrowly converges to m and* $\int_X d(x, x_0) m_n(dx) \to \int_X d(x, x_0) m(dx)$.

iii) (m_n) *narrowly converges to m and* $\lim_{R \to +\infty} \sup_n \int_{B_R(x_0)^c} d(x, x_0) m_n(dx) = 0$.

Sketch of Proof $(i) \Rightarrow (ii)$. Let us assume that $\mathbf{d}_1(m_n, m) \to 0$. Then, for any Lipschitz continuous map ϕ, we have $\int \phi m_n(dx) \to \int \phi m(dx)$ by definition of $\mathbf{d}_1$. In particular, if we chose $\phi(x) = d(x, x_0)$, we have $\int_X d(x, x_0) m_n(dx) \to \int_X d(x, x_0) m(dx)$. We now prove the weak-* convergence of (m_n). Let $\phi : X \to \mathbb{R}$ be continuous and bounded and let (ϕ_k) be the sequence defined in Lemma 1.4.

Then

$$\int \phi (m_n - m)(dx) = \int \phi_k (m_n - m)(dx) + \int (\phi - \phi_k)(m_n - m)(dx).$$

Fix $\varepsilon > 0$. As $(\int_X d(x, x_0)m_n(dx))$ converges and $m \in \mathcal{P}_1(X)$, we can find $R > 0$ large such that

$$\sup_n m_n (X \backslash B_R(x_0)) + m(X \backslash B_R(x_0)) \le \varepsilon.$$

On the other hand, we can find k large enough such that $\|\phi_k - \phi\|_{L^\infty(B_R(x_0))} \le \varepsilon$, by local uniform convergence of (ϕ_k). Finally, if n is large enough, we have $|\int \phi_k(m_n - m)(dx)| \le \varepsilon$, by the convergence of (m_n) to m in $\mathbf{d}_1$. So

$$\left| \int \phi (m_n - m)(dx) \right| \le \left| \int \phi_k (m_n - m)(dx) \right| + \left| \int_{X \backslash B_R(x_0)} (\phi - \phi_k)(m_n - m)(dx) \right|$$

$$+ \left| \int_{B_R(x_0)} (\phi - \phi_k)(m_n - m)(dx) \right|$$

$$\le \left| \int \phi_k (m_n - m)(dx)) \right|$$

$$+ (\|\phi_k\|_\infty + \|\phi\|_\infty)(m_n (X \backslash B_R(x_0))$$

$$+ m(X \backslash B_R(x_0))) + 2\|\phi_k - \phi\|_{L^\infty(B_R(x_0))}$$

$$\le \varepsilon + 2\|\phi\|_\infty \varepsilon + 2\varepsilon.$$

This shows the weak-* convergence of (m_n) to m.

$(ii) \Rightarrow (iii)$. Let us assume that (m_n) narrowly converges to m and $\int_X d(x, x_0)m_n(dx) \rightarrow \int_X d(x, x_0)m(dx)$. We have to check that $\lim_{R \to +\infty} \sup_n \int_{B_R(x_0)^c} d(x, x_0)m_n(dx) = 0$. For this we argue by contradiction, assuming that there is $\varepsilon > 0$ and a subsequence still denoted (m_n) and $R_n \to +\infty$ such that

$$\int_{B_{R_n}(x_0)^c} d(x, x_0)m_n(dx) \ge \varepsilon.$$

Then, for any $M > 0$ and any n large enough so that $R_n \ge M$,

$$\int_X d(x, x_0)m_n(dx) = \int_X (d(x, x_0) \wedge M)m_n(dx) + \int_{B_M(x_0)^c} d(x, x_0)m_n(dx) - M \int_{B_M(x_0)^c} m_n(dx)$$

$$\ge \int_X (d(x, x_0) \wedge M)m_n(dx) + \varepsilon - M m_n(B_M(x_0)^c).$$

We let $n \to +\infty$ in the above inequality to get, as $(\int_X d(x, x_0) m_n(dx))$ converges to $\int_X d(x, x_0) m(dx)$ and (m_n) converges to m narrowly,

$$\int_X d(x, x_0) m(dx) \geq \int_X (d(x, x_0) \wedge M) m(dx) + \varepsilon - M m(\overline{B_M(x_0)^c}).$$

As $\int_X d(x, x_0) m(dx)$ is finite, the last term in the right-hand side tends to 0 as M tends to infinity while the first one tends to $\int_X d(x, x_0) m(dx)$ by monotone convergence: this leads to a contradiction.

(ii) $\Rightarrow$ (iii). Let us assume that (m_n) weakly-* converges to m and that

$$\lim_{R \to +\infty} \sup_n \int_{B_R(x_0)^c} d(x, x_0) m_n(dx) = 0. \text{ Fix } \varepsilon > 0. \text{ In view of the last condition,}$$

we can find $R > 0$ large enough such that

$$\sup_n \int_{B_R(x_0)^c} d(x, x_0) m_n(dx) \leq \varepsilon \quad \text{and} \quad \int_{B_R(x_0)^c} d(x, x_0) m(dx) \leq \varepsilon.$$

As the sequence (m_n) converges, it is tight by Prokhorov theorem, and we can find a compact subset K of X such that

$$\sup_n \int_{K^c} m_n(dx) \leq R^{-1} \varepsilon \quad \text{and} \quad \int_{K^c} m(dx) \leq R^{-1} \varepsilon.$$

Let $\mathcal{K}_0$ be the set of 1-Lipschitz continuous maps on X which vanish at x_0. Note that, for any $\phi \in \mathcal{K}_0$, we have

$$|\phi(x)| = |\phi(x) - \phi(x_0)| \leq d(x, x_0).$$

Therefore

$$\mathbf{d}_1(m_n, m) = \sup_{\phi \in \mathcal{K}_0} \int_X \phi(x)(m_n - m)(dx)$$

$$\leq \sup_{\phi \in \mathcal{K}_0} \left[\int_K \phi(x)(m_n - m)(dx) + \int_{B_R(x_0) \setminus K} d(x, x_0)(m_n + m)(dx) \right.$$

$$\left. + \int_{B_R(x_0)^c} d(x, x_0)(m_n + m)(dx) \right] \leq \sup_{\phi \in \mathcal{K}_0} \left[\int_K \phi(x)(m_n - m)(dx) \right]$$

$$+ R(m_n + m_n)(K^c) + 2\varepsilon \leq \sup_{\phi \in \mathcal{K}_0} \left[\int_K \phi(x)(m_n - m)(dx) \right] + 4\varepsilon.$$

By Ascoli-Arzelà, there exists $\phi_n \in \mathcal{K}_0$ optimal in the right-hand side. In addition, we can assume that (ϕ_n) converges uniformly, up to a subsequence, to some 1-Lipschitz continuous map $\phi : K \to \mathbb{R}$. We can extend ϕ_n and ϕ to X by setting

$$\tilde{\phi}_n(x) = \sup_{y \in K}[\phi_n(y) - d(y, x_0)], \qquad \tilde{\phi}(x) = \sup_{y \in K}[\phi(y) - d(y, x_0)].$$

Then one easily checks that $(\tilde{\phi}_n)$ converges uniformly to $\tilde{\phi}$ in X, so that, by weak-* convergence of (m_n) to m we have:

$$\lim_n \int_X \tilde{\phi}_n(x)(m_n - m)(dx) = 0.$$

As the $(\tilde{\phi}_n)$ are 1-Lipschitz continuous and coincide with ϕ_n on K, we have, arguing as above,

$$\mathbf{d}_1(m_n, m) \leq \int_K \phi_n(x)(m_n - m)(dx) + 4\varepsilon \leq \int_X \tilde{\phi}_n(x)(m_n - m)(dx) + 6\varepsilon.$$

Letting $n \to +\infty$ in this inequality implies $\mathbf{d}_1(m_n, m) \to 0$. More precisely, we have proved that this holds for at least a subsequence of m_n. But since this argument applies to m_n as well as to any of its subsequences, a standard argument allows us to conclude the desired result. □

In the case where $X = \mathbb{R}^d$, we repeatedly use the following compactness criterium:

Lemma 1.5 *Let $r > 1$ and $\mathcal{K} \subset \mathcal{P}_1(\mathbb{R}^d)$ be such that*

$$\sup_{\mu \in \mathcal{K}} \int_{\mathbb{R}^d} |x|^r \mu(dx) < +\infty .$$

Then $\mathcal{K}$ is relatively compact for the $\mathbf{d}_1$ distance.

Note that bounded subsets of $\mathcal{P}_1(\mathbb{R}^d)$ are not relatively compact for the $\mathbf{d}_1$ distance. For instance, in dimension $d = 1$, the sequence of measures $\mu_n = \frac{n-1}{n}\delta_0 + \frac{1}{n}\delta_n$ satisfies $\mathbf{d}_2(\mu_n, \delta_0) = 1$ for any $n \geq 1$ but μ_n narrowly converges to δ_0.

Proof of Lemma 1.5 Let $\varepsilon > 0$ and $R > 0$ sufficiently large. We have for any $\mu \in \mathcal{K}$:

$$\mu(\mathbb{R}^d \setminus B_R(0)) \leq \int_{\mathbb{R}^d \setminus B_R(0)} \frac{|x|^r}{R^r} \mu(dx) \leq \frac{C}{R^r} < \varepsilon , \qquad (1.9)$$

where $C = \sup_{\mu \in \mathcal{K}} \int_{\mathbb{R}^d} |x|^r \mu(dx) < +\infty$. So $\mathcal{K}$ is tight.

Let now (μ_n) be a sequence in $\mathcal{K}$. From the previous step we know that (μ_n) is tight and therefore there is a subsequence, again denoted (μ_n), which narrowly converges to some μ. By (1.9) and (iii) in Proposition 1.1 the convergence also holds for the distance $\mathbf{d}_1$.   □

1.2.2.2 The $\mathbf{d}_2$ Distance

Here we assume for simplicity that $X = \mathbb{R}^d$. Another useful distance on the space of measures is the Wasserstein distance $\mathbf{d}_2$. It is defined on the space $\mathcal{P}_2(\mathbb{R}^d)$ of Borel probability measures m with a finite second order moment (i.e., $\int_{\mathbb{R}^d} |x|^2 m(dx) < +\infty$) by

$$\mathbf{d}_2(m_1, m_2) := \inf_\pi \left(\int_{\mathbb{R}^d \times \mathbb{R}^d} |x - y|^2 \pi(x, y) \right)^{1/2},$$

where the infimum is taken over the Borel probability measures π on $\mathbb{R}^d \times \mathbb{R}^d$ with first marginal given by m_1 and second marginal by m_2:

$$\int_{\mathbb{R}^d \times \mathbb{R}^d} \phi(x)\pi(dx, dy) = \int_{\mathbb{R}^d} \phi(x)m_1(dx),$$

$$\int_{\mathbb{R}^d \times \mathbb{R}^d} \phi(y)\pi(dx, dy) = \int_{\mathbb{R}^d} \phi(y)m_2(dy) \quad \forall \phi \in C_b^0(\mathbb{R}^d).$$

Given a "sufficiently rich" probability space $(\Omega, \mathcal{F}, \mathbb{P})$, the distance can be defined equivalently by

$$\mathbf{d}_2(m_1, m_2) = \inf_{X,Y} \left(\mathbb{E}\left[|X - Y|^2 \right] \right)^{1/2},$$

where the infimum is taken over random variables X, Y over Ω with law m_1 and m_2 respectively.

1.2.3 *Mean Field Limits*

We complete this preliminary part by the analysis of large particle systems. Classical references on this topic are the monographs or texbooks by Sznitman [177], Spohn [176] and Golse [111].

We consider system of N-particles (where $N \in \mathbb{N}^*$ is a large number) and we want to understand the behavior of the system as the number N tends to infinity. We work with the following system: for $i = 1, \dots, N$,

$$\begin{cases} dX_t^i = b(X_t^i, m_{X_t}^N)dt + dB_t^i, \qquad m_{X_t}^N := \frac{1}{N}\sum_{j=1}^N \delta_{X_t^j} \\ X_0^i = Z^i \end{cases} \tag{1.10}$$

where the (B^i) are independent Brownian motions, the Z^i are i.i.d. random variables in $\mathbb{R}^d$ which are also independent of the (B^i). The map $b : \mathbb{R}^d \times \mathcal{P}_1(\mathbb{R}^d) \to \mathbb{R}^d$ is assumed to be globally Lipschitz continuous. Note that, under these assumptions, the solution (X^i) to (1.10) exists and is unique, since this is an ordinary system of SDEs with Lipschitz continuous drift. A key point is that, because the (Z^i) have the same law and the equations satisfied by the X^i are symmetric, the X^i have the same law (they are actually "exchangeable").

We want to understand the limit of the (X^i) as $N \to +\infty$. The heuristic idea is that, as N is large, the (X^i) become more and more independent, so that they become almost i.i.d. The law of large numbers then implies that

$$\frac{1}{N}\sum_{j=1}^N b(X_t^i, X_t^j) \approx \tilde{\mathbb{E}}\left[b(X_t^i, \tilde{X}_t^i)\right] = \int_{\mathbb{R}^d} b(X_t^i, y)\mathbb{P}_{X_t^i}(dy),$$

where $\tilde{X}_t^i$ is an independent copy of X_t^i and $\tilde{\mathbb{E}}$ is the expectation with respect to this independent copy. Therefore we expect the X^i to be close to the solution $\bar{X}^i$ to the McKean-Vlasov equation

$$\begin{cases} d\bar{X}_t^i = b(\bar{X}_t^i, \mathcal{L}(\bar{X}_t^i))dt + dB_t^i, \\ \bar{X}_0^i = Z^i \end{cases} \tag{1.11}$$

This is exactly what we are going to show. For doing so, we proceed in 3 steps: firstly, we generalize the law of large numbers by considering the convergence of empirical measures (the Glivenko-Cantelli law of large numbers), secondly we prove the existence and the uniqueness of a solution to the McKean-Vlasov equation (1.11) and, thirdly, we establish the convergence.

1.2.3.1 The Glivenko-Cantelli Law of Large Numbers

Here we consider (X_n) a sequence of i.i.d. random variables on a fixed probability space $(\Omega, \mathcal{F}, \mathbb{P})$, with $\mathbb{E}[|X_1|] < +\infty$. We denote by m the law of X_1. The law of large numbers states that, a.s. and in L^1,

$$\lim_{N \to +\infty} \frac{1}{N}\sum_{n=1}^N X_n = \mathbb{E}[X_1].$$

Our aim is to show that a slightly stronger convergence holds: let

$$m_X^N := \frac{1}{N} \sum_{n=1}^{N} \delta_{X_n}$$

Note that m_X^N is a random measure, in the sense that m_X^N is a.s. a measure and that, for any Borel set $A \subset X$, $m_X^N(A)$ is a random variable. The following result is (sometimes) known as the Glivenko-Cantelli Theorem.

Theorem 1.1 *If* $\mathbb{E}[|X_1|] < +\infty$, *then, a.s. and in* L^1,

$$\lim_{N \to +\infty} \mathbf{d}_1(m_X^N, m) = 0.$$

Remark 1.2 It is often useful to quantify the convergence speed in the law of large numbers. Such results can be found in the text books [171] or, in a sharper form, in [68, Theorem 5.8], see also the references therein.

Sketch of Proof Let $\phi \in C_b^0(X)$. Then, by the law of large numbers,

$$\int_{\mathbb{R}^d} \phi(x) m_X^N(dx) = \frac{1}{N} \sum_{n=1}^{N} \phi(X_n) = \mathbb{E}[\phi(X_1)] \qquad a.s.$$

By a separability argument, it is not difficult to check that the set of zero probability in the above convergence can be chosen independent of ϕ. So (m_X^N) converge weakly-* to m a.s. Note also that

$$\int_{\mathbb{R}^d} d(x, x_0) m_X^N(dx) = \frac{1}{N} \sum_{n=1}^{N} d(X_n, x_0)$$

where the random variables $(d(X_n, x_0))$ are i.i.d. and in L^1. By the law of large numbers we have

$$\int_{\mathbb{R}^d} d(x, x_0) m_X^N(dx) \to \int_{\mathbb{R}^d} d(x, x_0) m(dx) \qquad a.s.$$

By Proposition 1.1, (m_X^N) converges a.s. in $\mathbf{d}_1$ to m. It remains to show that this convergence also holds in expectation. For this we note that

$$\mathbf{d}_1(m_X^N, m) = \sup_{\phi} \int_{\mathbb{R}^d} \phi(m_X^N - m)(dx) \leq \sup_{\phi} \frac{1}{N} \sum_{i=1}^{N} \phi(X_i) - \int_{\mathbb{R}^d} \phi(x) m(dx),$$

where the supremum is taken over the 1-Lipschitz continuous maps ϕ with $\phi(0) = 0$. So

$$\mathbf{d}_1(m_X^N, m) \le \frac{1}{N} \sum_{i=1}^{N} |X_i| + \int_{\mathbb{R}^d} |x| m(dx).$$

As the right-hand side converges in L^1, $\mathbf{d}_1(m_X^N, m)$ is uniformly integrable which implies its convergence in expectation to 0. $\square$

1.2.3.2   The Well-Posedness of the McKean-Vlasov Equation

Theorem 1.2 *Let us assume that* $b : \mathbb{R}^d \times \mathcal{P}_1(\mathbb{R}^d) \to \mathbb{R}^d$ *is globally Lipschitz continuous and let* $Z \in L^2(\Omega)$. *Then the McKean-Vlasov equation*

$$\begin{cases} dX_t = b(X_t, \mathcal{L}(X_t))dt + dB_t \\ X_0 = Z \end{cases}$$

has a unique solution, i.e., a progressively measurable process such that $\mathbb{E}\left[\int_0^T |X_s|^2 ds\right] < +\infty$ *for any* $T > 0$.

Remark 1.3 By Itô's formula, the law m_t of a solution X_t solves in the sense of distributions the McKean-Vlasov equation

$$\begin{cases} \partial_t m_t - \frac{1}{2}\Delta m_t + \text{div}(m_t b(x, m_t)) = 0 & \text{in } (0, T) \times \mathbb{R}^d \\ m_0 = \mathcal{L}(Z) & \text{in } \mathbb{R}^d. \end{cases}$$

One can show (and we will admit) that this equation has a unique solution, which proves the uniqueness in law of the process X.

Proof Let $\alpha > 0$ to be chosen later and E be the set of progressively measurable processes (X_t) such that

$$\|X\|_E := \mathbb{E}\left[\int_0^\infty e^{-\alpha t} |X_t| dt\right] < +\infty.$$

Then $(E, \|\cdot\|_E)$ is a Banach space. On E we define the map Φ by

$$\Phi(X)_t = Z + \int_0^t b(X_s, \mathcal{L}(X_s))ds + B_t, \qquad t \ge 0.$$

Let us check that the map Φ is well defined from E to E. Note first that $\Phi(X)$ is indeed progressively measurable. By the L-Lipschitz continuity of b (for some $L > 0$),

$$|\Phi(X)_t| \le |Z| + \int_0^t |b(X_s, \mathcal{L}(X_s))| ds + |B_t|$$

$$\le |Z| + t|b(0, \delta_0)| + L \int_0^t (|X_s| + \mathbf{d}_1(\mathcal{L}(X_s), \delta_0)) ds + |B_t|,$$

where one can easily check that $\mathbf{d}_1(\mathcal{L}(X_s), \delta_0) = \mathbb{E}\left[\int_0^t |X_s| ds\right]$. So

$$\mathbb{E}\left[\int_0^{+\infty} e^{-\alpha t} |\Phi(X)_t| dt\right] \le \alpha^{-1}\mathbb{E}[|Z|] + \alpha^{-2}|b(0, \delta_0)|$$

$$+ 2L\mathbb{E}\left[\int_0^{+\infty} e^{-\alpha t} \int_0^t |X_s| ds dt\right] + \int_0^{+\infty} e^{-\alpha t} \mathbb{E}[|B_t|] dt$$

$$= \alpha^{-1}\mathbb{E}[|Z|] + \alpha^{-2}|b(0, \delta_0)|$$

$$+ \frac{2L}{\alpha}\mathbb{E}\left[\int_0^{+\infty} e^{-\alpha s} |X_s| ds\right] + C_d \int_0^{+\infty} t^{1/2} e^{-\alpha t} dt,$$

where C_d depends only on dimension. This proves that $\Phi(X)$ belongs to E.

Let us finally check that Φ is a contraction. We have, if $X, Y \in E$,

$$|\Phi(X)_t - \Phi(Y)_t| \le \int_0^t |b(X_s, \mathcal{L}(X_s)) - b(Y_s, \mathcal{L}(Y_s))| dt$$

$$\le Lip(b)\left(\int_0^t \mathbf{d}_1(\mathbb{P}_{X_s}, \mathbb{P}_{Y_s}) dt + \int_0^t |X_s - Y_s| dt\right).$$

Recall that $\mathbf{d}_1(\mathbb{P}_{X_s}, \mathbb{P}_{Y_s}) \le \mathbb{E}[|X_s - Y_s|]$. So multiplying by $e^{-\alpha t}$ and taking expectation, we obtain:

$$\|\Phi(X) - \Phi(Y)\|_E = \mathbb{E}\left[\int_0^{+\infty} e^{-\alpha t} |\Phi(X)_s - \Phi(Y)_s| dt\right]$$

$$\le 2Lip(b) \int_0^{+\infty} e^{-\alpha t} \int_0^t \mathbb{E}[|X_s - Y_s|] ds dt$$

$$\le \frac{2Lip(b)}{\alpha} \|X - Y\|_E.$$

If we choose $\alpha > 2Lip(b)$, then Φ is a contraction in the Banach space E and therefore has a unique fixed point. It is easy to check that this fixed point is the unique solution to our problem. $\qquad\square$

1.2.3.3 The Mean Field Limit

Let (X^i) be the solution to the particle system (1.10) and $(\bar{X}^i)$ be the solution to (1.11). Let us note that, as the (B^i) and the (Z^i) are independent with the same law, the $(\bar{X}^i_t)$ are i.i.d. for any $t \geq 0$.

Theorem 1.3 *We have, for any $T > 0$,*

$$\lim_{N \to +\infty} \sup_{i=1,\dots,N} \mathbb{E} \left[\sup_{t \in [0,T]} |X^i_t - \bar{X}^i_t| \right] = 0.$$

Remark: a similar result holds when there is a non constant volatility term σ in front of the Brownian motion. The proof is then slightly more intricate.

Proof We consider

$$X^i_t - \bar{X}^i_t = \int_0^t \left(b(X^i_t, m^N_{X_t}) - b(\bar{X}^i_t, \mathcal{L}(\bar{X}^i_t)) \right) dt.$$

By the uniqueness in law of the solution to the McKean-Vlasov equation we can denote by $m(t) := \mathcal{L}(\bar{X}^i_t)$ (it is independent of i). Then, setting $m^N_{\bar{X}_t} := \dfrac{1}{N} \sum_{j=1}^N \delta_{\bar{X}^j_t}$ and using the triangle inequality, we have

$$|X^i_t - \bar{X}^i_t| \leq \int_0^t \left| b(X^i_t, m^N_{X_t}) - b(\bar{X}^i_s, m^N_{\bar{X}_s}) \right| ds + \int_0^t \left| b(\bar{X}^i_s, m^N_{\bar{X}_s}) - b(\bar{X}^i_s, m(s)) \right| ds$$

$$\leq Lip(b) \int_0^t (|X^i_s - \bar{X}^i_s| + \mathbf{d}_1(m^N_{X_t}, m^N_{\bar{X}_s})) ds + Lip(b) \int_0^t \mathbf{d}_1(m^N_{\bar{X}_s}, m(s)) ds$$

$$\leq Lip(b) \int_0^t (|X^i_s - \bar{X}^i_s| + \frac{1}{N} \sum_{j=1}^N |X^j_s - \bar{X}^j_s|) ds + Lip(b) \int_0^t \mathbf{d}_1(m^N_{\bar{X}_s}, m(s)) ds,$$

$$\tag{1.12}$$

since

$$\mathbf{d}_1(m^N_{X_t}, m^N_{\bar{X}_s}) \leq \frac{1}{N} \sum_{j=1}^N |X^j_s - \bar{X}^j_s|.$$

Summing over $i = 1, \dots, N$, we get

$$\frac{1}{N} \sum_{i=1}^N |X^i_t - \bar{X}^i_t| \leq 2 Lip(b) \int_0^t \frac{1}{N} \sum_{j=1}^N |X^j_s - \bar{X}^j_s| ds + Lip(b) \int_0^t \mathbf{d}_1(m^N_{\bar{X}_s}, m(s)) ds.$$

Using Gronwall Lemma, we find, for any $T > 0$, and for some constant C_T depending on $Lip(b)$,

$$\sup_{t \in [0,T]} \frac{1}{N} \sum_{i=1}^{N} |X_t^i - \bar{X}_t^i| \leq C_T \int_0^T \mathbf{d}_1(m_{\bar{X}_s}^N, m(s)) ds, \tag{1.13}$$

where C_T depends on T and $Lip(b)$ (but not on N). Then we can come back to (1.12), use first Gronwall Lemma and then (1.13) to get, for any $T > 0$, and for some (new) constant C_T depending on $Lip(b)$ and which might change from line to line,

$$\sup_{t \in [0,T]} |X_t^i - \bar{X}_t^i| \leq C_T \int_0^T (\frac{1}{N} \sum_{j=1}^{N} |X_s^j - \bar{X}_s^j| + \mathbf{d}_1(m_{\bar{X}_s}^N, m(s))) ds$$

$$\leq C_T \int_0^T \mathbf{d}_1(m_{\bar{X}_s}^N, m(s)) ds.$$

We now take expectation to obtain

$$\mathbb{E}\left[\sup_{t \in [0,T]} |X_t^i - \bar{X}_t^i| \right] \leq C_T \int_0^T \mathbb{E}\left[\mathbf{d}_1(m_{\bar{X}_s}^N, m(s)) ds \right].$$

One can finally check exactly as in the proof of Theorem 1.1 that the right-hand side tends to 0.   □

1.3   The Mean Field Game System

In this section, we focus on the mean field game system of PDEs (henceforth, MFG system) introduced by J.-M. Lasry and P.-L. Lions [144, 145]. It takes the form of a backward Hamilton-Jacobi equation coupled with a forward Kolmogorov equation

$$\begin{cases} -\partial_t u - \varepsilon \Delta u + H(x, Du, m) = 0, \\ \partial_t m - \varepsilon \Delta m - \text{div}(m H_p(x, Du, m) = 0, \\ m(0) = m_0, \ u(T) = G(x, m(T)). \end{cases} \tag{1.14}$$

The Hamilton-Jacobi equation formalizes the individual optimization problem and is solved by the value function of each agent, while the Kolmogorov equation describes the evolution of the population density.

We will first derive in a heuristic way the MFG system (1.14). Then we will discuss several PDE methods used to obtain the existence and uniqueness of solutions, in both the diffusive case ($\varepsilon > 0$) and the deterministic case ($\varepsilon = 0$).

In order to give a more clear and complete presentation of the PDE approach, we will mostly focus on the simplest form of the system (where the cost of control is separate from the mean-field dependent cost):

$$
\begin{cases}
-\partial_t u - \varepsilon \Delta u + H(x, Du) = F(x, m) \\
u(T) = G(x, m(T)) \\
\partial_t m - \varepsilon \Delta m - \operatorname{div}(m\, H_p(x, Du)) = 0 \\
m(0) = m_0,
\end{cases}
\tag{1.15}
$$

where we will distinguish two kind of regimes, depending on the case of smoothing couplings F, G (operators on the space of measures) rather than on the case of local couplings (functions defined on the density of absolutely continuous measures). Sample results of existence and uniqueness will be given in both cases.

For simplicity, we will restrict the analysis of system (1.15) to the periodic case. This means that x belongs to the d-dimensional torus $\mathbb{T}^d := \mathbb{R}^d/\mathbb{Z}^d$, and all x-dependent functions are $\mathbb{Z}^d$-periodic in space. We denote by $\mathcal{P}(\mathbb{T}^d)$ the set of Borel probability measures on $\mathbb{T}^d$, endowed as before with the Monge-Kantorovich distance $\mathbf{d}_1$:

$$
\mathbf{d}_1(m, m') = \sup_{\phi} \int_{\mathbb{T}^d} \phi\, d(m - m') \qquad \forall m, m' \in \mathcal{P}(\mathbb{T}^d),
$$

where the supremum is taken over all 1-Lipschitz continuous maps $\phi : \mathbb{T}^d \to \mathbb{R}$. This distance metricizes the narrow topology on $\mathcal{P}(\mathbb{T}^d)$. Recall that $\mathcal{P}(\mathbb{T}^d)$ is a compact metric space. For $T > 0$, we set $Q_T := (0, T) \times \mathbb{T}^d$.

1.3.1   Heuristic Derivation of the MFG System

We describe here the simplest class of mean field games, when the state space is $\mathbb{R}^d$. In this control problem with infinitely many agents, each small agent controls his/her own dynamics:

$$
X_s = x + \int_t^s b(X_r, \alpha_r, m(r))dr + \int_t^s \sigma(X_r, \alpha_r, m(r))dB_r,
\tag{1.16}
$$

where X lives in $\mathbb{R}^d$, α is the control (taking its values in a fixed set A) and B is a given M-dimensional Brownian motion. The difference with Sect. 1.2.1 is the dependence of the drift with respect to the distribution $(m(t))$ of all the players. This (time dependent) distribution $(m(t))$ belongs to the set $\mathcal{P}(\mathbb{R}^d)$ and is, at this stage, supposed to be given: one should think at $(m(t))$ as the *anticipation* made by the agents on their future time dependent distribution. The coefficients $b : \mathbb{R}^d \times A \times$

$\mathcal{P}(\mathbb{R}^d) \to \mathbb{R}^d$ and $\sigma : \mathbb{R}^d \times A \times \mathcal{P}(\mathbb{R}^d) \to \mathbb{R}^{d \times M}$ are assumed to be smooth enough for the solution (X_t) to exist.

The cost of a small player is given by

$$J(t, x, \alpha) = \mathbb{E}\left[\int_t^T L(X_s, \alpha_s, m(s))ds + G(X_T, m(T))\right]. \tag{1.17}$$

Here $T > 0$ is the finite horizon of the problem, $L : \mathbb{R}^d \times A \times \mathcal{P}(\mathbb{R}^d) \to \mathbb{R}$ and $G : \mathbb{R}^d \times \mathcal{P}(\mathbb{R}^d) \to \mathbb{R}$ are given continuous maps.

If we define the value function u as

$$u(t, x) = \inf_\alpha J(t, x, \alpha),$$

then, at least in a formal way, u solves the Hamilton-Jacobi equation

$$\begin{cases} -\partial_t u(t, x) + H(x, Du(t, x), D^2u(t, x), m(t)) = 0 & \text{in } (0, T) \times \mathbb{R}^d \\ u(T, x) = G(x, m(T)) & \text{in } \mathbb{R}^d \end{cases}$$

where the Hamiltonian $H : \mathbb{R}^d \times \mathbb{R}^d \times \mathbb{R}^{d \times d} \times \mathcal{P}(\mathbb{R}^d) \to \mathbb{R}$ is defined by

$$H(x, p, M, m) := \sup_{a \in A}\left[-L(x, a, m) - p \cdot b(x, a, m) - \frac{1}{2}\text{Tr}(\sigma\sigma^*(x, a, m)M)\right].$$

Let us now introduce $\alpha^*(t, x) \in A$ as a maximum point in the definition of H when $p = Du(t, x)$ and $M = D^2u(t, x)$. Namely

$$H(x, Du(t, x), D^2u(t, x), m(t)) = -L(x, \alpha^*(t, x), m(t)) - Du(t, x) \cdot b(x, \alpha^*(t, x), m(t))$$

$$- \frac{1}{2}\text{Tr}(\sigma\sigma^*(x, \alpha^*(t, x))D^2u(t, x), m(t)).$$

$$\tag{1.18}$$

Recall from Sect. 1.2.1 that α^* is the optimal feedback for the problem. However, we stress that here u and α^* depend on the time-dependent family of measures $(m(t))$.

We now discuss the evolution of the population density. For this we make the two following assumptions. Firstly we assume that all the agents control the same system (1.16) (although not necessarily starting from the same initial position) and minimize the same cost J. As a consequence, the dynamics at optimum of each player is given by

$$dX_s^* = b(X_s^*, \alpha^*(s, X_s^*), m(s))ds + \sigma(X_s^*, \alpha^*(s, X_s^*), m(s))dB_s.$$

Secondly, we assume that the initial position of the agents and the noise driving their dynamics are independent: in particular, there is no "common noise" impacting all the players. The initial distribution of the agents at time $t = 0$ is denoted by

$m_0 \in \mathcal{P}(\mathbb{R}^d)$. From the analysis of the mean field limit of Sect. 1.2.3 (in the simple case where the coefficients do not depend on the other agents) the actual distribution $(\tilde{m}(s))$ of all agents at time s is simply given by the law of (X_s^*) with $\mathcal{L}(X_0^*) = m_0$.

Let us now write the equation satisfied by $(\tilde{m}(s))$. By Itô's formula, we have, for any smooth map $\phi : [0, T) \times \mathbb{R}^d \to \mathbb{R}$ with a compact support:

$$0 = \mathbb{E}\big[\phi(T, X_T^*)\big] = \mathbb{E}\big[\phi(0, X_0^*)\big] + \int_0^T \mathbb{E}\Big[\partial_t \phi(s, X_s^*) + b(X_s^*, \alpha^*(s, X_s^*), m(s)) \cdot D\phi(s, X_s^*)$$

$$+ \frac{1}{2}\mathrm{Tr}(\sigma\sigma^*(X_s^*, \alpha^*(s, X_s^*), m(s))D^2\phi(s, X_s^*))\Big]ds$$

$$= \int_{\mathbb{R}^d} \phi(0, x)m_0(dx) + \int_0^T \int_{\mathbb{R}^d} \Big[\partial_t \phi(s, x) + b(x, \alpha^*(s, x), m(s)) \cdot D\phi(s, x)$$

$$+ \frac{1}{2}\mathrm{Tr}(\sigma\sigma^*(x, \alpha^*(s, x), m(s))D^2\phi(s, x))\Big]\tilde{m}(t, dx)ds$$

After integration by parts, we obtain that $(\tilde{m}(t))$ satisfies, in the sense of distributions,

$$\begin{cases} \partial_t \tilde{m} - \dfrac{1}{2}\sum_{ij} D_{ij}^2(\tilde{m}(t, x)a_{ij}(x, \alpha^*(t, x), \\ m(t))) + \mathrm{div}(\tilde{m}(t, x)b(x, \alpha^*(t, x), m(t))) = 0 \quad \text{in } (0, T) \times \mathbb{R}^d, \\ \tilde{m}(0) = m_0 \qquad \text{in } \mathbb{R}^d, \end{cases}$$

where $a = \sigma\sigma^*$.

At equilibrium, one expects the anticipation $(m(t))$ made by the agents to be correct: $\tilde{m}(t) = m(t)$. Collecting the above equations leads to the MFG system:

$$\begin{cases} -\partial_t u(t, x) + H(x, Du(t, x), D^2u(t, x), m(t)) = 0 \quad \text{in } (0, T) \times \mathbb{R}^d, \\ \partial_t m - \dfrac{1}{2}\sum_{ij} D_{ij}^2(m(t, x)a_{ij}(x, \alpha^*(t, x), m(t)) + \mathrm{div}(m(t, x)b(x, \alpha^*(t, x), \\ m(s))) = 0 \quad \text{in } (0, T) \times \mathbb{R}^d, \\ m(0) = m_0, \ u(T, x) = G(x, m(T)) \qquad \text{in } \mathbb{R}^d, \end{cases}$$

where α^* is given by (1.18) and $a = \sigma\sigma^*$.

In order to simplify a little this system, let us assume that $M = d$ and $\sigma = \sqrt{2\varepsilon}I_d$ (where now ε is a constant). We set (warning! abuse of notation!)

$$H(x, p, m) := \sup_{a \in A}[-L(x, a, m) - p \cdot b(x, a, m)]$$

and note that, by Lemma 1.1, under suitable assumptions one has (see (1.6))

$$H_p(x, Du(t, x), m(t)) = -b(x, \alpha^*(t, x), m(t)).$$

In this case the MFG system becomes

$$\begin{cases} -\partial_t u(t, x) - \varepsilon \Delta u(t, x) + H(x, Du(t, x), m(t)) = 0 & \text{in } (0, T) \times \mathbb{R}^d, \\ \partial_t m - \varepsilon \Delta m(t, x) - \text{div}(m(t, x) H_p(x, Du(t, x), m(t))) = 0 & \text{in } (0, T) \times \mathbb{R}^d, \\ m(0) = m_0, \ u(T, x) = G(x, m(T)) & \text{in } \mathbb{R}^d, \end{cases}$$

This system will be the main object of analysis of this chapter. Note that it is not a standard PDE system, since the first equation is backward in time, while the second one is forward in time. As this analysis is not too easy, it will be more convenient to work with periodic boundary condition (namely on the d-dimensional torus $\mathbb{T}^d = \mathbb{R}^d / \mathbb{Z}^d$).

1.3.2 Second Order MFG System with Smoothing Couplings

We start with the analysis of the MFG system (1.15). Hereafter, we assume that x belongs to the d-dimensional torus $\mathbb{T}^d$. In this section we assume that the coupling functions are smoothing operators defined on the set $C^0([0, T], \mathcal{P}(\mathbb{T}^d))$. To stress that the couplings are operators, we will write their action as $F[m]$ and $G[m]$, so the system will be written as

$$\begin{cases} -\partial_t u - \varepsilon \Delta u + H(t, x, Du) = F[m] \\ u(T) = G[m(T)] \\ \partial_t m - \varepsilon \Delta m - \text{div}(m \, H_p(t, x, Du)) = 0 \\ m(0) = m_0 \,. \end{cases} \tag{1.19}$$

Definition 1.1 We say that a pair (u, m) is a *classical* solution to (1.19) if

(i) $m \in C^0([0, T], \mathcal{P}(\mathbb{T}^d))$ and $m(0) = m_0$; $u \in C(\overline{Q_T})$ and $u(T) = G[m(T)]$
(ii) u, m are continuous functions in $(0, T) \times \mathbb{T}^d$, of class C^2 in space and C^1 in time, and the two equations are satisfied pointwise for $x \in \mathbb{T}^d$ and $t \in (0, T)$.

Let us stress that the above definition only requires u, m to be smooth for $t \in (0, T)$, which allows m_0 to be a general probability measure. Of course, the smoothness can extend up to $t = 0$ and/or $t = T$ in case m_0 and/or $G[\cdot]$ are sufficiently smooth. We also notice that the above definition requires $H_p(x, p)$ to be differentiable, in order for m to be a classical solution. It is often convenient to use a weaker notion as well: we will simply say that (u, m) is a solution of (1.19) (without using the adjective *classical*) if (u, m) satisfy point (i) in the above Definition, m and Du are locally bounded and if both equations are satisfied in the sense of distributions in $(0, T)$, i.e. against test functions $\varphi \in C^1((0, T) \times \mathbb{T}^d)$ with compact support in $(0, T)$.

The smoothing character of the couplings F, G is assumed in the following conditions:

$$F : C^0([0, T], \mathcal{P}(\mathbb{T}^d)) \to C^0(\overline{Q_T})$$

is continuous with range into a bounded set of $L^\infty(0, T; W^{1,\infty}(\mathbb{T}^d))$ (1.20)

and similarly

$$G : \mathcal{P}(\mathbb{T}^d) \cap L^1(\mathbb{T}^d) \to C^0(\mathbb{T}^d)$$

is continuous with range into a bounded set of $W^{1,\infty}(\mathbb{T}^d)$.   (1.21)

We notice that functions $F(t, x, m)$ which are (locally) Lipschitz continuous on $Q_T \times \mathcal{P}(\mathbb{T}^d)$ naturally provide with corresponding operators given by $F[m] = F(t, x, m(t))$; the above assumption (1.20) is satisfied if, for instance, F is Lipschitz in x uniformly for $m \in \mathcal{P}(\mathbb{T}^d)$. Examples of such smoothing operators are easily obtained by convolution.

As is now well-known in the theory, a special case occurs if the operators F, G are *monotone*. This can be understood as an extension of the standard monotonicity for L^2 operators (indeed, F and G are defined in $L^2(Q_T)$ and $L^2(\mathbb{T}^d)$ respectively). For instance, F is said to be monotone if

$$\int_0^T \int_{\mathbb{T}^d} (F[m_1] - F[m_2]) d(m_1 - m_2) \geq 0 \qquad \forall m_1, m_2 \in C^0([0, T], \mathcal{P}(\mathbb{T}^d))$$

and a similar definition applies to G. Let us observe that the monotonicity condition on F, G is satisfied for a restricted class of convolution operators, but one can take for instance $F[m] = f(k \star m) \star k$, where f is a nondecreasing function and $k \geq 0$ a smooth symmetric kernel.

We start by giving one of the early results by J.-M. Lasry and P.-L. Lions.

Theorem 1.4 ([144, 145]) *Let F, G satisfy conditions (1.20), (1.21). Assume that $H \in C^1(Q_T \times \mathbb{R}^d)$ is convex with respect to p and satisfies at least one of the two following assumptions:*

$$\exists\ c_0 > 0 : \quad |H_p(t, x, p)| \leq c_0(1 + |p|) \quad \forall(t, x, p) \in Q_T \times \mathbb{R}^d \qquad (1.22)$$

$$\exists\ c_1 > 0 : \quad H_x(t, x, p) \cdot p \geq -c_1(1 + |p|^2) \quad \forall(t, x, p) \in Q_T \times \mathbb{R}^d \qquad (1.23)$$

Then, for every $m_0 \in \mathcal{P}(\mathbb{T}^d)$, there exist $u \in L^\infty(0, T; W^{1,\infty}(\mathbb{T}^d)) \cap C^0(\overline{Q_T})$ and $m \in C^0([0, T], \mathcal{P}(\mathbb{T}^d))$ such that (u, m) is a solution to (1.19).

In addition, let F, G be monotone operators. If one of the two following conditions hold:

$$\begin{cases} \int_0^T \int_{\mathbb{T}^d} (F[m_1] - F[m_2]) d(m_1 - m_2) = 0 \Rightarrow F[m_1] = F[m_2] \\ \int_{\mathbb{T}^d} (G[m_1] - G[m_2]) d(m_1 - m_2) = 0 \Rightarrow G[m_1] = G[m_2] \end{cases} \qquad (1.24)$$

$$H(t, x, p_1) - H(t, x, p_2) - H_p(t, x, p_2)(p_1 - p_2) = 0$$

$$\Rightarrow H_p(t, x, p_1) = H_p(t, x, p_2) \quad \forall p_1, p_2 \in \mathbb{R}^d \qquad (1.25)$$

then (u, m) is unique in the above class.

Finally, if in addition $H_p \in C^1(Q_T \times \mathbb{R}^d)$ and F is a bounded map in the space of Hölder continuous functions, then (u, m) is a classical solution.

Proof We start by assuming that m_0 is Hölder continuous in $\mathbb{T}^d$. We set

$$X := \{ m \in C^0([0, T], L^1(\mathbb{T}^d)) : m \geq 0, \int_{\mathbb{T}^d} m(t) = 1 \ \forall t \in [0, T] \},$$

and we define the following operator: for $\mu \in X$, if u_μ denotes the unique solution to

$$\begin{cases} -\partial_t u_\mu - \varepsilon \Delta u_\mu + H(t, x, Du_\mu) = F[\mu] \\ u_\mu(T) = G[\mu(T)], \end{cases}$$

then we set $m := \Phi(\mu)$ as the solution to

$$\begin{cases} \partial_t m - \varepsilon \Delta m - \mathrm{div}(m H_p(t, x, Du_\mu)) = 0, \\ m(0) = m_0. \end{cases}$$

We observe that, for given $\mu \in X$, $F[\mu]$ belongs to $L^\infty(0, T; W^{1,\infty}(\mathbb{T}^d))$ and $G[\mu]$ is Lipschitz as well. If condition (1.22) is satisfied, then H has at most quadratic growth since $|H(t, x, p)| \leq c(1+|p|^2)$ for some constant $c > 0$. Hence the classical parabolic theory applies (see [142, Chapter V, Thm 3.1]); there exists a constant $K > 0$ and $\alpha \in (0, 1)$ such that $Du_\mu \in C^\alpha(Q_T)$ and

$$\|Du_\mu\|_\infty \leq K. \qquad (1.26)$$

More precisely, the constant K is independent of μ due to the assumptions on the range of F, G in (1.20)–(1.21).

By contrast, if condition (1.23) holds true, then H has not necessarily natural growth; however, a gradient estimate follows by using the classical Bernstein's method. This means that we look at the equation satisfied by $w := |Du|^2$. Assuming u to be smooth, a direct computation gives

$$\partial_t w - \varepsilon \Delta w = -2|D^2 u|^2 - 2Du \cdot D\left(H(t, x, Du) - F[m]\right)$$
$$\leq -H_p \cdot Dw - 2H_x(t, x, Du) \cdot Du + 2DuDF[m]$$
$$\leq -H_p \cdot Dw + 2c_1(1 + |Du|^2) + |Du|^2 + \|F[m]\|^2_{W^{1,\infty}}$$

where we used both assumptions (1.20) and (1.23). Since $\|F[m]\|_{W^{1,\infty}}$ is bounded uniformly with respect to m, we conclude that there exists $C > 0$ such that

$$\partial_t w - \varepsilon \Delta w + H_p \cdot Dw \leq C(1 + w).$$

At time $t = T$ we have $\|w(T)\|_\infty \leq \|DG[m]\|^2_\infty \leq C$, so we deduce, by maximum principle, that

$$\|Du\|^2_\infty = \|w\|_\infty \leq C_T \tag{1.27}$$

for some constant C_T depending on T, c_1, F, G. Therefore, (1.26) holds true under condition (1.23) as well.

Eventually, we conclude that $H_p(t, x, Du_\mu)$ is uniformly bounded for $\mu \in X$. By parabolic regularity (see e.g. [142, Chapter V, Thms 1.1 and 2.1]), this implies that m is uniformly bounded in $C^\alpha(\overline{Q_T})$, for some $\alpha \in (0, 1)$. In particular, the operator Φ has bounded range in $C^\alpha(\overline{Q_T})$, so the range of Φ is a compact subset of X. The continuity of Φ is straightforward: if $\mu_n \to \mu$, we have $F[\mu_n] \to F[\mu]$ in $C(\overline{Q_T})$, so u_n converges uniformly to the corresponding solution u_μ, while Du_n converges a.e. to Du_μ, hence $H_p(t, x, Du_n) \to H_p(t, x, Du_\mu)$ in $L^p(Q_T)$ for every $p > 1$, which entails the convergence of m_n towards $m = \Phi(\mu)$. By Schauder's fixed point theorem (see e.g. [110]) applied to Φ, we deduce the existence of m such that $m = \Phi(m)$, which means a solution (u, m) of (1.19).

For general $m_0 \in \mathcal{P}(\mathbb{T}^d)$, we can proceed by approximation. Given a sequence of smooth functions m_{0n} converging to m_0 in $\mathcal{P}(\mathbb{T}^d)$, the corresponding solutions u_n will satisfy (1.26) uniformly thanks to (1.20)–(1.21). Using as before the parabolic regularity one gets that Du_n is relatively compact in $C^0([a, b] \times \mathbb{T}^d)$ for all compact subsets $[a, b] \subset (0, T)$. Hence $H_p(t, x, Du_n)$ converges in $L^p(Q_T)$ for every $p < \infty$. By standard stability of Fokker-Planck equations, this implies the compactness of m_n in $C^0([0, T]; \mathcal{P}(\mathbb{T}^d))$. In particular, we deduce both the initial and the terminal condition (due to the continuity of G). Finally, the limit couple (u, m) satisfies $u \in L^\infty(0, T; W^{1,\infty}(\mathbb{T}^d))$, $m \in C^0([0, T]; \mathcal{P}(\mathbb{T}^d))$ and is a solution of (1.19). In fact, by the parabolic regularity recalled before, this solution satisfies $m, Du \in C^\alpha(Q_T)$ for some $\alpha \in (0, 1)$. If F is a bounded map in the space of Hölder continuous functions, then we bootstrap the regularity once more. We have that $F[m]$ is Hölder continuous, so is $H(t, x, Du)$ and therefore by Schauder regularity (see e.g. [142, Chapter IV]) u belongs to $C^{1+\frac{\alpha}{2}, 2+\alpha}(Q_T)$ for some $\alpha \in (0, 1)$. If H_p is C^1, this implies that $div(H_p(t, x, Du))$ is Hölder continuous as well, and we conclude that m is also a classical solution in $(0, T)$.

Uniqueness: Let (u_1, m_1) and (u_2, m_2) be solutions of (1.19) such that $u_i \in L^\infty(0, T; W^{1,\infty}(\mathbb{T}^d)) \cap C(\overline{Q_T})$ and $m_i \in C^0([0, T]; \mathcal{P}(\mathbb{T}^d))$. As we already used above, the m_i are locally bounded and Hölder continuous; therefore, $m_1 - m_2$ can be justified as test function in the equation of $u_1 - u_2$ (and viceversa) in any interval (a, b) compactly contained in $(0, T)$. It follows that

$$-\frac{d}{dt} \int_{\mathbb{T}^d} (u_1 - u_2)(m_1 - m_2) = \int_{\mathbb{T}^d} (F[m_1] - F[m_2])(m_1 - m_2)$$

$$+ \int_{\mathbb{T}^d} m_1 \left\{ H(t, x, Du_2) - H(t, x, Du_1) - H_p(t, x, Du_1)(Du_2 - Du_1) \right\}$$

$$+ \int_{\mathbb{T}^d} m_2 \left\{ H(t, x, Du_1) - H(t, x, Du_2) - H_p(t, x, Du_2)(Du_1 - Du_2) \right\}$$

$$(1.28)$$

where the equality is meant in the weak sense in $(0, T)$. By convexity of H and monotonicity of F, it follows that $\int_{\mathbb{T}^d} (u_1 - u_2)(m_1 - m_2)$ is non increasing in time. Moreover, this quantity is continuous in $[0, T]$ because $u_i \in C(\overline{Q_T})$ and $m_i \in C^0([0, T]; \mathcal{P}(\mathbb{T}^d))$. By monotonicity of G, this quantity is nonnegative at $t = T$, however it vanishes for $t = 0$. We deduce that it vanishes for all $t \in [0, T]$. In particular, the previous equality implies that all terms in the right-hand side are equal to zero. If condition (1.24) holds true, this implies that $F[m_1] = F[m_2]$ and $G[m_1(T)] = G[m_2(T)]$; hence, by uniqueness of the parabolic equation (namely, by maximum principle), we deduce that $u_1 = u_2$. This implies $H_p(t, x, Du_1) = H_p(t, x, Du_2)$, and for the Fokker-Planck equation this implies that $m_1 = m_2$. Indeed, given a bounded drift $b \in L^\infty(Q_T)$, one can easily verify with a duality argument that if $\mu \in C^0([0, T]; \mathcal{P}(\mathbb{T}^d))$ is a weak solution of the equation $\partial_t \mu - \Delta \mu - div(b \mu) = 0$ and $\mu(0) = 0$, then $\mu \equiv 0$. Alternatively, if (1.25) holds true, then we first obtain that $H_p(t, x, Du_1) = H_p(t, x, Du_2)$, hence we deduce that $m_1 = m_2$ and we conclude by uniqueness of u.   □

Remark 1.4 Uniqueness of the solution of (1.19) is not expected in general if Lasry-Lions' monotonicity condition fails. This lack of uniqueness is well-documented in the literature: see for instance [25, 40, 68, 149]. By contrast, it is relatively easy to check that uniqueness holds if the horizon is "short" or if the functions H and G do not "depend too much" on m, see e.g. [25].

The existence part of the above result can easily be extended to more general MFG systems, in which the Hamiltonian has no separate structure:

$$\begin{cases} -\partial_t u - \varepsilon \Delta u + H(t, x, Du, m(t)) = 0 & \text{in } (0, T) \times \mathbb{T}^d \\ \partial_t m - \varepsilon \Delta m - div\left(m\, H_p(t, x, Du, m(t))\right) = 0 & \text{in } (0, T) \times \mathbb{T}^d \\ m(0) = m_0, \ u(T, x) = G[m(T)] & \text{in } \mathbb{T}^d \end{cases} \quad (1.29)$$

The notion of classical solution is given as before. A general existence result in this direction sounds as follows.

Theorem 1.5 *Assume that $H : Q_T \times \mathbb{R}^d \times \mathcal{P}(\mathbb{T}^d) \to \mathbb{R}$ is a continuous function, differentiable with respect to p, and such that both H and H_p are C^1 continuous on $Q_T \times \mathbb{R}^d \times \mathcal{P}(\mathbb{T}^d)$, and in addition H satisfies the growth condition*

$$H_x(t, x, p, m) \cdot p \geq -C_0(1+|p|^2) \qquad \forall (t, x, p, m) \in Q_T \times \mathbb{R}^d \times \mathcal{P}(\mathbb{T}^d) \tag{1.30}$$

for some constant $C_0 > 0$. Assume that G satisfies (1.21) and that $m_0 \in \mathcal{P}(\mathbb{T}^d)$. Then there is at least one classical solution to (1.29).

Remark 1.5 Of course, the solution found in the above Theorem is smooth up to $t = 0$ (respectively $t = T$) if, for some $\beta \in (0, 1)$, $m_0 \in C^{2+\beta}(\mathbb{T}^d)$ (respectively, $G[m]$ is bounded in $C^{2+\beta}(\mathbb{T}^d)$ uniformly with respect to $m \in \mathcal{P}(\mathbb{T}^d)$).

The proof is relatively easy and relies on gradient estimates for Hamilton-Jacobi equations (as already used in Theorem 1.4 above) and on the following estimate on the McKean-Vlasov equation

$$\begin{cases} \partial_t m - \varepsilon \Delta m - \operatorname{div}(m \, b(t, x, m(t))) = 0 & \text{in } (0, T) \times \mathbb{T}^d \\ m(0) = m_0 \end{cases} \tag{1.31}$$

To this purpose, it is convenient to introduce the following stochastic differential equation (SDE)

$$\begin{cases} dX_t = b(t, X_t, \mathcal{L}(X_t))dt + \sqrt{2\varepsilon} \, dB_t, & t \in [0, T] \\ X_0 = Z_0 \end{cases} \tag{1.32}$$

where (B_t) is a standard d-dimensional Brownian motion over some probability space $(\Omega, \mathcal{A}, \mathbb{P})$ and where the initial condition $Z_0 \in L^1(\Omega)$ is random and independent of (B_t).

We assume that the vector field $b : [0, T] \times \mathbb{T}^d \times \mathcal{P}(\mathbb{T}^d) \to \mathbb{R}^d$ is continuous in time and Lipschitz continuous in (x, m) uniformly in t. Under the above condition on b, we have proved in Sect. 1.2.3.2 that there is a unique solution to (1.32). This solution is closely related to equation (1.31).

Lemma 1.6 *Under the above condition on b, if $\mathcal{L}(Z_0) = m_0$, then $m(t) := \mathcal{L}(X_t)$ is a weak solution of (1.31) and satisfies*

$$\mathbf{d}_1(m(t), m(s)) \leq c_0(1 + \|b\|_\infty)|t - s|^{\frac{1}{2}} \qquad \forall s, t \in [0, T] \tag{1.33}$$

for some constant $c_0 = c_0(T)$ independent of $\varepsilon \in (0, 1]$.

Proof The fact that $m(t) := \mathcal{L}(X_t)$ is a weak solution of (1.31) is a straightforward consequence of Itô's formula: if $\varphi : Q_T \to \mathbb{R}$ is smooth, then

$$\varphi(t, X_t) = \varphi(0, Z_0)$$
$$+ \int_0^t [\varphi_t(s, X_s) + D\varphi(s, X_s) \cdot b(s, X_s, m(s)) + \varepsilon \Delta \varphi(s, X_s)] \, ds$$
$$+ \int_0^t D\varphi(s, X_s) \cdot dB_s.$$

Taking the expectation on both sides of the equality, we have, since $\mathbb{E}\left[\int_0^t D\varphi(s, X_s) \cdot dB_s\right] = 0$,

$$\mathbb{E}[\varphi(t, X_t)] = \mathbb{E}\left[\varphi(0, Z_0) + \int_0^t [\varphi_t(s, X_s) + D\varphi(s, X_s) \cdot b(s, X_s, m(s))\right.$$
$$\left. + \varepsilon \Delta \varphi(s, X_s)] \, ds\right].$$

So by definition of $m(t)$, we get

$$\int_{\mathbb{R}^d} \varphi(t, x) m(t, dx)$$
$$= \int_{\mathbb{R}^d} \varphi(0, x) m_0(dx)$$
$$+ \int_0^t \int_{\mathbb{R}^d} [\varphi_t(s, x) + D\varphi(s, x) \cdot b(s, x, m(s)) + \varepsilon \Delta \varphi(s, x)] m(s, dx) \, ds$$

hence m is a weak solution to (1.31), provided we check that m is continuous in time. This is the aim of the next estimate. Let $\phi : \mathbb{T}^d \to \mathbb{R}$ be 1-Lipschitz continuous and take, for instance, $s < t$. Then, using (1.8), we have

$$\mathbf{d}_1(m(t), m(s)) \leq \mathbb{E}[|X_t - X_s|] \leq \mathbb{E}\left[\int_s^t |b(\tau, X_\tau, m(\tau))| \, d\tau + \sqrt{2\varepsilon} \, |B_t - B_s|\right]$$
$$\leq \|b\|_\infty (t - s) + \sqrt{2\varepsilon(t - s)} \qquad (1.34)$$

which yields (1.33).   □

Remark 1.6 Observe that not only the estimate (1.33) is independent of the diffusion coefficient ε, but actually the precise form (1.34) shows that, when $\varepsilon \to 0$, the map $m(t)$ becomes Lipschitz in the time variable.

We further notice that an estimate also follows, similarly as in (1.34), for the Wasserstein distance. Indeed, recalling that $\mathbf{d}_2(m_1, m_2) = \inf_{X,Y} \left(\mathbb{E}[|X - Y|^2]\right)^{1/2}$, proceeding similarly as in (1.34) yields

$$\mathbf{d}_2(m(t), m(s)) \leq \sqrt{|t-s|}\, \mathbb{E}\left[\int_s^t |b(\tau, X_\tau, m(\tau))|^2 \, d\tau\right]^{\frac{1}{2}} + o(1) \quad \text{as } \varepsilon \to 0$$

$$\leq \sqrt{|t-s|} \left(\int_0^T \int_{\mathbb{T}^d} |b|^2 m \, dx dt\right)^{\frac{1}{2}} + o(1) \quad \text{as } \varepsilon \to 0.$$

We prove now Theorem 1.5.

Proof of Theorem 1.5 For a large constant C_1 to be chosen below, let $\mathcal{C}$ be the set of maps $\mu \in C^0([0, T], \mathcal{P}(\mathbb{T}^d))$ such that

$$\sup_{s \neq t} \frac{\mathbf{d}_1(\mu(s), \mu(t))}{|t-s|^{\frac{1}{2}}} \leq C_1. \tag{1.35}$$

Then $\mathcal{C}$ is a closed convex subset of $C^0([0, T], \mathcal{P}(\mathbb{T}^d))$. It is actually compact thanks to Ascoli's Theorem and the compactness of the set $\mathcal{P}(\mathbb{T}^d)$. To any $\mu \in \mathcal{C}$ we associate $m = \Psi(\mu) \in \mathcal{C}$ in the following way. Let u be the solution to

$$\begin{cases} -\partial_t u - \varepsilon \Delta u + H(t, x, Du, \mu(t)) = 0 & \text{in } (0, T) \times \mathbb{T}^d \\ u(T) = G[\mu(T)] & \text{in } \mathbb{T}^d \end{cases} \tag{1.36}$$

Then we define $m = \Psi(\mu)$ as the solution of the Fokker-Planck equation

$$\begin{cases} \partial_t m - \varepsilon \Delta m - \operatorname{div}\left(m\, H_p(t, x, Du, \mu(t))\right) = 0 & \text{in } (0, T) \times \mathbb{T}^d \\ m(0) = m_0 & \text{in } \mathbb{T}^d. \end{cases} \tag{1.37}$$

Let us check that Ψ is well-defined and continuous. We start by assuming that H is globally Lipschitz continuous. Then, by standard parabolic theory (see e.g. [142, Chapter V, Thm 6.1]), equation (1.36) has a unique classical solution u. Moreover, u is of class $C^{1+\alpha/2, 2+\alpha}(Q_T)$ where the constant α do not depend on μ. In addition, the Bernstein gradient estimate (1.27) holds exactly as in Theorem 1.4, which means that

$$\|Du\|_\infty \leq K$$

for some constant K only depending on T, $\|DG[\mu]\|_\infty$ and on the constant C_0 in (1.30). Due to (1.21), the constant K is therefore independent of μ. We see now that the global Lipschitz condition on H can be dropped: indeed, it is enough to replace $H(t, x, p, m)$ with $\tilde{H}(t, x, p, m) = \zeta(p) H(t, x, p, m) + (1 - \zeta(p))|p|$ where $\zeta : \mathbb{R}^d \to [0, 1]$ is a smooth function such that $\zeta(p) \equiv 1$ for $|p| \leq 2K$ and $\zeta(p) \equiv 0$ for $|p| > 2K + 1$. Thanks to the gradient estimate, solving the problem for $\tilde{H}$ is the same as for H.

Next we turn to the Fokker-Planck equation (1.37). Since the drift $b(t, x) :=$ $-H_p(t, x, Du(x), \mu(t))$ belongs to $L^\infty(Q_T)$, there is a unique solution $m \in C^0([0, T]; \mathcal{P}(\mathbb{T}^d))$; moreover, since b is bounded independently of μ, say by a constant C_2, from Lemma 1.6 we have the following estimates on m:

$$\mathbf{d}_1(m(t), m(s)) \le c_0(1 + \|H_p(\cdot, Du, \mu)\|_\infty)|t - s|^{\frac{1}{2}} \le c_0(1 + C_2)|t - s|^{\frac{1}{2}} \qquad \forall s, t \in [0, T].$$

So if we choose C_1 so large that $C_1 \ge c_0(1 + C_2)$, then m belongs to $\mathcal{C}$. Moreover, if we write the equation in the form

$$\partial_t m - \varepsilon \Delta m - Dm \cdot H_p(t, x, Du, \mu(t)) - m \operatorname{div} H_p(t, x, Du, \mu(t)) = 0$$

then we observe that m is a classical solution in $(0, T)$. Indeed, since $u \in C^{1+\alpha/2, 2+\alpha}(Q_T)$ and $t \to \mu(t)$ is Holder continuous, the maps $(t, x) \to H_p(t, x, Du, \mu(t))$ and $(t, x) \to \operatorname{div} H_p(t, x, Du, \mu(t))$ belong to $C^\alpha(Q_T)$, so that the solution m belongs to $C^{1+\alpha/2, 2+\alpha}(Q_T)$ [142, Chapter IV, Thm 10.1].

We have just proved that the mapping $\Psi : \mu \to m = \Psi(\mu)$ is well-defined from $\mathcal{C}$ into itself. The continuity of Ψ can be proved exactly as in Theorem 1.4. We conclude by Schauder fixed point Theorem that the continuous map $\mu \to m = \Psi(\mu)$ has a fixed point $\bar{\mu}$ in $\mathcal{C}$. Let $\bar{u}$ be associated to $\bar{\mu}$ as above. Then $(\bar{u}, \bar{\mu})$ is a solution of our system (1.29). In addition $(\bar{u}, \bar{\mu})$ is a classical solution in view of the above estimates.   □

Let us mention that there are no general criteria for the uniqueness of solutions to (1.29) in arbitrary time horizon T, except for the Lasry-Lions' monotonicity condition (1.24) for the case of separate Hamiltonian treated in Theorem 1.4. In case of local dependence of $H(t, x, p, m)$ with respect to the density m of the measure, a structure condition on H ensuring the uniqueness was given by P.-L. Lions in [149] (Lessons 5-12/11 2010) and will be discussed later in Theorem 1.13, in the subsection devoted to local couplings.

Otherwise, the uniqueness of solutions to (1.29) can be proved for short time horizon, e.g. using directly the Banach fixed point theorem for contraction mappings, in order to produce both existence and uniqueness of solutions (as in the papers by Caines, Huang and Malhamé [132], under a smallness assumption on the coefficients or on the time interval).

1.3.3 Application to Games with Finitely Many Players

In this subsection, we show how to apply the previous results on the MFG system to study a N-player differential game in which the number N of players is "large".

1.3.3.1 The N-Player Game

The dynamic of player i (where $i \in \{1, \ldots, N\}$) is given by

$$dX_t^i = b^i(X_t^1, \ldots, X_t^N, \alpha_t^1, \ldots, \alpha_t^N)dt + \sqrt{2}dB_t^i, \qquad X_0^i = Z^i,$$

where (B_t^i) is a d-dimensional Brownian motion.[1] The initial condition X_0^i for this system is also random and has for law $\tilde{m}_0 \in \mathcal{P}_1(\mathbb{R}^d)$, and we assume that all Z^i and all the Brownian motions (B_t^i) $(i = 1, \ldots, N)$ are independent. Player i can choose his bounded control α^i with values in $\mathbb{R}^d$ and adapted to the filtration $(\mathcal{F}_t = \sigma\{X_0^j, B_s^j, s \leq t, j = 1, \ldots, N\})$. We make the structure assumption that the drift b^i of player i depends only on his/her own control and position and on the distribution of the other players. Namely:

$$b^i(x^1, \ldots, x^N, \alpha^1, \ldots, \alpha^N) = b(x^i, \alpha^i, \pi \sharp m_{\mathbf{x}}^{N,i}),$$

$$\text{where } \mathbf{x} = (x^1, \ldots, x^N) \in (\mathbb{R}^d)^N \text{ and } m_{\mathbf{x}}^{N,i} = \sum_{j \neq i} \delta_{x^j},$$

where $b : \mathbb{T}^d \times \mathbb{R}^d \times \mathcal{P}(\mathbb{T}^d) \to \mathbb{R}^d$ is a globally Lipschitz continuous map. We have denoted by $\pi : \mathbb{R}^d \to \mathbb{T}^d$ the canonical projection and by $\pi \sharp \tilde{m}_0$ the image of the measure $\tilde{m}_0$ by the map π. The fact that the players interact through the projection over $\mathbb{T}^d$ of the empirical measure $m_{\mathbf{x}}^{N,i}$ is only a simplifying assumption related to the fact that we have so far led our analysis of the MFG system on the torus. Indeed, here we systematically see maps defined on $\mathbb{T}^d$ as $\mathbb{Z}^d$-periodic maps on $\mathbb{R}^d$.

The cost of player i is then given by

$$\mathcal{J}_i^N(\alpha^1, \ldots, \alpha^N) = \mathcal{J}_i^N(\alpha^i, (\alpha^j)_{j \neq i})$$

$$= \mathbb{E}\left[\int_0^T L^i(X_t^1, \ldots, X_t^N, \alpha_t^1, \ldots, \alpha_t^N)dt + G^i(X_T^1, \ldots, X_T^N)\right].$$

Here again we make the structure assumption that the running cost L^i of player i depends only on his/her own control and position and on the distribution of the other players' positions, while the terminal cost depends only on his/her position and on the distribution of the other players' positions:

$$L^i(x^1, \ldots, x^N, \alpha^1, \ldots, \alpha^N) = L(x^i, \alpha^i, \pi \sharp m_{\mathbf{x}}^{N,i}), \qquad G^i(x^1, \ldots, x^N) = G(x^i, \pi \sharp m_{\mathbf{x}}^{N,i}),$$

---

[1] In order to avoid the (possible but) cumbersome definition of stochastic processes on the torus but, at the same time, be able to use the results of the previous parts, we work here with diffusions in $\mathbb{R}^d$ with periodic coefficients and assume that the mean field dependence of the data is always through the projection of the measures over the torus.

where $L : \mathbb{T}^d \times \mathbb{R}^d \times \mathcal{P}(\mathbb{T}^d) \to \mathbb{R}$ and $G : \mathbb{T}^d \times \mathcal{P}(\mathbb{T}^d) \to \mathbb{R}$ are continuous maps, with

$$|L(x, \alpha, m) - L(x', \alpha, m')| + |G(x, m) - G(x', m')| \le K(|x - x'| + \mathbf{d}_1(m, m')),$$

the constant K being independent of α. In addition, we need a coercivity assumption on L with respect to α:

$$L(x, \alpha, m) \ge C^{-1}|\alpha| - C, \tag{1.38}$$

where C is independent of $(x, m) \in \mathbb{T}^d \times \mathcal{P}(\mathbb{T}^d)$. These assumptions are a little strong in practice, but allow us to avoid several (very) technical points in the proofs.

In that setting, a natural notion of equilibrium is the following.

Definition 1.2 We say that a family $(\bar{\alpha}^1, \ldots, \bar{\alpha}^N)$ of bounded open-loop controls is an ε-Nash equilibrium of the N-player game (where $\varepsilon > 0$) if, for any $i \in \{1, \ldots, N\}$ and any bounded open-loop control α^i,

$$\mathcal{J}_i^N(\bar{\alpha}^i, (\bar{\alpha}^j)_{j \ne i}) \le \mathcal{J}_i^N(\alpha^i, (\bar{\alpha}^j)_{j \ne i}) + \varepsilon.$$

1.3.3.2 The MFG System and the N-Player Game

Our aim is to understand to what extent the MFG system can provide an ε-Nash equilibrium of the N-player game, at least if N is large enough. For this, we set

$$H(x, p, m) = \sup_{\alpha \in \mathbb{R}^d} \{-b(x, \alpha, m) \cdot p - L(x, \alpha, m)\}$$

and we assume that H, G and $m_0 := \pi \sharp \tilde{m}_0$ satisfy the assumptions of Theorem 1.5.

Hereafter, we fix (u, m) a classical solution to (1.29) (here with $\varepsilon = 1$). Following the arguments of Sect. 1.3.1, we recall the interpretation of the MFG system. In the mean-field approach, a generic player controls the solution to the SDE

$$X_t = X_0 + \int_0^t b(X_s, \alpha_s, m(s))ds + \sqrt{2}B_s,$$

and faces the minimization problem

$$\inf_\alpha \mathcal{J}(\alpha) \quad \text{where} \quad \mathcal{J}(\alpha) = \mathbb{E}\left[\int_0^T L(X_s, \alpha_s, m(s)) \, ds + G(X_T, m(T))\right].$$

In the above dynamics we assume that X_0 is a fixed random initial condition with law $m_0 \in \mathcal{P}(\mathbb{T}^d)$ and the control α is adapted to some filtration $(\mathcal{F}_t)$. We also assume that (B_t) is a d-dimensional Brownian motion adapted to $(\mathcal{F}_t)$ and that X_0 and (B_t) are independent.

Then, given a solution (u, m) of (1.29), $u(0)$ is the optimal value and the feedback strategy α^* such that $b(x, \alpha^*(t, x), m(t)) := -H_p(x, Du(t, x), m(t))$ is optimal for the single player. Namely:

Lemma 1.7 *Let $(\bar{X}_t)$ be the solution of the stochastic differential equation*

$$\begin{cases} d\bar{X}_t = b(\bar{X}_t, \alpha^*(t, \bar{X}_t), m(t))dt + \sqrt{2}dB_t \\ \bar{X}_0 = X_0 \end{cases}$$

and set $\bar{\alpha}_t = \alpha^*(t, \bar{X}_t)$. Then

$$\inf_\alpha \mathcal{J}(\alpha) = \mathcal{J}(\bar{\alpha}) = \int_{\mathbb{T}^d} u(0, x)\, m_0(dx) .$$

Our goal now is to show that the strategy given by the mean field game is almost optimal for the N-player problem. We assume that the feedback $\alpha^*(t, x)$ defined above is continuous in (t, x) and globally Lipschitz continuous in x uniformly in t. With the feedback strategy α^* one can associate the open-loop control $\bar{\alpha}^i$ obtained by solving the system of SDEs:

$$d\bar{X}_t^i = b(\bar{X}_t^i, \alpha^*(t, \bar{X}_t^i), m_{\mathbf{X}_t}^{N,i})dt + \sqrt{2}dB_t^i, \; X_0^i = Z^i \qquad (\text{where } \mathbf{X}_t = (X_t^1, \ldots, X_t^N)),$$

$$(1.39)$$

and setting $\bar{\alpha}_t^i = \alpha^*(t, \bar{X}_t^i)$. We are going to show that the controls $\bar{\alpha}^i$ realize an approximate Nash equilibrium for the N-player game.

Theorem 1.6 *Assume that (Z^i) are i.i.d. random variables on $\mathbb{R}^d$ such that $\mathbb{E}[|Z^1|^q] < +\infty$ for some $q > 4$. There exists a constant $C > 0$ such that, for any $N \in \mathbb{N}^*$, the symmetric strategy $(\bar{\alpha}^1, \ldots, \bar{\alpha}^N)$ is a $C\varepsilon_N$-Nash equilibrium in the game $\mathcal{J}_1^N, \ldots, \mathcal{J}_N^N$ where*

$$\varepsilon_N := \begin{cases} N^{-1/2} & \text{if } d < 4 \\ N^{-1/2}\ln(N) & \text{if } d = 4 \\ N^{-2/d} & \text{if } d > 4 \end{cases}$$

Namely, for any $i \in \{1, \ldots, N\}$ and for any control α^i adapted to the filtration $(\mathcal{F}_t)$,

$$\mathcal{J}_i^N(\bar{\alpha}^i, (\bar{\alpha}^j)_{j\neq i}) \leq \mathcal{J}_i^N(\alpha^i, (\bar{\alpha}^j)_{j\neq i}) + C\varepsilon_N.$$

The Lipschitz continuity assumption on H and G with respect to m allows us to quantify the error. If H and G are just continuous with respect to m, one can only say that, for any $\varepsilon > 0$, there exists N_0 such that the symmetric strategy $(\bar{\alpha}^1, \ldots, \bar{\alpha}^N)$ is an ε-Nash equilibrium in the game $\mathcal{J}_1^N, \ldots, \mathcal{J}_N^N$ for any $N \geq N_0$.

Before starting the proof, we need the following result on product measures which can be found in [68] (Theorem 5.8. See also the references therein):

Lemma 1.8 *Assume that* (Z_i) *are i.i.d. random variables on* $\mathbb{R}^d$ *of law* μ *such that* $\mathbb{E}[|Z_0|^q] < +\infty$ *for some* $q > 4$. *Then there is a constant* C, *depending only on* d, q *and* $\mathbb{E}[|Z_0|^q]$, *such that*

$$\mathbb{E}\left[\mathbf{d}_2(m_Z^N, \mu)\right] \le \begin{cases} CN^{-1/2} & \text{if } d < 4 \\ CN^{-1/2} \ln(N) & \text{if } d = 4 \\ CN^{-2/d} & \text{if } d > 4 \end{cases}$$

Proof of Theorem 1.6 From the symmetry of the problem, it is enough to show that

$$\mathcal{J}_1^N(\bar\alpha^1, (\bar\alpha^j)_{j\ge2}) \le \mathcal{J}_1^N(\alpha^1, (\bar\alpha^j)_{j\ne2}) + C\varepsilon_N \tag{1.40}$$

for any control α, as soon as N is large enough. We note for this that the map $\tilde b(t, x, m) := b(x, \alpha^*(t, x), \pi \sharp m)$ is globally Lipschitz continuous in (x, m) uniformly in t thanks to our assumptions on b and α^*. Following the proof of Theorem 1.3 in Sect. 1.2.3, we have therefore that

$$\mathbb{E}\left[\sup_{t\in[0,T]} \frac{1}{N}\sum_{i=1}^{N}|X_t^i - \bar X_t^i|\right] + \mathbb{E}\left[\sup_{t\in[0,T]} |X_t^i - \bar X_t^i|\right] \le C_T \int_0^T \mathbb{E}\left[\mathbf{d}_1(m_{\bar X_s}^N, m(s))ds\right], \tag{1.41}$$

where $\bar X_t = (\bar X^1, \dots, \bar X^N)$ solves

$$d\bar X_t^i = b(\bar X_t^i, \alpha^*(t, \bar X_t^i), \pi \sharp \mathcal{L}(\bar X_t^i))dt + \sqrt{2}dB_t^i, \qquad \bar X_0^i = Z^i.$$

By uniqueness in law of the solution of the McKean-Vlasov equation, we have that the $\bar X_s^i$ are i.i.d. with a law $\tilde m(s)$, where $\tilde m(s)$ solves

$$\partial_t \tilde m - \Delta \tilde m - \text{div}(\tilde m b(x, \alpha^*(t, x), \pi\sharp\tilde m(t))) = 0, \qquad \tilde m(0) = \tilde m_0.$$

In view of the assumption on b, it is easy to check that

$$\mathbb{E}[\sup_{t\in[0,T]} |\bar X_t^1|^q] \le C(1 + \mathbb{E}[|Z^1|^q]) < +\infty.$$

Therefore using Lemma 1.8, we have

$$\mathbb{E}\left[\sup_{t\in[0,T]} \mathbf{d}_1(m_{\bar X_t}^{N,1}, \tilde m(t))\right] \le \mathbb{E}\left[\sup_{t\in[0,T]} \mathbf{d}_2(m_{\bar X_t}^{N,1}, \tilde m(t))\right] \le C\varepsilon_N.$$

Note also that, by the uniqueness of the solution of the McKean-Vlasov equation (this time in $\mathbb{T}^d$), we have that $\pi \sharp \tilde{m} = m$ since both flows solve the same equation. Hence

$$\mathbb{E}\left[\sup_{t \in [0,T]} \mathbf{d}_1(\pi \sharp m_{\bar{X}_t}^{N,1}, m(t))\right] \leq \mathbb{E}\left[\sup_{t \in [0,T]} \mathbf{d}_1(m_{\bar{X}_t}^{N,1}, \tilde{m}(t))\right] \leq C\varepsilon_N.$$

Using (1.41) we obtain therefore

$$\mathbb{E}\left[\sup_{t \in [0,T]} \frac{1}{N}\sum_{i=1}^N |X_t^i - \bar{X}_t^i|\right] + \mathbb{E}\left[\sup_{t \in [0,T]} |X_t^i - \bar{X}_t^i|\right] \leq C\varepsilon_N.$$

In particular, by Lemma 1.7 and the local Lipschitz continuity of L and G, we get

$$\mathcal{J}_1^N(\bar{\alpha}^1, (\bar{\alpha}^j)_{j \geq 2}) = \mathbb{E}\left[\int_0^T L(X_s^1, \bar{\alpha}_s^1, \pi \sharp m_{X_s}^{N,i}) \, ds + G(X_T^1, \pi \sharp m_{X_T}^{N,i})\right]$$

$$\leq \mathbb{E}\left[\int_0^T L(\bar{X}_s^1, \bar{\alpha}_s^1, m(s)) \, ds + G(\bar{X}_T^1, m(T))\right] + C\varepsilon_N$$

$$\leq \int_{\mathbb{T}^d} u(0, x)m_0(dx)dx + C\varepsilon_N. \tag{1.42}$$

Let now α^1 be a bounded control adapted to the filtration $(\mathcal{F}_t)$ and (Y_t^i) be the solution to

$$dY_t^1 = b(Y_t^1, \alpha_t^1, m_{\mathbf{Y_t}}^{N,1})dt + \sqrt{2}dB_t^1, \ Y_0^1 = Z^1,$$

and

$$dY_t^i = b(Y_t^i, \bar{\alpha}_t^i, m_{\mathbf{Y_t}}^{N,i})dt + \sqrt{2}dB_t^i, \ Y_0^i = Z^i.$$

We first note that we can restrict our analysis to the case where $\mathbb{E}[\int_0^T |\alpha_s^1|ds] \leq A$, for A large enough. Indeed, if $\mathbb{E}[\int_0^T |\alpha_s^1|ds] > A$, we have by assumption (1.38) and inequality (1.42), as soon as A is large enough (independent of N) and N is large enough:

$$\mathcal{J}_1^N(\alpha^1, (\bar{\alpha}^j)_{j \geq 2}) \geq C^{-1}\mathbb{E}\left[\int_0^T |\alpha_s^1|ds\right] - CT \geq C^{-1}A - CT$$

$$\geq \int_{\mathbb{T}^d} u(0, x)m_0(dx)dx + 1 \geq \mathcal{J}_1^N(\bar{\alpha}^1, (\bar{\alpha}^j)_{j \geq 2}) + 1 - C\varepsilon_N$$

$$\geq \mathcal{J}_1^N(\bar{\alpha}^1, (\bar{\alpha}^j)_{j \geq 2}) - (C/2)\varepsilon_N.$$

From now on we assume that $\int_0^T |\alpha_s^1| ds \leq A$. Let us first estimate $\mathbf{d}_1(m_{\mathbf{Y}_s}^{N,1}, m_{\mathbf{X}_s}^{N,1}))$. Note that we have, by Lipschitz continuity of b,

$$|Y_t^1 - X_t^1| \leq C \int_0^t (|Y_s^1 - X_s^1| + |\alpha_s^1 - \bar{\alpha}_s^1| + \mathbf{d}_1(m_{\mathbf{Y}_s}^{N,1}, m_{\mathbf{X}_s}^{N,1}))ds$$

$$\leq C \int_0^t (|Y_s^1 - X_s^1| + |\alpha_s^1 - \bar{\alpha}_s^1| + \frac{1}{N-1} \sum_{j=2}^N |Y_s^j - X_s^j|)ds$$

while for $i \in \{2, \ldots, N\}$ we have, arguing in the same way,

$$|Y_t^i - X_t^i| \leq C \int_0^t (|Y_s^i - X_s^i| + \frac{1}{N-1} \sum_{j \neq i}^N |Y_s^j - X_s^j|)ds.$$

So

$$\frac{1}{N} \sum_{i=1}^N |Y_t^i - X_t^i| \leq C \int_0^t (\frac{1}{N} |\alpha_s^1 - \bar{\alpha}_s^1| + \frac{1}{N} \sum_{i=1}^N |Y_s^i - X_s^i|)ds.$$

By Gronwall lemma we obtain therefore

$$\frac{1}{N} \sum_{i=1}^N |Y_t^i - X_t^i| \leq \frac{C}{N} \int_0^t |\alpha_s^1 - \bar{\alpha}_s^1| ds.$$

So

$$\sup_{t \in [0,T]} \mathbf{d}_1(m_{\mathbf{Y}_s}^{N,1}, m_{\mathbf{X}_s}^{N,1})) \leq \sup_{t \in [0,T]} \frac{1}{N-1} \sum_{i=2}^N |Y_t^i - X_t^i|$$

$$\leq \sup_{t \in [0,T]} \frac{1}{N-1} \sum_{i=1}^N |Y_t^i - X_t^i| \leq \frac{C}{N} \int_0^T |\alpha_s^1 - \bar{\alpha}_s^1| ds.$$

As $\mathbb{E}[\int_0^T |\alpha_s^1 - \bar{\alpha}_s^1| ds] \leq A$, this shows that

$$\mathbb{E}\left[\sup_{t \in [0,T]} \mathbf{d}_1(\pi \sharp m_{\mathbf{Y}_s}^{N,1}, m(s)) \right] \leq \mathbb{E}\left[\sup_{t \in [0,T]} \mathbf{d}_1(m_{\mathbf{Y}_s}^{N,1}, m_{\mathbf{X}_s}^{N,1}) \right]$$

$$+ \left[\sup_{t \in [0,T]} \mathbf{d}_1(\pi \sharp m_{\mathbf{X}_s}^{N,1}, m(s)) \right]$$

$$\leq \frac{CA}{N} + C\varepsilon_N \leq C_A \varepsilon_N,$$

where C_A depends also on A. Therefore, using again the Lipschitz continuity of L and G with respect to m, we get

$$
\begin{aligned}
\mathcal{J}_1^N(\alpha^1, (\bar{\alpha}^j)_{j\neq 2}) &= \mathbb{E}\left[\int_0^T L(Y_s^1, \alpha_s^1, \pi \sharp m_{\mathbf{Y}_s}^{N,i}) \, ds + G(Y_T^1, \pi \sharp m_{\mathbf{Y}_t}^{N,i})\right] \\
&\geq \mathbb{E}\left[\int_0^T L(Y_s^1, \alpha_s^1, m(s)) \, ds + G(X_T^1, m(T))\right] - C_A \varepsilon_N \\
&\geq \int_{\mathbb{T}^d} u(0, x) m_0(dx) - C_A \varepsilon_N,
\end{aligned}
$$

where the last inequality comes from the optimality of $\bar{\alpha}$ in Lemma 1.7. Recalling (1.42) proves the result. $\square$

We conclude this subsection by recalling that the use of the MFG system to obtain ε-*Nash equilibria* (Theorem 1.6) has been initiated—in a slightly different framework—in a series of papers due to Caines, Huang and Malhamé: see in particular [131] (for linear dynamics) and [132] (for nonlinear dynamics). In these papers, the dependence with respect to the empirical measure of dynamics and payoff occurs through an average, so that the CLT implies that the error term is of order $N^{-1/2}$. The genuinely non linear version of the result given above is a variation on a result by Carmona and Delarue [67] (see also [68], Section 6 in Vol. II). Many variations and extensions of these results have followed since then: we refer to [68] and the references therein.

We discuss below, in Sect. 1.4.4, the reverse statement: to what extent the MFG system pops up as the limit of Nash equilibria. Let us just underline at this stage that this latter problem is much more challenging.

1.3.4 The Vanishing Viscosity Limit and the First Order System with Smoothing Couplings

We now analyze the vanishing viscosity limit for system (1.19) and the corresponding existence and uniqueness of solutions for the deterministic problem

$$
\begin{cases}
-\partial_t u + H(t, x, Du) = F[m] \\
u(T) = G[m(T)] \\
\partial_t m - \text{div}(m \, H_p(t, x, Du)) = 0 \\
m(0) = m_0.
\end{cases}
\tag{1.43}
$$

To this purpose we strengthen the assumptions on F, G, H. Namely, we assume that

$F : C^0([0, T]; \mathcal{P}(\mathbb{T}^d)) \to C^0(\overline{Q_T})$ is continuous with range into a bounded set of

$\quad L^\infty(0, T; W^{2,\infty}(\mathbb{T}^d))$ and F is a bounded map from $C^\alpha([0, T]; \mathcal{P}(\mathbb{T}^d))$

$\quad$ into $C^\alpha(\overline{Q_T})$, for any $\alpha \in (0, 1)$.

$$(1.44)$$

Similarly, we assume that

$G : \mathcal{P}(\mathbb{T}^d) \to C^0(\mathbb{T}^d)$ is continuous with range into a bounded set of $W^{2,\infty}(\mathbb{T}^d)$.
$$(1.45)$$

Moreover we assume that $H \in C^2(\overline{Q_T} \times \mathbb{R}^d)$ and satisfies

$$\exists\, c_0 > 0 : \quad c_0^{-1} I_d \le H_{pp}(t, x, p) \le c_0 I_d \qquad \forall (t, x, p) \in Q_T \times \mathbb{R}^d \quad (1.46)$$

and one between (1.23) or the following condition:

$$\exists\, c_1 > 0 : \quad |H_{xx}(t, x, p)| \le c_1(1 + |p|^2),$$
$$|H_{xp}(t, x, p)| \le c_1(1 + |p|) \qquad \forall (t, x, p) \in Q_T \times \mathbb{R}^d \quad (1.47)$$

Under the above smoothing conditions on the couplings F, G, it will be possible to consider u as a viscosity solution of the Hamilton-Jacobi equation and to make use of several regularity results already known from the standard viscosity solutions' theory. Hence, the notion of solution which is the most suitable here is the following one.

Definition 1.3 A couple (u, m) is a solution to (1.43) if $u \in C^0(\overline{Q_T}) \cap L^\infty(0, T, W^{1,\infty}(\mathbb{T}^d))$, $m \in C^0([0, T]; \mathcal{P}(\mathbb{T}^d))$, u is a viscosity solution of the Hamilton-Jacobi equation, with $u(T) = G[m(T)]$, and m is a distributional solution of the continuity equation such that $m(0) = m_0$.

Assumptions (1.44) and (1.45), together with the uniform convexity of the Hamiltonian ((1.46)), are crucial here in order to guarantee an estimate of semi-concavity for the function u. This is usually a fundamental regularity property of solutions of first order equations, but this is most relevant here because of the properties inherited by the drift term $H_p(t, x, Du)$ in the continuity equation. Let us recall the definition and some properties of semi-concavity. Proofs and references can be found, for instance, in the monograph [48].

Definition 1.4 A map $w : \mathbb{R}^d \to \mathbb{R}$ is semi-concave if there is some $C > 0$ such that one of the following equivalent conditions is satisfied:

1. the map $x \to w(x) - \frac{C}{2}|x|^2$ is concave in $\mathbb{R}^d$,

2. $w(\lambda x + (1-\lambda)y) \geq \lambda w(x) + (1-\lambda)w(y) - C\lambda(1-\lambda)|x-y|^2$ for any $x, y \in \mathbb{R}^d$, $\lambda \in [0, 1]$,
3. $D^2 w \leq C\, I_d$ in the sense of distributions,
4. $(p-q)\cdot(x-y) \leq C|x-y|^2$ for any $x, y \in \mathbb{R}^d$, $p \in D^+ w(x)$ and $q \in D^+ w(y)$, where $D^+ w$ denotes the super-differential of w, namely

$$D^+ w(x) = \left\{ p \in \mathbb{R}^d \; ; \; \limsup_{y \to x} \frac{w(y) - w(x) - (p, y-x)}{|y-x|} \leq 0 \right\}.$$

We will use later the following main consequences of semi-concavity (see e.g. [48]).

Lemma 1.9 *Let* $w : \mathbb{R}^d \to \mathbb{R}$ *be semi-concave. Then* w *is locally Lipschitz continuous in* $\mathbb{R}^d$, *and it is differentiable at* x *if and only if* $D^+ w(x)$ *is a singleton.*

Moreover $D^+ w(x)$ *is the closed convex hull of the set* $D^* w(x)$ *of reachable gradients defined by*

$$D^* w(x) = \left\{ p \in \mathbb{R}^d \; : \; \exists x_n \to x \text{ such that } Dw(x_n) \text{ exists and converges to } p \right\}.$$

In particular, for any $x \in \mathbb{R}^d$, $D^+ w(x)$ *is a compact, convex and non empty subset of* $\mathbb{R}^d$.

Finally, if (w_n) *is a sequence of uniformly semi-concave maps on* $\mathbb{R}^d$ *which pointwisely converges to a map* $w : \mathbb{R}^d \to \mathbb{R}$, *then the convergence is locally uniform,* $Dw_n(x)$ *converges to* $Dw(x)$ *for a.e.* $x \in \mathbb{R}^d$ *and* w *is semi-concave. Moreover, for any* $x_n \to x$ *and any* $p_n \in D^+ w_n(x_n)$, *the set of cluster points of* (p_n) *is contained in* $D^+ w(x)$.

The following theorem is given in [145], and some details also appeared in [55]. Here we give a slightly more general version of the result.

Theorem 1.7 *Let* $m_0 \in L^\infty(\mathbb{T}^d)$. *Assume (1.44)–(1.46) and that at least one between the conditions (1.23) and (1.47) holds true. Let* $(u^\varepsilon, m^\varepsilon)$ *be a solution of (1.19). Then there exists a subsequence, not relabeled, and a couple* $(u, m) \in W^{1,\infty}(Q_T) \times L^\infty(Q_T)$ *such that*

$$u^\varepsilon \to u \qquad \text{in } C(\overline{Q_T}), \qquad m^\varepsilon \to m \qquad \text{in } L^\infty(Q_T) - weak^*,$$

and (u, m) *is a solution of (1.43) in the sense of Definition 1.3.*

Proof

Step 1. (Bounds for $u^\varepsilon, m^\varepsilon$) Let us recall that, by Theorem 1.4, u^ε and m^ε are classical solutions in $(0, T)$. First of all, by maximum principle and (1.44)–(1.45), it follows that u^ε is uniformly bounded in Q_T. The next key point consists in proving a semi concavity estimate for u^ε. To this purpose, let ξ be a direction in $\mathbb{R}^d$. We drop the index ε for simplicity and we set $u_{\xi\xi}(t, x) = D^2 u(t, x)\xi \cdot \xi$.

Then we look at the equation satisfied by $w := u_{\xi\xi} + \lambda(u + M)^2$ where λ, M are positive constants to be fixed later. Straightforward computations give the following:

$$-\partial_t w - \varepsilon \Delta w + H_p(t, x, Du) \cdot Dw + H_{\xi\xi}(t, x, Du)$$

$$+ 2H_{\xi p}(t, x, Du) \cdot Du_\xi + H_{pp}(t, x, Du)Du_\xi \cdot Du_\xi$$

$$= (F[m])_{\xi\xi} - 2\lambda(u + M)(H(t, x, Du) - F[m]) - 2\lambda\varepsilon|Du|^2.$$

We choose $M = \|u\|_\infty + 1$, and we use the coercivity of H which satisfies, from (1.46), $H(t, x, p) \geq \frac{1}{2}c_0^{-1}|p|^2 - c$ for some constant $c > 0$. Therefore we estimate

$$-\partial_t w - \varepsilon \Delta w + H_p(t, x, Du) \cdot Dw + H_{\xi\xi}(t, x, Du) + 2H_{\xi p}(t, x, Du) \cdot Du_\xi$$

$$+ H_{pp}(t, x, Du)Du_\xi \cdot Du_\xi$$

$$\leq (F[m])_{\xi\xi} - \lambda c_0^{-1}|Du|^2 + c\lambda(1 + \|u\|_\infty)(1 + \|F[m]\|_\infty).$$

Now we estimate the terms with the second derivatives of H, using condition (1.46) and one between (1.23) and (1.47). To this purpose, we notice that, if (1.23) holds true, then we already know that Du^ε is uniformly bounded in Q_T, see (1.27) in Theorem 1.4. Then the bounds assumed in (1.47) come for free because H is a C^2 function and the arguments (t, x, Du) live in compact sets. Therefore, we can proceed using (1.47) in both cases. Thanks to Young's inequality, we estimate

$$H_{\xi\xi}(t, x, Du) + 2H_{\xi p}(t, x, Du) \cdot Du_\xi + H_{pp}(t, x, Du)Du_\xi \cdot Du_\xi$$

$$\geq \frac{1}{2}c_0^{-1}|Du_\xi|^2 - c(1 + |Du|^2)$$

hence we deduce that

$$-\partial_t w - \varepsilon \Delta w + H_p(t, x, Du) \cdot Dw + \frac{1}{2}c_0^{-1}|Du_\xi|^2$$

$$\leq c(1 + |Du|^2) + \|D_{xx}^2 F[m]\|_\infty - \lambda c_0^{-1}|Du|^2 + c\lambda(1 + \|F[m]\|_\infty)$$

where we used that $\|u\|_\infty$ is bounded and we denote by c any generic constant independent of ε. The terms given by $F[m]$ are uniformly bounded due to (1.44). Thus, by choosing λ sufficiently large we deduce that, at an internal maximum point (t, x) of w, we have $|Du_\xi|^2 \leq C$ for a constant C independent of ε. Since $|Du_\xi| \geq |u_{\xi\xi}| \geq |w| - c\|u\|_\infty^2$, this gives an upper bound at the maximum point of $w(t, x)$, whenever it is attained for $t < T$. By the way, if the maximum of w is reached at T, then $\max w \leq \|G[m]\|_{W^{2,\infty}} + c\|u\|_\infty^2$. We conclude an estimate for max w, and therefore an upper bound for $u_{\xi\xi}$. The bound being independent of ξ, we have obtained so far that

$$D^2 u^\varepsilon (t, x) \le C \qquad \forall (t, x) \in Q_T$$

for a constant C independent of ε. Since u^ε is $\mathbb{Z}^d$-periodic, this also implies a uniform bound for $\|Du^\varepsilon\|_\infty$.

At this stage, let us observe that the above estimate has been obtained as if $u_{\xi\xi}$ was a smooth function, but this is a minor point: indeed, since $u \in C^{1,2}(Q_T)$ and $H \in C^2$, we have $w \in C^0(Q_T)$ and the above computation shows that $-\partial_t w - \varepsilon \Delta w + H_p(x, Du) \cdot Dw$ is itself a continuous function; so the estimate for w follows applying the maximum principle for continuous solutions.

Now we easily deduce an upper bound on m^ε as well. Indeed, m^ε satisfies, for some constant $K > 0$

$$\partial_t m - \varepsilon \Delta m - Dm \cdot H_p(t, x, Du) = m \operatorname{Tr}\left(H_{pp}(t, x, Du) D^2 u\right) \le K m$$

thanks to the semi concavity estimate and the upper bound of H_{pp} given in (1.46). We deduce that m^ε satisfies

$$\|m^\varepsilon\|_\infty \le e^{Kt} \|m_0\|_\infty . \tag{1.48}$$

Step 2. (Compactness) From the previous step we know that Du^ε is uniformly bounded. Lemma 1.6 then implies that the map $t \to m^\varepsilon(t)$ is Hölder continuous in $\mathcal{P}(\mathbb{T}^d)$, uniformly in ε. This implies that m^ε is relatively compact in $C^0([0, T], \mathcal{P}(\mathbb{T}^d))$. Moreover, from the uniform bound (1.48), m^ε is also relatively compact in the weak$-*$ topology of $L^\infty(Q_T)$. Therefore, up to subsequences m^ε converges in L^∞-weak* and in $C^0([0, T], P(\mathbb{T}^d))$ towards some $m \in L^\infty(Q_T) \cap C^0([0, T], P(\mathbb{T}^d))$. In particular, $m(0) = m_0$. In order to pass to the limit in the equation of m^ε, we observe that the uniform semi concavity bound implies that, up to subsequences, $H_p(t, x, Du^\varepsilon)$ almost everywhere converges towards $H_p(t, x, Du)$ (see Lemma 1.9), and then it converges strongly in $L^1(Q_T)$ by Lebesgue theorem. This allows us to pass to the limit in the product $m^\varepsilon H_p(t, x, Du^\varepsilon)$ and deduce that m is a distributional solution of the continuity equation.

We conclude now with the compactness of u^ε. Since Du^ε is bounded, we only need to check the uniform continuity of u^ε in time, which is done with a standard time-translation argument. First we observe that, as $G[m^\varepsilon(T)] \in W^{2,\infty}(\mathbb{T}^d)$, the maps $w^+(x, t) = G[m^\varepsilon(T)] + C_1(T - t)$ and $w^-(x, t) = G[m^\varepsilon(T)] - C_1(T - t)$ are, respectively, a super and sub solution of the equation of u^ε, for C_1 sufficiently large (but not depending of ε). Hence, by comparison principle

$$\|u^\varepsilon(t) - G[m^\varepsilon(T)]\|_\infty \le C_1(T - t) . \tag{1.49}$$

For $h > 0$, we consider $u_h^\varepsilon(x, t) = u^\varepsilon(x, t - h)$ in (h, T), which satisfies

$$-\partial_t u_h^\varepsilon - \varepsilon \Delta u_h^\varepsilon + H(t - h, x, Du_h^\varepsilon) = F[m^\varepsilon](t - h).$$

Because of the uniform Hölder regularity of the map $t \to m^\varepsilon(t)$ in $P(\mathbb{T}^d)$ and the assumption (1.44) (with $\alpha = \frac{1}{2}$), we have

$$\sup_{t \in [h,T]} \left\| F[m^\varepsilon](t - h) - F[m^\varepsilon](t) \right\|_\infty \leq C \sqrt{h}$$

and since H is locally Lipschitz and Du^ε is uniformly bounded we also have

$$\sup_{t \in [h,T]} |H(t - h, x, Du_h^\varepsilon) - H(t, x, Du_h^\varepsilon)| \leq C h.$$

For the terminal condition we also have, using (1.49),

$$\|u_h^\varepsilon(T) - u^\varepsilon(T)\|_\infty = \|u^\varepsilon(T - h) - G[m^\varepsilon(T)]\|_\infty \leq C_1 h.$$

By the L^∞ stability (say, the comparison principle) we deduce that

$$\|u^\varepsilon(t - h) - u^\varepsilon(t)\|_\infty \leq C(T - t) \sqrt{h} + C_1 h.$$

This proves the equi-continuity of u^ε in time, and so we conclude that u^ε is relatively compact in $C^0(\overline{Q_T})$.

By continuity assumptions on F and G, we know that $G[m^\varepsilon(T)]$ converges to $G[m(T)]$ in $C^0(\mathbb{T}^d)$, and $F[m^\varepsilon]$ converges to $F[m]$ in $C^0(\overline{Q_T})$. It is now possible to apply the classical stability results for viscosity solutions and we deduce, as $\varepsilon \to 0$, that u is a viscosity solution of the HJ equation $-\partial_t u + H(t, x, Du) = F[m]$, with $u(T) = G[m(T)]$. Note that, as $H(t, x, Du)$ and $F[m]$ are bounded, by standard results in viscosity solutions' theory it turns out that u is Lipschitz continuous in time as well. $\square$

Let us now turn to the question of uniqueness of solutions to (1.43). On one hand, the uniqueness will again rely on the monotonicity argument introduced by Lasry and Lions; on another hand, the new difficulty lies in showing the uniqueness of m from the continuity equation; this step is highly non trivial and we will detail it later.

Theorem 1.8 *Let $m_0 \in L^\infty(\mathbb{T}^d)$, and let H and F satisfy the conditions of Theorem 1.7. In addition, assume that F, G are monotone (nondecreasing) operators, i.e.*

$$\int_0^T \int_{\mathbb{T}^d} (F[m_1] - F[m_2]) d(m_1 - m_2) \geq 0 \qquad \forall m_1, m_2 \in C^0([0, T], \mathcal{P}(\mathbb{T}^d))$$

and

$$\int_{\mathbb{T}^d} (G[m_1] - G[m_2])d(m_1 - m_2) \geq 0 \qquad \forall m_1, m_2 \in \mathcal{P}(\mathbb{T}^d).$$

Then (1.43) *admits a unique solution (in the sense of Definition 1.3)* (u, m) *such that* $m \in L^{\infty}(Q_T)$.

Proof We first observe that the Lasry-Lions monotonicity argument works perfectly in the setting of solutions given above. Indeed, let (u^1, m^1) and (u^2, m^2) be two solutions of (1.43) in the sense of Definition 1.3, with the additional property that $m^1, m^2 \in L^{\infty}(Q_T)$. We recall that for $m = m^i$, we have the weak formulation

$$\int_0^T \int_{\mathbb{T}^d} \left(-m \, \partial_t \varphi + m \, H_p(t, x, Du) \cdot D\varphi \right) = 0 \qquad \forall \varphi \in C_c^1((0, T) \times \mathbb{T}^d).$$

(1.50)

Since $m \in L^{\infty}(Q_T)$, by an approximation argument it is easy to extend this formulation to hold for every $\varphi \in W^{1,\infty}(Q_T)$ with compact support in $(0, T)$. We recall here that $u = u^i$ belongs to $L^{\infty}(0, T, W^{1,\infty}(\mathbb{T}^d))$ by definition and then, by properties of viscosity solutions, it is also Lipschitz in time. Hence $u \in W^{1,\infty}(Q_T)$. In particular, u is almost everywhere differentiable in Q_T and, by definition of viscosity solutions, it satisfies

$$-\partial_t u + H(t, x, Du) = F[m] \qquad \text{a.e. in } Q_T.$$

(1.51)

Let here $\xi = \xi(t)$ be a function in $W^{1,\infty}(0, T)$ with compact support. Using (1.50) with $m = m^i$ and $\varphi = u^j \xi$ and (1.51) for u^j, $i, j = 1, 2$, we obtain the usual equality (1.28) in the weak form

$$\int_0^T \int_{\mathbb{T}^d} (u_1 - u_2)(m_1 - m_2)\partial_t \xi$$

$$= \int_0^T \int_{\mathbb{T}^d} \xi \left(F[m_1] - F[m_2] \right)(m_1 - m_2)$$

$$+ \int_0^T \int_{\mathbb{T}^d} \xi \, m_1 \left\{ H(t, x, Du_2) - H(t, x, Du_1) - H_p(t, x, Du_1)(Du_2 - Du_1) \right\}$$

$$+ \int_0^T \int_{\mathbb{T}^d} \xi \, m_2 \left\{ H(t, x, Du_1) - H(t, x, Du_2) - H_p(t, x, Du_2)(Du_1 - Du_2) \right\}.$$

(1.52)

Now we take $\xi = \xi_\varepsilon(t)$ such that ξ is supported in $(\varepsilon, T - \varepsilon)$, $\xi \equiv 1$ for $t \in (2\varepsilon, T - 2\varepsilon)$ and ξ is linear in $(\varepsilon, 2\varepsilon)$ and in $(T - 2\varepsilon, T - \varepsilon)$. Of course we have

$\xi_\varepsilon \to 1$ and all integrals in the right-hand side of (1.52) converge by Lebesgue theorem. The boundary layers terms give

$$\int_0^T \int_{\mathbb{T}^d} (u_1 - u_2)(m_1 - m_2)\partial_t \xi_\varepsilon = \frac{1}{\varepsilon} \int_\varepsilon^{2\varepsilon} \int_{\mathbb{T}^d} (u_1 - u_2)(m_1 - m_2)$$

$$- \frac{1}{\varepsilon} \int_{T-2\varepsilon}^{T-\varepsilon} \int_{\mathbb{T}^d} (u_1 - u_2)(m_1 - m_2)$$

where we can pass to the limit because $m^i \in C^0([0, T], \mathcal{P}(\mathbb{T}^d))$ and $u^i \in C(\overline{Q_T})$. Therefore letting $\varepsilon \to 0$ in (1.52), and using the same initial condition for m^1, m^2 we conclude that

$$\int_{\mathbb{T}^d} (G[m_1(T)] - G[m_2(T)]) d(m_1(T) - m_2(T)) + \int_0^T \int_{\mathbb{T}^d} (F[m_1] - F[m_2])(m_1 - m_2)$$

$$+ \int_0^T \int_{\mathbb{T}^d} m_1 \{H(t, x, Du_2) - H(t, x, Du_1) - H_p(t, x, Du_1)(Du_2 - Du_1)\}$$

$$+ \int_0^T \int_{\mathbb{T}^d} m_2 \{H(t, x, Du_1) - H(t, x, Du_2) - H_p(t, x, Du_2)(Du_1 - Du_2)\} = 0.$$

Thanks to the monotonicity condition on F, G, and to the strict convexity of H, given by (1.46), this implies that $Du^1 = Du^2$ a.e. in $\{m^1 > 0\} \cup \{m^2 > 0\}$. In particular, m^1 and m^2 solve the same Kolmogorov equation: m^1 and m^2 are both solutions to

$$\partial_t m - \operatorname{div}(m H_p(t, x, Du^1(t, x))) = 0, \qquad m(0) = m_0.$$

We admit for a while the (difficult) fact that this entails the equality $m^1 = m^2$ (see Lemma 1.10 below). Then u^1 and u^2 are two viscosity solutions of the same equation with the same terminal condition; by comparison, they are therefore equal. □

In order to complete the above proof, we are left with the main task, which is the content of the following result.

Lemma 1.10 *Assume that $u \in C(\overline{Q_T})$ is a viscosity solution to*

$$- \partial_t u + H(t, x, Du) = F(t, x), \qquad u(T, x) = u_T(x), \qquad (1.53)$$

where $H : [0, T] \times \mathbb{T}^d \times \mathbb{R}^d \to \mathbb{R}$ satisfies the conditions of Theorem 1.7 and $F \in C(\overline{Q_T}) \cap L^\infty(0, T; W^{2,\infty}(\mathbb{T}^d))$, $u_T \in W^{2,\infty}(\mathbb{T}^d)$.
Then, for any $m_0 \in L^\infty(\mathbb{T}^d)$, the transport equation

$$\partial_t m - \operatorname{div}(m H_p(t, x, Du)) = 0, \qquad m(0, x) = m_0(x) \qquad (1.54)$$

possesses at most one weak solution in L^∞.

The proof of the Lemma is delicate and is the aim of the rest of the section. The difficulty comes from the fact that the vector field $H_p(t, x, Du)$ is not smooth: it is actually discontinuous in general. The analysis of transport equations with non smooth vector fields has attracted a lot of attention since the Di Perna-Lions seminal paper [98]. We rely here on Ambrosio's approach [15, 16], in particular for the "superposition principle" (see Theorem 1.9 below). A key point will be played by the semi concavity property of u. In particular, this implies that $H_p(t, x, Du)$ has bounded variation; nevertheless, this does not seem to be enough to apply previous results on the continuity equation, where the vector field is usually supposed to have a non singular divergence. We will overcome this problem by using the optimal control representation of $H_p(t, x, Du)$ and the related properties of the characteristic curves.

Let us first point out some basic properties of the solution u of (1.53). Henceforth, for simplicity (and without loss of generality) we assume that $F = 0$ in (1.53), which is always possible up to defining a new Hamiltonian $\tilde{H} = H - F$.

We already know that u is unique and Lipschitz continuous, and it is obtained by viscous approximation. Therefore, one can check (exactly as in Theorem 1.7) that u is semiconcave in space for any t, with a modulus bounded independently of t. Moreover, we will extensively use the fact that u can be represented as the value function of a problem of calculus of variations:

$$u(t, x) = \inf_{\gamma:\ \gamma(t)=x} \int_t^T L(s, \gamma(s), \dot{\gamma}(s))ds + u_T(\gamma(T)) \tag{1.55}$$

where $L(t, x, \xi) = \sup_{p \in \mathbb{R}^d}[-\xi \cdot p - H(t, x, p)]$ and where $\gamma \in AC([0, T]; \mathbb{T}^d)$ are absolutely continuous curves in $[0, T]$. For $(t, x) \in [0, T) \times \mathbb{T}^d$ we denote by $\mathcal{A}(t, x)$ the set of optimal trajectories for the control problem (1.55). One easily checks that, under the above assumptions on H, such set is nonempty, and that, if $(t_n, x_n) \to (t, x)$ and $\gamma_n \in \mathcal{A}(t_n, x_n)$, then, up to some subsequence, γ_n weakly converges in H^1 to some $\gamma \in \mathcal{A}(t, x)$.

We need to analyze precisely the connection between the differentiability of u with respect to the x variable and the uniqueness of the minimizer in (1.55). The following properties are well-known in the theory of optimal control and Hamilton-Jacobi equations (see e.g. [48, Chapter 6]), but we will give the proofs for the reader's convenience.

Lemma 1.11 (Regularity of u Along Optimal Solutions) *Let $(t, x) \in [0, T] \times \mathbb{T}^d$ and $\gamma \in \mathcal{A}(t, x)$. Then*

1. *(Uniqueness of the optimal control along optimal trajectories) for any $s \in (t, T]$, the restriction of γ to $[s, T]$ is the unique element of $\mathcal{A}(s, \gamma(s))$.*
2. *(Uniqueness of the optimal trajectories) $Du(t, x)$ exists if and only if $\mathcal{A}(t, x)$ is reduced to a singleton. In this case, $\dot{\gamma}(t) = -H_p(t, x, Du(t, x))$ where $\mathcal{A}(t, x) = \{\gamma\}$.*

Remark 1.7 In particular, if we combine the above statements, we see that, for any $\gamma \in \mathcal{A}(t, x)$, $u(s, \cdot)$ is always differentiable at $\gamma(s)$ for $s \in (t, T)$, with $\dot{\gamma}(s) = -H_p(s, \gamma(s), Du(s, \gamma(s)))$.

Proof We recall that, since H is C^2 and strictly convex in p, then L is also C^2 and strictly convex in ξ, which ensures the regularity of the minimizers. So if $\gamma \in \mathcal{A}(t, x)$, then γ is of class C^2 on $[t, T]$ and satisfies the Euler-Lagrange equation

$$\frac{d}{dt} L_\xi(s, \gamma(s), \dot{\gamma}(s)) = L_x(s, \gamma(s), \dot{\gamma}(s)) \qquad \forall s \in [t, T] \tag{1.56}$$

with the trasversality condition

$$Du_T(\gamma(T)) = -L_\xi(T, \gamma(T), \dot{\gamma}(T)). \tag{1.57}$$

Let $\gamma_1 \in \mathcal{A}(s, \gamma(s))$. For any $h > 0$ small we build some $\gamma_h \in \mathcal{A}(t, x)$ in the following way:

$$\gamma_h(\tau) = \begin{cases} \gamma(\tau) & \text{if } \tau \in [t, s-h) \\ \gamma(s-h) + (\tau - (s-h))\dfrac{\gamma_1(s+h) - \gamma(s-h)}{2h} & \text{if } \tau \in [s-h, s+h) \\ \gamma_1(\tau) & \text{if } \tau \in [s+h, T] \end{cases}$$

Since $\gamma_{|[s,T]}$ and γ_1 are optimal for $u(s, \gamma(s))$, the concatenation γ_0 of $\gamma_{|[t,s]}$ and γ_1 is also optimal for $u(t, x)$. So, comparing the payoff for γ_0 (which is optimal) and the payoff for γ_h we have

$$\int_{s-h}^{s} L(\tau, \gamma(\tau), \dot{\gamma}(\tau))d\tau + \int_{s}^{s+h} L(\tau, \gamma_1(\tau), \dot{\gamma}_1(\tau))d\tau$$

$$-\int_{s-h}^{s+h} L(\tau, \gamma_h(\tau), \frac{\gamma_1(s+h) - \gamma(s-h)}{2h})d\tau \le 0.$$

We divide this inequality by h and let $h \to 0^+$ to get

$$L(s, \gamma(s), \dot{\gamma}(s)) + L(s, \gamma(s), \dot{\gamma}_1(s)) - 2L(s, \gamma(s), \frac{1}{2}(\dot{\gamma}(s) + \dot{\gamma}_1(s))) \le 0$$

since $\lim_{h \to 0,\, s \in [s-h, s+h]} \gamma_h(s) = \gamma(s) = \gamma_1(s)$. By strict convexity of L with respect to the last variable, we conclude that $\dot{\gamma}(s) = \dot{\gamma}_1(s)$. Since we also have $\gamma(s) = \gamma_1(s)$, and since both $\gamma(\cdot)$ and $\gamma_1(\cdot)$ satisfy on the time interval $[s, T]$ the second order equation (1.56), we conclude that $\gamma(\tau) = \gamma_1(\tau)$ on $[s, T]$. This means that the optimal solution for $u(s, \gamma(s))$ is unique.

Next we show that, if $Du(t, x)$ exists, then $\mathcal{A}(t, x)$ is reduced to a singleton and $\dot{\gamma}(t) = -H_p(t, x, Du(t, x))$ where $\mathcal{A}(t, x) = \{\gamma\}$. Indeed, let $\gamma \in \mathcal{A}(t, x)$. Then, for any $v \in \mathbb{R}^d$,

$$u(t, x + v) \le \int_t^T L(s, \gamma(s) + v, \dot{\gamma}(s))ds + u_T(\gamma(T) + v) .$$

Since equality holds for $v = 0$ and since left- and right-hand sides are differentiable with respect to v at $v = 0$ we get by (1.56)–(1.57):

$$Du(t, x) = \int_t^T L_x(s, \gamma(s), \dot{\gamma}(s))ds + Du_T(\gamma(T))$$
$$= \int_t^T \frac{d}{dt} L_\xi(s, \gamma(s), \dot{\gamma}(s)) + Du_T(\gamma(T)) = -L_\xi(t, x, \dot{\gamma}(t)) .$$

By definition of L, this means that $\dot{\gamma}(t) = -H_p(t, x, Du(t, x))$ and therefore $\gamma(\cdot)$ is the unique solution of the Euler-Lagrange equation with initial conditions $\gamma(t) = x$ and $\dot{\gamma}(t) = -H_p(t, x, Du(t, x))$. This shows the claim.

Conversely, let us prove that, if $\mathcal{A}(t, x)$ is a singleton, then $u(t, \cdot)$ is differentiable at x. For this we note that, if p belongs to $D^*u(t, x)$ (the set of reachable gradients of the map $u(t, \cdot)$), then the solution to (1.56), with initial conditions $\gamma(t) = x$, $\dot{\gamma}(t) = -H_p(t, x, p)$, is optimal. Indeed, by definition of p, there is a sequence $x_n \to x$ such that $u(t, \cdot)$ is differentiable at x_n and $Du(t, x_n) \to p$. Now, since $u(t, \cdot)$ is differentiable at x_n, we know by what proved before that the unique solution $\gamma_n(\cdot)$ to (1.56) with initial conditions $\gamma_n(t) = x_n$, $\dot{\gamma}_n(t) = -H_p(t, x_n, Du(t, x_n))$, is optimal. Passing to the limit as $n \to +\infty$ implies (by the stability of optimal trajectories), that $\gamma(\cdot)$, which is the uniform limit of the $\gamma_n(\cdot)$, is also optimal.

Now, from our assumptions, there is a unique optimal curve in $\mathcal{A}(t, x)$. Therefore $D^*u(t, x)$ has to be reduced to a singleton, which implies, since $u(t, \cdot)$ is semi-concave, that $u(t, \cdot)$ is differentiable at x (Lemma 1.9). $\qquad \square$

We now turn the attention to the solutions of the differential equation

$$\begin{cases} \dot{\gamma}(s) = -H_p(s, \gamma(s), Du(s, \gamma(s))) & \text{a.e. in } [t, T] \\ \gamma(t) = x . \end{cases} \qquad (1.58)$$

Here we fix a Borel representative of $Du(t, x)$ (e.g. a measurable selection of $D^+u(t, x)$), so that the vector field $H_p(t, x, Du(t, x))$ is defined everywhere in Q_T. In what follows, we say that γ is a solution to (1.58) if $\gamma \in AC([0, T]; \mathbb{T}^d)$, if $u(s, \cdot)$ is differentiable at $\gamma(s)$ for a.e. $s \in (t, T)$ and if

$$\gamma(s) = x - \int_t^s H_p(\tau, \gamma(\tau), Du(\tau, \gamma(\tau)))d\tau \qquad \forall s \in [t, T] .$$

We already know (see Remark 1.7) that, if $\gamma \in \mathcal{A}(t, x)$, then γ is a solution to (1.58); now we show that the converse is also true.

Lemma 1.12 (Optimal Synthesis) *Let $(t, x) \in [0, T) \times \mathbb{T}^d$ and $\gamma(\cdot)$ be a solution to (1.58). Then the trajectory γ is optimal for $u(t, x)$.*

In particular, if $u(t, \cdot)$ is differentiable at x, then equation (1.58) has a unique solution, corresponding to the optimal trajectory.

Proof We first note that $\gamma(\cdot)$ is Lipschitz continuous because so is u. Let $s \in (t, T)$ be such that equation (1.58) holds (in particular $u(s, \cdot)$ is differentiable at $\gamma(s)$) and the Lipschitz continuous map $s \to u(s, \gamma(s))$ has a derivative at s. Since u is Lipschitz continuous, Lebourg's mean value Theorem [94, Th. 2.3.7] states that, for any $h > 0$ small, there is some $(s_h, y_h) \in [(s, \gamma(s)), (s + h, \gamma(s + h))]$ and some $(\xi_t^h, \xi_x^h) \in \mathrm{Co}D^*_{t,x}u(s_h, y_h)$ with

$$u(s + h, \gamma(s + h)) - u(s, \gamma(s)) = \xi_t^h h + \xi_x^h \cdot (\gamma(s + h) - \gamma(s)) , \qquad (1.59)$$

(where $\mathrm{Co}D^*_{t,x}u(s, y)$ stands for the closure of the convex hull of the set of reachable gradients $D^*_{t,x}u(s, y)$). From Carathéodory Theorem, there are $(\lambda^{h,i}, \xi_t^{h,i}, \xi_x^{h,i})_{i=1,\dots,d+2}$ such that $\lambda^{h,i} \geq 0$, $\sum_i \lambda^{h,i} = 1$, with $(\xi_t^{h,i}, \xi_x^{h,i}) \in D^*_{t,x}u(s_h, y_h)$ and

$$(\xi_t^h, \xi_x^h) = \sum_i \lambda^{h,i}(\xi_t^{h,i}, \xi_x^{h,i}) .$$

Note that the $\xi_x^{h,i}$ converge to $Du(s, \gamma(s))$ as $h \to 0$ because, from Lemma 1.9, any cluster point of the $\xi_x^{h,i}$ must belong to $D^+u(s, \gamma(s))$, which is reduced to $Du(s, \gamma(s))$ since $u(s, \cdot)$ is differentiable at $\gamma(s)$. In particular, $\xi_x^h = \sum_i \lambda^{h,i}\xi_x^{h,i}$ converges to $Du(s, \gamma(s))$ as $h \to 0$.

Since u is a viscosity solution of (1.53) and $(\xi_t^{h,i}, \xi_x^{h,i}) \in D^*_{t,x}u(s_h, y_h)$, we have

$$-\xi_t^{h,i} + H(s_h, y_h, \xi_x^{h,i}) = 0 .$$

Therefore $\xi_t^h = \sum_i \lambda^{h,i}\xi_t^{h,i} = \sum_i \lambda^{h,i} H(s_h, y_h, \xi_x^{h,i})$ converges to $H(s, \gamma(s), Du(s, \gamma(s))$ as $h \to 0$.

Then, dividing (1.59) by h and letting $h \to 0^+$ we get

$$\frac{d}{ds}u(s, \gamma(s)) = H(s, \gamma(s), Du(s, \gamma(s)) + Du(s, \gamma(s)) \cdot \dot{\gamma}(s) .$$

Since $\dot{\gamma}(s) = -H_p(s, \gamma(s), Du(s, \gamma(s)))$, this implies that

$$\frac{d}{ds}u(s, \gamma(s)) = -L(s, \gamma(s), \dot{\gamma}(s)) \quad \text{a.e. in } (t, T) .$$

Integrating the above inequality over $[t, T]$ we finally obtain, since $u(T, y) = u_T(y)$,

$$u(x, t) = \int_t^T L(s, \gamma(s), \dot{\gamma}(s)) \, ds + u_T(\gamma(T)) \, .$$

which means that γ is optimal. The last statement of the Lemma is a direct consequence of Lemma 1.11-(2). □

The next step is a key result by Ambrosio (the so-called superposition principle) on the probabilistic representation of weak solutions to the continuity equation

$$\partial_t \mu + \text{div}(\mu b(t, x)) = 0 \, . \tag{1.60}$$

For this let us define for any $t \in [0, T]$ the map $e_t : C^0([0, T], \mathbb{T}^d) \to \mathbb{T}^d$ by $e_t(\gamma) = \gamma(t)$ for $\gamma \in C^0([0, T], \mathbb{T}^d)$.

Theorem 1.9 ([15]) *Let $b : [0, T] \times \mathbb{T}^d \to \mathbb{R}^d$ be a given Borel vector field and μ be a solution to (1.60) such that $\int_{Q_T} |b|^2 \, d\mu < \infty$. Then there exists a Borel probability measure η on $C^0([0, T], \mathbb{T}^d)$ such that $\mu(t) = e_t \sharp \eta$ for any t and, for η-a.e. $\gamma \in C^0([0, T], \mathbb{T}^d)$, γ is a solution to the ODE*

$$\begin{cases} \dot{\gamma}(s) = b(s, \gamma(s)) & \text{a.e. in } [0, T] \\ \gamma(0) = x \, . \end{cases} \tag{1.61}$$

We will also need the notion of disintegration of a measure and the following well-known disintegration theorem, see for instance [18, Thm 5.3.1].

Theorem 1.10 *Let X and Y be two Polish spaces and λ be a Borel probability measure on $X \times Y$. Let us set $\mu = \pi_X \sharp \lambda$, where π_X is the standard projection from $X \times Y$ onto X. Then there exists a μ-almost everywhere uniquely determined family of Borel probability measures (λ_x) on Y such that*

1. *the function $x \mapsto \lambda_x$ is Borel measurable, in the sense that $x \mapsto \lambda_x(B)$ is a Borel-measurable function for each Borel-measurable set $B \subset Y$,*
2. *for every Borel-measurable function $f : X \times Y \to [0, +\infty]$,*

$$\int_{X \times Y} f(x, y) d\lambda(x, y) = \int_X \int_Y f(x, y) \, d\lambda_x(y) d\mu(x).$$

We are finally ready to prove the uniqueness result:

Proof of Lemma 1.10 Let m be a solution of the transport equation (1.54). We set $\Gamma := C^0([0, T], \mathbb{T}^d)$. From Ambrosio superposition principle, there exists a Borel probability measure η on Γ such that $m(t) = e_t \sharp \eta$ for any t and, for η-a.e. $\gamma \in \Gamma$,

γ is a solution to the ODE $\dot{\gamma} = -H_p(t, \gamma(t), Du(t, \gamma(t)))$. We notice that, since $m \in L^1(Q_T)$, for any subset $E \subset Q_T$ of zero measure we have

$$\int_0^T \int_\Gamma 1_{\{\gamma(t) \in E\}} d\eta = \int_0^T \int_{\mathbb{T}^d} 1_E \, dm_t = 0$$

which means that $\gamma(t) \in E^c$ for a.e. $t \in (0, T)$ and η-a.e. $\gamma \in \Gamma$. In particular, since u is a.e. differentiable, this implies that $u(t, \cdot)$ is differentiable at $\gamma(t)$ for a.e. $t \in (0, T)$ and η-a.e. $\gamma \in \Gamma$. As $m_0 = e_0 \sharp \eta$, we can disintegrate the measure η into $\eta = \int_{\mathbb{T}^d} \eta_x dm_0(x)$, where $\gamma(0) = x$ for η_x-a.e. γ and m_0-a.e. $x \in \mathbb{T}^d$. Therefore, since m_0 is absolutely continuous, for m_0-a.e. $x \in \mathbb{T}^d$, η_x-a.e. map γ is a solution to the ODE starting from x. By Lemma 1.12 we know that such a solution γ is optimal for the calculus of variation problem (1.55). As, moreover, for a.e. $x \in \mathbb{T}^d$ the solution of this problem is reduced to a singleton $\{\bar{\gamma}_x\}$, we can conclude that $d\eta_x(\gamma) = \delta_{\bar{\gamma}_x}$ for m_0-a.e. $x \in \mathbb{T}^d$. Hence, for any continuous map $\phi : \mathbb{T}^d \to \mathbb{R}$, one has

$$\int_{\mathbb{T}^d} \phi(x) m(t, x)) dx = \int_{\mathbb{T}^d} \phi(\bar{\gamma}_x(t)) m_0(x) dx$$

which defines m uniquely. $\square$

1.3.5 Second Order MFG System with Local Couplings

We now consider the case that the coupling functions F, G depend on the local density of the measure $m(t, x)$. Thus we assume that $F \in C^0(\overline{Q}_T \times \mathbb{R})$ and $G \in C^0(\mathbb{T}^d \times \mathbb{R})$ and we consider the system

$$\begin{cases} -\partial_t u - \varepsilon \Delta u + H(t, x, Du) = F(t, x, m(t, x)) \\ u(T) = G(x, m(T, x)) \\ \partial_t m - \varepsilon \Delta m - \mathrm{div}(m \, H_p(t, x, Du)) = 0 \\ m(0) = m_0 \,. \end{cases} \tag{1.62}$$

We assume that both nonlinearities are bounded below:

$$\exists \, c_0 \in \mathbb{R} : \quad F(t, x, m) \geq c_0, \qquad G(x, m) \geq c_0 \quad \forall (t, x, m) \in \overline{Q}_T \times \mathbb{R}_+ \tag{1.63}$$

where $\mathbb{R}_+ = [0, \infty)$. We observe that F, G could be allowed to be measurable with respect to t and x, and bounded when the real variable m lies in compact sets. However, we simplify here the presentation by assuming continuity with respect to all variables.

1.3.5.1 Existence and Uniqueness of Solutions

For local couplings, there are typically two cases where the existence of solutions can be readily proved, namely whenever F, G are bounded, or $H(t, x, p)$ is globally Lipschitz in p. We give a sample result in this latter case. We warn the reader that, in the study of system (1.62) with local couplings, the notion of solution to be used may strongly depend on the regularity of (u, m) which is available. As a general framework, both equations will be understood in distributional sense, and a basic notion of weak solution will be discussed later.

In this first result that we give, assuming m_0 and $H_p(t, x, p)$ to be bounded, the function m turns out to be globally bounded and regular for $t > 0$. Then (u, m) is a solution of (1.62) in the sense that $m \in L^\infty(Q_T) \cap L^2(0, T; H^1(\mathbb{T}^d))$ is a weak solution of the Fokker-Planck equation, with $m \in C^0([0, T], L^1(\mathbb{T}^d))$ and $m(0) = m_0$, whereas $u \in C(\overline{Q_T}) \cap L^2(0, T; H^1(\mathbb{T}^d))$, with $u(T) = G(x, m(T))$, and is a weak solution of the first equation.

Theorem 1.11 *Let $m_0 \in L^\infty(\mathbb{T}^d)$, $m_0 \geq 0$ with $\int_{\mathbb{T}^d} m_0 = 1$. Assume that $H(t, x, p)$ is a Carathéodory function such that H is convex and differentiable with respect to p and satisfies*

$$\exists\, \beta > 0 : \quad |H_p(t, x, p)| \leq \beta \qquad \forall (t, x, p) \in Q_T \times \mathbb{R}^d . \tag{1.64}$$

Then there exists a solution (u, m) to (1.62) with $Du, m \in C^\alpha(Q_T)$ for some $\alpha \in (0, 1)$. If $F(t, x, m)$ is a locally Hölder continuous function and $H(t, x, p)$, $H_p(t, x, p)$ are of class C^1, then (u, m) is a classical solution in $(0, T)$.

Finally, if $F(t, x, \cdot)$, $G(x, \cdot)$ are nondecreasing, then the solution is unique.

Proof For simplicity, we fix the diffusion coefficient $\varepsilon = 1$. We set

$$K = \{m \in C^0([0, T]; L^2(\mathbb{T}^d)) \cap L^\infty(Q_T) : \|m\|_\infty \leq L\} \tag{1.65}$$

where L will be fixed later. For any $\mu \in K$, defining $u_\mu \in L^2(0, T; H^1(\mathbb{T}^d))$ the (unique) bounded solution to

$$\begin{cases} -\partial_t u_\mu - \Delta u_\mu + H(t, x, Du_\mu) = F(t, x, \mu) \\ u_\mu(T) = G(x, \mu(T)), \end{cases}$$

one sets $m := \Phi(\mu)$ as the solution to

$$\begin{cases} \partial_t m - \Delta m - \operatorname{div}(m H_p(t, x, Du_\mu)) = 0, \\ m(0) = m_0. \end{cases}$$

Due to the global bound on H_p, there exists $L > 0$, depending only on β and $\|m_0\|_\infty$, such that $\|m\|_\infty \leq L$. This fixes the value of L in (1.65), so that K is an invariant convex subset of $C^0([0, T]; L^2(\mathbb{T}^d))$. Continuity and compactness of Φ are an easy exercise, so Schauder's fixed point theorem applies which yields a solution. By parabolic regularity for Fokker-Planck equations with bounded drift, m is $C^\alpha(Q_T)$ for some $\alpha > 0$, and so is Du from the first equation. Finally, if the nonlinearity F preserves the Hölder regularity of m, and if H, H_p are of class C^1, then the Schauder's theory can be applied exactly as in Theorem 1.4, so u and m will belong to $C^{1+\frac{\alpha}{2}, 2+\alpha}(Q_T)$ for some $\alpha \in (0, 1)$ and they will be classical solutions.

The uniqueness follows by the time monotonicity estimate (1.28), which still holds for any two possible solutions $(u_1, m_1), (u_2, m_2)$ of system (1.62), because they are bounded. The convexity of H and the monotonicity of F, G imply that $F(t, x, m_1) = F(t, x, m_2)$ and $G(x, m_1(T)) = G(x, m_2(T))$. This readily yields $u_1 = u_2$ by standard uniqueness of the Bellman equation with Lipschitz Hamiltonian and bounded solutions. Since $H_p(t, x, Du_1) = H_p(t, x, Du_2)$, from the Fokker-Planck equation we deduce $m_1 = m_2$.   □

Let us stress that the global Lipschitz bound for H implies a global L^∞ bound for m and Du, which is independent of the time horizon T as well. We will come back to that in Sect. 1.3.6.

Remark 1.8 The existence of solutions would still hold assuming the minimal condition that the initial distribution $m_0 \in \mathcal{P}(\mathbb{T}^d)$. The proof remains essentially the same up to using the smoothing effect in the Fokker-Planck equation, where $\|m(t)\|_\infty \leq Ct^{-\frac{d}{2}}$ for some C only depending on the constant β in (1.64). However, it is unclear how to prove uniqueness when m_0 is just a probability measure, unless some restriction is assumed on the growth of the coupling F. Of course one can combine the growth of F with respect to m and the integrability assumption of m_0 in order to get uniqueness results for some class of unbounded initial data, but this is not surprising.

Remark 1.9 The monotonicity condition on F and G can be slightly relaxed, depending on the diffusive coefficient ε and on $\|m_0\|_\infty$. In particular, if H satisfies (1.64) and is locally uniformly convex with respect to p, there exists a positive value γ, depending on H, F, ε and $\|m_0\|_\infty$, such that (1.62) admits a unique solution whenever $F(x, m) + \gamma m$ is nondecreasing in m. The value of γ tends to zero if $\|m_0\|_\infty \to \infty$ or if $\varepsilon \to 0$. Indeed, this is an effect of diffusivity, which could be understood in the theory as the impact of the independent noise in the players' dynamics against a mild aggregation cost. This phenomenon was observed first in [149] and recently addressed in [91] in relation with the long-time stabilization of the system.

When the Hamiltonian has not linear growth in the gradient, the existence and uniqueness of solutions with local couplings is no longer a trivial issue. The main problem is that solutions can hardly be proved to be smooth unless the growth of the coupling functions F, G or the growth of the Hamiltonian are restricted (see Remark 1.12 below).

On one hand, unbounded solutions of the Bellman equation may be not unique. On another hand, if the drift $H_p(t, x, Du)$ has not enough integrability, the standard parabolic estimates (including boundedness and strict positivity of the solution) are not available for the Fokker-Planck equation. This kind of questions are discussed in [165], where a theory of existence and uniqueness of weak solutions is developed using arguments from renormalized solutions and L^1-theory. We give a sample result of this type, assuming here that the Hamiltonian $H(t, x, p)$ satisfies the following coercivity and growth conditions in $Q_T \times \mathbb{R}^d$:

$$H(t, x, p) \geq \alpha|p|^2 - \gamma \tag{1.66}$$

$$|H_p(t, x, p)| \leq \beta (1 + |p|) \tag{1.67}$$

$$H_p(t, x, p) \cdot p - H(t, x, p) \geq \alpha |p|^2 - \gamma \tag{1.68}$$

for some constants $\alpha, \beta, \gamma > 0$.

We stress that, under conditions (1.66)–(1.68), and for couplings F, G with general growth, the existence of smooth solutions is not known.

Definition 1.5 Assume (1.66)–(1.68). A couple (u, m) is a *weak solution* to system (1.62) if

- $F(t, x, m) \in L^1(Q_T)$, $G(x, m(T)) \in L^1(\mathbb{T}^d)$ and $u \in L^2(0, T; H^1(\mathbb{T}^d))$ is a distributional solution of

$$\begin{cases} -\partial_t u - \varepsilon \Delta u + H(t, x, Du) = F(t, x, m(t, x)) \\ u(T) = G(x, m(T, x)) \end{cases}$$

- $m \in C^0([0, T]; L^1(\mathbb{T}^d))$, $m |Du|^2 \in L^1(Q_T)$ and m is a distributional solution of

$$\begin{cases} \partial_t m - \varepsilon \Delta m - \operatorname{div}(m H_p(t, x, Du)) = 0 \\ m(0) = m_0. \end{cases}$$

Let us stress that the terminal condition for u is understood in $L^1(\mathbb{T}^d)$, because $u \in C^0([0, T]; L^1(\mathbb{T}^d))$ as a consequence of the equation itself.

The following result is essentially taken from [165], although the uniqueness statement that we give below generalizes the original result, by establishing that the uniqueness of u always holds m-almost everywhere. This seems to be the most general well-posedness result available so far for system (1.62), in terms of the conditions allowed on H and F, G. Later we discuss the issue of smooth solutions, some special cases, and several related results, including other possible approaches to weak solutions.

Theorem 1.12 *[[165]] Assume that $H(t, x, p)$ is convex in p and satisfies conditions (1.66)–(1.68), and that F, G satisfy (1.63) and $G(x, \cdot)$ is nondecreasing. Then, for any $m_0 \in L^\infty(\mathbb{T}^d)$, there exists a weak solution to (1.62).*

If we assume in addition that $F(x, \cdot)$ is nondecreasing, then $F(x, m) = F(x, \tilde{m})$ and $G(x, m(T)) = G(x, \tilde{m}(T))$ for any two couples of weak solutions $(u, m), (\tilde{u}, \tilde{m})$. Moreover, if at least one of the following two assumptions holds:

(i) $F(x, \cdot)$ is increasing
(ii) $H(t, x, p) - H(t, x, q) - H_p(t, x, q) \cdot (p - q) = 0 \Rightarrow H_p(t, x, p) = H_p(t, x, q) \quad \forall p, q \in \mathbb{R}^d$

then $m = \tilde{m}$ and $u = \tilde{u}$ m-almost everywhere.

In particular, there is at most one weak solution (u, m) with $m > 0$ and, if $m_0 > 0$ and $\log(m_0) \in L^1(\mathbb{T}^d)$, there exists one and only one weak solution.

Remark 1.10 We stress that if (u, m) is a weak solution such that $u, m \in L^\infty(Q_T)$, then both u and m belong to $L^2(0, T; H^1(\mathbb{T}^d))$ and the two equations hold in the usual formulation of finite energy solutions, e.g. against test functions $\varphi \in L^2(0, T; H^1(\mathbb{T}^d)) \cap L^\infty(Q_T)$ with $\partial_t \varphi \in L^2(0, T; H^1(\mathbb{T}^d)') + L^1(Q_T)$. This fact can be deduced, for instance, from the characterization that weak solutions in the sense of Definition 1.5 are also renormalized solutions (see [165, Lemma 4.2]).

In addition, if (u, m) are bounded weak solutions, further results in the literature can be applied: since $F(x, m)$ is bounded and H has at most quadratic growth, it turns out that Du is also bounded for $t < T$, which is enough to ensure that $m \in C^\alpha(Q_T)$ and $m(t) > 0$ for $t > 0$. In other words, bounded weak solutions are regularized with standard bootstrap arguments. In particular, under the assumptions of Theorem 1.12, *bounded weak solutions are unique.*

The existence part of Theorem 1.12 requires many technical tools which we will only sketch here, referring to [165] for the details. It is instructive first to recall the basic a priori estimates of the system (1.62), which explain the natural framework of *weak solutions.* We stress that the estimates below are independent of the diffusion constant ε.

Lemma 1.13 *Assume that (u, m) are bounded weak solutions to system (1.62) and F, G are continuous functions satisfying (1.63). There exists a constant K, independent on ε, such that*

$$\int_0^T \int_{\mathbb{T}^d} m\{H_p(t, x, Du)Du - H(t, x, Du)\} + \int_0^T \int_{\mathbb{T}^d} H(t, x, Du)$$

$$+ \int_0^T \int_{\mathbb{T}^d} F(t, x, m)m + \int_{\mathbb{T}^d} G(x, m)m \leq K. \tag{1.69}$$

The constant K depends on $\|m_0\|_\infty, T, \|H(t, x, 0)\|_\infty, c_0$ and $\displaystyle\sup_{m \leq 2\|m_0\|_\infty} [F(t, x, m) + G(x, m)]$.

Proof We omit the dependence on t of the nonlinearities, which plays no role. Since u and m are bounded, they can be used as test functions in the usual weak formulations of both equations. This yields the energy equality

$$\int_{\mathbb{T}^d} G(x, m(T))m(T) + \int_0^T \int_{\mathbb{T}^d} F(x, m)m$$
$$+ \int_0^T \int_{\mathbb{T}^d} m\left[H_p(x, Du) \cdot Du - H(x, Du)\right] = \int_{\mathbb{T}^d} m_0 u(0) \quad (1.70)$$

which implies

$$\int_{\mathbb{T}^d} G(x, m(T))m(T) + \int_0^T \int_{\mathbb{T}^d} F(x, m)m$$
$$+ \int_0^T \int_{\mathbb{T}^d} m\left[H_p(x, Du) \cdot Du - H(x, Du)\right] \le \|m_0\|_\infty \int_{\mathbb{T}^d} u(0)_+$$
$$\le \|m_0\|_\infty \left\{ \int_0^T \int_{\mathbb{T}^d} (F(x, m)) + \int_{\mathbb{T}^d} G(x, m(T)) - \int_0^T \int_{\mathbb{T}^d} H(x, Du) \right\}$$
$$+ \|m_0\|_\infty \int_{\mathbb{T}^d} u(0)_-$$

where we used that $\int_{\mathbb{T}^d} u(0) = \int_0^T \int_{\mathbb{T}^d} F(x, m) + \int_{\mathbb{T}^d} G(x, m(T)) - \int_0^T \int_{\mathbb{T}^d} H(x, Du)$.

Now we estimate the right-hand side of the previous inequality. From assumption (1.63) and the maximum principle, we have that u is bounded below by a constant depending on c_0 and the L^∞- bound of $H(x, 0)$, so last term is bounded. We also have $F(x, m) \le \frac{1}{2\|m_0\|_\infty} F(x, m)m + C$, for some constant C depending on $\sup_{m \le 2\|m_0\|_\infty} F(x, m)$. Similarly we estimate $G(x, m)$. Therefore, we conclude that (1.69) holds true. $\qquad \square$

Proof of Theorem 1.12 (Sketch) Without loss of generality, we fix the diffusion coefficient $\varepsilon = 1$.

Existence To start with, one can build a sequence of smooth solutions, e.g. by defining $F^n(t, x, m) = \rho^n \star F(t, \cdot, \rho^n \star m))(x)$, $G^n(x, m) = \rho^n \star G(\cdot, \rho^n \star m))(x)$, where $\star$ denotes the convolution with respect to the spatial variable and ρ^n is a standard symmetric mollifier, i.e. $\rho^n(x) = n^N \rho(nx)$ for a nonnegative function $\rho \in C_c^\infty(\mathbb{R}^d)$ such that $\int_{\mathbb{R}^d} \rho(x)dx = 1$.

The existence of a bounded solution u^n, m^n is given, for instance, by Theorem 1.4. From assumption (1.63) and the maximum principle, we have that u^n is bounded from below. Due to the a priori estimates (1.69), applied to (u^n, m^n), and thanks to (1.68) and (1.66), we have that

u^n is bounded in $L^2((0, T); H^1(\mathbb{T}^d)$, and $m^n |Du^n|^2$ is bounded in $L^1(Q_T)$.

In addition, we have that

$$F(t, x, m^n), G(x, m^n(T)) \quad \text{are bounded and equi-integrable in } L^1(Q_T)$$

$$\text{and } L^1(\mathbb{T}^d), \text{ respectively.} \tag{1.71}$$

The heart of the existence proof consists then in considering both the stability properties of the viscous HJ equation

$$- \partial_t u^n - \Delta u^n + H(t, x, Du^n) = f^n \tag{1.72}$$

for some f^n converging in $L^1(Q_T)$, and the compactness of the Fokker-Planck equation

$$\partial_t m^n - \Delta m^n - \text{div}(m^n b^n) = 0 \tag{1.73}$$

where $m^n |b^n|^2$ is bounded (or eventually, converging) in $L^1(Q_T)$.

Indeed, as a first step one uses (1.73) to show that m^n is relatively compact in $L^1(Q_T)$, as well as in $C^0([0, T]; W^{-1,q}(\mathbb{T}^d))$ for some dual Sobolev space $W^{-1,q}(\mathbb{T}^d)$, and for every t we have that $m^n(t)$ is relatively compact in $\mathcal{P}(\mathbb{T}^d)$. Using the extra estimates (1.71), $m^n(T)$ is relatively compact in the weak L^1 topology and $F(t, x, m^n)$ is compact in $L^1(Q_T)$. If we turn the attention to the Bellman equation (1.72), the L^1 convergence of f^n is enough to ensure that u^n, Du^n are relatively compact in $L^1(Q_T)$ and, thanks to existing results of the L^1-theory for divergence form operators, one concludes that $u^n \to u$ which solves

$$-\partial_t u - \Delta u + H(t, x, Du) = F(t, x, m).$$

The convergence of Du^n now implies that $m^n H_p(t, x, Du^n) \to m H_p(t, x, Du)$ in $L^1(Q_T)$ and m can be proved to be a weak solution of the limit equation. The proof of the existence would be concluded if not for the coupling in the terminal condition $u(T)$; in fact, to establish that $u(T) = G(x, m(T))$ some extra work is needed, and this can be achieved by using the monotonicity of $G(x, \cdot)$. For the full proof of this stability argument, we refer to [165][Thm 4.9].

Uniqueness To shortness notations, we omit here the dependence on t of the nonlinearities H, F. A key point for uniqueness is to establish that both u and m are renormalized solutions of their respective equations (see [165, Lemma 4.2]). This means that if (u, m) is any weak solution, then u satisfies

$$- \partial_t S_h(u) - \Delta S_h(u) + S_h'(u)H(x, Du) = F(x, m)S_h'(u) - S_h''(u)|Du|^2 \tag{1.74}$$

where $S_h(r)$ is the sequence of functions (an approximation of the identity function) defined as

$$S_h(r) = h \, S\left(\frac{r}{h}\right), \text{ where } S(r) = \int_0^r S'(r)dr, \quad S'(r) = \begin{cases} 1 & \text{if } |s| \le 1 \\ 2 - |s| & \text{if } 1 < |s| \le 2 \\ 0 & \text{if } |s| > 2 \end{cases}$$

(1.75)

Notice that $S_h(r) \to r$ as $h \to \infty$ and since S'_h has compact support the *renormalized equation* (1.74) is restricted to a set where u is bounded. Similarly m is also a renormalized solution, in particular it satisfies

$$\partial_t S_n(m) - \Delta S_n(m) - \text{div}(S'_n(m)m \, H_p(x, Du)) = \omega_n,$$

$$\text{for some } \omega_n \xrightarrow{n \to \infty} 0 \text{ in } L^1(Q_T).$$

(1.76)

We recall that the renormalized formulations are proved to hold for all weak solutions, since $F(x, m) \in L^1(Q_T)$ and since $m|Du|^2 \in L^1(Q_T)$. In addition, it is proved in [165, Lemma 4.6] that, for any couple of weak solutions (u, m) and $(\tilde{u}, \tilde{m})$, the following *crossed regularity* holds: $m|D\tilde{u}|^2, \tilde{m}|Du|^2 \in L^1(Q_T)$. This is what is needed in order to perform first the Lasry-Lions' argument on the renormalized formulations, and then letting $n \to \infty$ and subsequently $h \to \infty$ to conclude the usual monotonicity inequality:

$$\int_0^T \int_{\mathbb{T}^d} (F(x, m) - F(x, \tilde{m}))(m - \tilde{m}) + \int_{\mathbb{T}^d} [G(x, m(T)) - G(x, \tilde{m}(T))][m(T) - \tilde{m}(T)]$$

$$+ \int_0^T \int_{\mathbb{T}^d} m \, [H(x, D\tilde{u}) - H(x, Du) - H_p(x, Du)(D\tilde{u} - Du)]$$

$$+ \int_0^T \int_{\mathbb{T}^d} \tilde{m} \, [H(x, Du) - H(x, D\tilde{u}) - H_p(x, D\tilde{u})(Du - D\tilde{u})] \le 0.$$

This implies, because $F(x, \cdot)$, $G(x, \cdot)$ are nondecreasing,

$$F(x, m) = F(x, \tilde{m}), \quad G(x, m(T)) = G(x, \tilde{m}(T))$$

(1.77)

and, from the convexity of $H(x, \cdot)$,

$$H(x, D\tilde{u}) - H(x, Du) = H_p(x, Du)(D\tilde{u} - Du) \quad \text{in } \{(t, x): \ m(t, x) > 0\}$$

$$H(x, Du) - H(x, D\tilde{u}) = H_p(x, D\tilde{u})(Du - D\tilde{u}) \quad \text{in } \{(t, x): \ \tilde{m}(t, x) > 0\}.$$

(1.78)

We warn the reader that (1.77) does not imply alone that $u = \tilde{u}$, because unbounded weak solutions to the Bellman equation may be not unique. So we need to use some extra information.

We first want to show that $m = \tilde{m}$. This is straightforward if $F(x, \cdot)$ is increasing. Otherwise, suppose that (1.25) holds true. Then we deduce that

$$m \, H_p(x, Du) = m \, H_p(x, D\tilde{u}) \qquad \tilde{m} \, H_p(x, Du) = \tilde{m} \, H_p(x, D\tilde{u}) \quad \text{a.e. in } Q_T.$$
(1.79)

We take now the difference of the renormalized equations of $m, \tilde{m}$, namely

$$\partial_t \, (S_n(m) - S_n(\tilde{m})) - \Delta \, (S_n(m) - S_n(\tilde{m}))$$
$$- \operatorname{div}(S_n'(m) m \, H_p(x, Du) - S_n'(\tilde{m}) \tilde{m} \, H_p(x, D\tilde{u})) = \omega_n - \tilde{\omega}_n$$

and we aim at showing that, roughly speaking, $\|m(t) - \tilde{m}(t)\|_{L^1(\mathbb{T}^d)}$ is time contractive. To do it rigorously, we consider the function $\Theta_\varepsilon(s) = \int_0^r \frac{T_\varepsilon(r)}{\varepsilon} dr$, with $T_\varepsilon(r) = \min(\varepsilon, r)$; then $\Theta_\varepsilon(s)$ approximates $|s|$ as $\varepsilon \to 0$. Using $\frac{T_\varepsilon(S_n(m) - S_n(\tilde{m}))}{\varepsilon}$ as test function in the previous equality we get

$$\int_{\mathbb{T}^d} \Theta_\varepsilon[S_n(m(t)) - S_n(\tilde{m}(t))]$$

$$+ \frac{1}{\varepsilon} \int_0^T \int_{\mathbb{T}^d} |DT_\varepsilon(S_n(m) - S_n(\tilde{m}))|^2 \le \|\omega_n\|_{L^1(Q_T)} + \|\tilde{\omega}_n\|_{L^1(Q_T)}$$

$$- \frac{1}{\varepsilon} \int_0^T \int_{\mathbb{T}^d} (S_n'(m) m \, H_p(x, Du) - S_n'(\tilde{m}) \tilde{m} \, H_p(x, D\tilde{u})) DT_\varepsilon(S_n(m) - S_n(\tilde{m}))$$

where we used that the test function is smaller than one. Thanks to Young's inequality we deduce

$$\int_{\mathbb{T}^d} \Theta_\varepsilon[S_n(m(t)) - S_n(\tilde{m}(t))] \le \frac{1}{4\varepsilon} \int_0^T \int_{\mathbb{T}^d} |S_n'(m) m \, H_p(x, Du)$$

$$- S_n'(\tilde{m}) \tilde{m} \, H_p(x, D\tilde{u})|^2 \, 1_{\{|S_n(m) - S_n(\tilde{m})| < \varepsilon\}}$$

$$+ \|\omega_n\|_{L^1(Q_T)} + \|\tilde{\omega}_n\|_{L^1(Q_T)}.$$
(1.80)

Now we use (1.79): if one between $m, \tilde{m}$ is positive, then $H_p(x, Du) = H_p(x, D\tilde{u})$, so

$$\int_0^T \int_{\mathbb{T}^d} |S_n'(m) m \, H_p(x, Du) - S_n'(\tilde{m}) \tilde{m} \, H_p(x, D\tilde{u})|^2 \, 1_{\{|S_n(m) - S_n(\tilde{m})| < \varepsilon\}}$$

$$= \int_0^T \int_{\mathbb{T}^d} |S_n'(m) m - S_n'(\tilde{m}) \tilde{m}|^2 \, |H_p(x, Du)|^2 \, 1_{\{|S_n(m) - S_n(\tilde{m})| < \varepsilon\}}$$

and since

$$|S'_n(m)m - S'_n(\tilde{m})\tilde{m}|^2 \, |H_p(x, Du)|^2 \, 1_{\{|S_n(m)-S_n(\tilde{m})|<\varepsilon\}} \leq C \varepsilon \, (m + \tilde{m})(1 + |Du|^2)$$

we can let $n \to \infty$ using Lebesgue's theorem since $m|Du|^2, \tilde{m}|Du|^2 \in L^1(Q_T)$. Therefore, letting $n \to \infty$, from (1.80) we obtain (for a.e. $\varepsilon > 0$):

$$
\begin{aligned}
\int_{\mathbb{T}^d} \Theta_\varepsilon[m(t) - \tilde{m}(t)] &\leq \frac{1}{4\varepsilon} \int_0^T \int_{\mathbb{T}^d} |m - \tilde{m}|^2 \, |H_p(x, Du)|^2 \, 1_{\{m-\tilde{m}|<\varepsilon\}} \\
&\leq \frac{1}{4} \int_0^T \int_{\mathbb{T}^d} |m - \tilde{m}| \, |H_p(x, Du)|^2 1_{\{m-\tilde{m}|<\varepsilon\}} .
\end{aligned}
\tag{1.81}
$$

Last term converges to zero as $\varepsilon \to 0$ (using again Lebesgue's theorem), whereas the first integral converges to $\|m(t) - \tilde{m}(t)\|_{L^1(\mathbb{T}^d)}$. Hence, by letting $\varepsilon \to 0$ we get $\|m(t) - \tilde{m}(t)\|_{L^1(\mathbb{T}^d)} = 0$. This concludes the proof of the uniqueness of m.

Now we show that u is unique m-a.e.; to this purpose, we are going to show that

$$\int_{\mathbb{T}^d} m(t)(u - \tilde{u})_+(t) \leq 0 \qquad \text{for a.e. } t < T. \tag{1.82}$$

To prove (1.82), we subtract the renormalized formulations (1.74) for $u, \tilde{u}$. By using the convexity of H, we have

$$
\begin{aligned}
&- \partial_t (S_h(u) - S_h(\tilde{u})) - \Delta(S_h(u) - S_h(\tilde{u})) + S'_h(u) H_p(x, D\tilde{u})D(u - \tilde{u}) \\
&+ (S'_h(u) - S'_h(\tilde{u}))H(x, D\tilde{u}) \\
&\qquad \leq F(x, m)(S'_h(u) - S'_h(\tilde{u})) - S''_h(u)|Du|^2 + S''_h(\tilde{u})|D\tilde{u}|^2 .
\end{aligned}
$$

We multiply this equation by $\varphi_\varepsilon := \frac{T_\varepsilon(S_h(u)-S_h(\tilde{u}))_+}{\varepsilon}$; denoting as before Θ_ε the primitive of $\frac{T_\varepsilon(t)}{t}$, using that $0 \leq \varphi_\varepsilon \leq 1$ we get, in weak sense,

$$
\begin{aligned}
&- \partial_t \Theta_\varepsilon[(S_h(u) - S_h(\tilde{u}))_+] - \Delta\Theta_\varepsilon[(S_h(u) - S_h(\tilde{u}))_+] + \varphi_\varepsilon S'_h(u) H_p(x, D\tilde{u})D(u - \tilde{u}) \\
&\qquad \leq |S'_h(u) - S'_h(\tilde{u})| \, |H(x, D\tilde{u}) + F(x, m)| + |S''_h(u)| \, |Du|^2 + |S''_h(\tilde{u})| \, |D\tilde{u}|^2 .
\end{aligned}
$$

Now we multiply by $S_n(m)$ this equation, we integrate in (t, T), we use that $u(T) = \tilde{u}(T)$ and (1.76). We obtain

$$
\begin{aligned}
&\int_{\mathbb{T}^d} S_n(m(t))\Theta_\varepsilon[(S_h(u(t)) - S_h(\tilde{u}(t)))_+] \\
&- \int_t^T \int_{\mathbb{T}^d} S'_n(m) \, m \, H_p(x, Du)\varphi_\varepsilon D(S_h(u) - S_h(\tilde{u}))
\end{aligned}
$$

$$+ \int_t^T \int_{\mathbb{T}^d} S_n(m) \, \varphi_\varepsilon S_h'(u) \, H_p(x, D\tilde{u}) D(u - \tilde{u})$$

$$\leq \int_t^T \int_{\mathbb{T}^d} S_n(m) \, |S_h'(u) - S_h'(\tilde{u})| \, |H(x, D\tilde{u}) + F(x, m)|$$

$$+ \int_t^T \int_{\mathbb{T}^d} S_n(m) [|S_h''(u)| \, |Du|^2 + |S_h''(\tilde{u})| \, |D\tilde{u}|^2]$$

$$+ \int_t^T \int_{\mathbb{T}^d} \Theta_\varepsilon [(S_h(u(t)) - S_h(\tilde{u}(t)))_+] \, \omega_n$$

where we used that $D\Theta_\varepsilon[(S_h(u(t)) - S_h(\tilde{u}(t)))_+] = \varphi_\varepsilon \, D(S_h(u) - S_h(\tilde{u}))$. Now we let $n, h \to \infty$, which is allowed using that $F(x, m)m, m|Du|^2, m|D\tilde{u}|^2 \in L^1(Q_T)$. First we let $n \to \infty$, so that we can use the L^1-convergence to zero of ω_n (whereas $S_h(u), S_h(\tilde{u})$ are bounded functions). Once n has gone to infinity, we let $h \to \infty$, so that $S_h' \to 1$; using dominated convergence in each term and Fatou's lemma in the first integral, we get

$$\int_{\mathbb{T}^d} m(t) \Theta_\varepsilon [(u(t) - \tilde{u}(t))_+] - \int_t^T \int_{\mathbb{T}^d} m H_p(x, Du) \varphi_\varepsilon D(u - \tilde{u})$$

$$+ \int_t^T \int_{\mathbb{T}^d} m \, \varphi_\varepsilon \, H_p(x, D\tilde{u}) D(u - \tilde{u}) \leq 0$$

for a.e. $t \in (0, T)$. Since $m \, H_p(x, Du) D(u - \tilde{u}) = m \, H_p(x, D\tilde{u}) D(u - \tilde{u})$ from (1.78) (where now $\tilde{m} = m$), we deduce that $\int_{\mathbb{T}^d} m(t) \Theta_\varepsilon [(u(t) - \tilde{u}(t))_+] \leq 0$. Letting $\varepsilon \to 0$ yields (1.82). Reversing the roles of $u, \tilde{u}$, we conclude that $u = \tilde{u}$ m-a.e.

Finally, it is proved in [165] that, if $\log m_0 \in L^1(\mathbb{T}^d)$, then we have $m > 0$ a.e., in which case we deduce that $u = \tilde{u}$ almost everywhere.   □

Several comments and remarks are in order as far as the previous result and MFG systems with local couplings are concerned.

Remark 1.11 (Extensions of Theorem 1.12)

(i) The result of Theorem 1.12 also holds with homogeneous Dirichlet or Neumann boundary conditions; this extension already appears in [165]. Let us stress that this is one of the main advantage for the use of renormalized solutions, which are well adapted to boundary conditions. Indeed, through the use of renormalization one wishes to approximate a weak solution with its own truncations, which often preserve natural boundary conditions. By contrast, the approximation of weak solutions through mollification introduces many technical problems when dealing with boundary conditions.

Results on the whole space $\mathbb{R}^d$ are also available in [166], assuming $m_0 \in L^1(\mathbb{T}^d) \cap L^\infty(\mathbb{T}^d)$; in that case u belongs to $L^\infty((0, T) \times \mathbb{R}^d) + L^\infty(0, T; L^1(\mathbb{R}^d))$ and $m \in L^\infty(0, T; L^1(\mathbb{R}^d))$.

(ii) Similar results hold by assuming the Hamiltonian coercive with q-growth, namely replacing $|p|^2$ with $|p|^q$ in (1.66), (1.68) and a $q-1$-growth for H_p, where $1 < q < 2$. However, general uniqueness results in this case have been proved so far only for the periodic case or for the whole space [165, 166].

(iii) The same results hold for more general diffusion coefficients, namely if the Laplacian is replaced by the divergence form operator $\operatorname{div}(A(t, x)D(\cdot))$ with $A(t, x) \in L^\infty(0, T; W^{1,\infty})$. In particular, this includes the case where Δu is replaced by $\operatorname{Tr}(\sigma(t, x)\sigma^*(t, x)D^2 u)$ for a bounded, Lipschitz and elliptic matrix $\sigma(t, \cdot)$, modeling diffusion processes associated to stochastic dynamics with Lipschitz diffusion coefficients.

Remark 1.12 (Smoothness of Solutions) Solutions of system (1.62) with local couplings can be proved to be more regular under growth restrictions on $H(t, x, p)$ and/or on $F(t, x, m)$.

An easy case occurs when

$$|H_p(t, x, p)| \leq C(1 + |p|^{q-1}) \qquad \text{with } q < \frac{d+2}{d+1}. \tag{1.83}$$

Indeed, since $F(t, x, m) \in L^1(Q_T)$ (regardless of the growth of F, see estimate (1.69)) then any weak solution u belongs to $L^s(0, T; W^{1,s}(\mathbb{T}))$ for any $s < \frac{d+2}{d+1}$. If (1.83) holds, this implies that $H_p(t, x, Du) \in L^r$ for some $r > d + 2$; in turn, by standard parabolic results, the solution of the Fokker-Planck equation becomes bounded in this case, and actually even Hölder continuous. One concludes that u is bounded as well, from the first equation, and actually Du is Hölder continuous as well. Smoothness up to C^2 regularity then follows according to the smoothness required on the coefficients.

A somewhat similar situation occurs if F has restricted growth, namely if

$$|F(t, x, m)| \leq C(1 + m^\gamma)$$

with $\gamma < \frac{2}{d}$; in this case estimate (1.69) implies that $F(t, x, m) \in L^r(Q_T)$ for some $r > \frac{2+d}{2}$, and the standard parabolic regularity immediately gives the boundedness, and then smoothness, of u, m.

The above two situations are straightforward applications, using parabolic regularity, of the a priori estimates (1.69); in particular they do not require any smoothness in the x-dependence of the nonlinearities, and directly apply to weak solutions in order to obtain their boundedness. We recall that *proving boundedness of weak solutions is enough to show that they are unique*, see Remark 1.10.

However, in order to get smooth solutions, one can go beyond the above conditions up to using refined estimates on the system. This was addressed first by P.-L. Lions, who showed that $F(x, m) \simeq m^\gamma$ with $\gamma < \frac{2}{d-2}$ was enough to ensure smoothness of solutions, standing on second order estimates which further exploit the monotonicity of the coupling F. This issue has been extensively investigated later in a series of papers by D. Gomes and co-workers (see e.g. [120, 121]; most results are encoded in the book [122]), coupling the second order estimates with regularity estimates for the Fokker-Planck equation obtained through the adjoint method introduced by L.C. Evans. In this series of contributions, some growth conditions on H and F have been given which allow to have smooth solutions, both for sub quadratic and for super quadratic Hamiltonians. They are specially important for the case that $H(x, p)$ grows superquadratically in p, because in that situation the approach through weak solutions as developed in Theorem 1.12 cannot be used. It must be said that the aforementioned regularity results usually require smoothness of the Hamiltonian and periodic setting, and the smoothness of solutions remains largely open under general growth assumptions.

Remark 1.13 (*Quadratic Hamiltonian and Hopf-Cole Reduction*) In the special case that $H(t, x, p) = \frac{1}{2}|p|^2 + b(t, x) \cdot p$, the system (1.62) can be transformed into a system of semi linear equations. By introducing the two new unknowns: $w = e^{-\frac{u}{2}}$ and $\varphi = m\, e^{\frac{u}{2}}$, then (1.62) (with $\varepsilon = 1$) is equivalent to the system

$$\begin{cases} -\partial_t w - \Delta w + b \cdot Dw + \frac{1}{2} w F(t, x, \varphi\, w) = 0 \\ \partial_t \varphi - \Delta \varphi - \operatorname{div}(b\, \varphi) + \frac{1}{2}\varphi F(t, x, \varphi\, w) = 0 \\ w(T) = e^{-G(x, \varphi(T)w(T))/2}, \quad \varphi(0) = \frac{m_0}{w(0)} \end{cases} \tag{1.84}$$

Notice that the system (1.84) appears to be simplified, compared to (1.62), but the initial-terminal conditions are both coupled. The initial condition at $t = 0$ makes sense because $w > 0$ by strong maximum principle. Still by maximum principle, the function φ is positive as well. Assuming $G(x, \cdot)$ to be nondecreasing, the condition $w(T) = e^{-G(x, \varphi(T)w(T))/2}$ defines $w(T)$ implicitly as a function of $\varphi(T)$; hence the final condition reads as $w(T) = \psi(x, \varphi(T))$ for a function ψ defined by the implicit relation $\psi(r) - e^{-\frac{1}{2}G(x, r\, \psi(r))} = 0$, for $r \geq 0$.

When $b = 0$ and G only depends on x, it is proved in [60] that weak solutions to (1.84) are bounded. The proof uses a Moser iteration scheme, and it can be easily verified that the proof still holds, without additional difficulty, for the case that $b \in L^\infty(0, T; W^{1,\infty}(\mathbb{T}^d))$ and $G(x, \cdot)$ is monotone. The equivalence between (1.84) and (1.62) (for this special H) is easy to verify for bounded solutions, since the maximum principle gives $\varphi, w > 0$, so $u = -2\log w$, $m = w\,\varphi$ defines (u, m) back from (1.84). Once solutions are shown to be bounded, then they are smooth $(m, Du \in C^\alpha(Q_T))$ by standard bootstrap arguments, and they are classical solutions in Q_T if F is locally Hölder continuous. Therefore, system (1.62) possesses regular, and even classical, solutions in the special case that $H(t, x, Du) = b(t, x) \cdot Du + \frac{1}{2}|Du|^2$, with b Lipschitz continuous in x, and this holds true without any growth restriction on the couplings F, G.

We mention here further related results on weak solutions and on systems with local coupling.

- Under general assumptions, essentially the conditions of Theorem 1.12 above, it has been proved that the discrete solutions of finite difference schemes, as defined in [2], converge to weak solutions as the numerical scheme approximates the continuous equation, i.e. as the mesh size tends to zero. This result is proved in [4] and provides an independent, alternative proof of the existence of weak solutions.

- A different notion of weak solution was introduced in [104] relying on the theory of motonone operators. In particular, if F, G are nondecreasing, then problem (1.62) can be rephrased as $\mathcal{A}(m, u) = 0$ where $\mathcal{A}$ is a monotone operator (on the couple (m, u)) defined as

$$\mathcal{A}(m, u) := \begin{pmatrix} \partial_t u + \Delta u + f(m) - H(x, Du) \\ \partial_t m - \Delta m - \operatorname{div}(m H_p(x, Du)) \end{pmatrix}$$

Since $\langle \mathcal{A}(m, u) - \mathcal{A}(\mu, v), (m - \mu, u - v) \rangle \geq 0$, where the duality is meant in distributional sense, $\mathcal{A}$ defines a monotone operator. Then the Minty-Browder theory of monotone operators suggests the possibility to define a notion of weak solution (u, m) as a couple satisfying

$$\langle \mathcal{A}(\varphi, v), (m - \varphi, u - v) \rangle \geq 0 \qquad \forall (\varphi, v) \in C^2(\overline{Q_T})^2. \tag{1.85}$$

This notion requires even less regularity on (m, u) than in Definition 1.5, and of course the existence of a couple (m, u) satisfying (1.85) is readily proved by weak stability and monotonicity, as in Minty-Browder's theory. However, the uniqueness of a solution of this kind is unclear, and has not been proved so far.

- The study of non monotone couplings $F(x, m)$ in (1.62) leads to different kind of questions and results. This direction has been mostly exploited for the stationary system [81, 87] and in special examples for the evolution case. We refer the reader to [92].

In a different direction, it is worth pointing out that the assumption that $F(x, m)$ be bounded from below could be relaxed by allowing $F(x, m) \to -\infty$ as $m \to 0^+$ as in the model case $F(x, m) \sim \log m$ for $m \to 0$. Results on this model can be found e.g. in [114, 127].

We conclude this section by mentioning the case of a general Hamiltonian function $H(t, x, Du, m)$, as in problem (1.29), where now $H : Q_T \times \mathbb{R}^d \times [0, +\infty) \to \mathbb{R}$ is a continuous function depending locally on the density m. In his courses at Collége de France, P.-L. Lions introduced structure conditions in order to have uniqueness of solutions (u, m) to the local MFG system:

$$\begin{cases} -\partial_t u - \varepsilon \Delta u + H(t, x, Du, m) = 0 & \text{in } (0, T) \times \mathbb{T}^d \\ \partial_t m - \varepsilon \Delta m - \operatorname{div}(m H_p(t, x, Du, m)) = 0 & \text{in } (0, T) \times \mathbb{T}^d \\ m(0) = m_0, \ u(x, T) = G(x, m(T))) & \text{in } \mathbb{T}^d \end{cases} \tag{1.86}$$

Assuming $H(t, x, p, m)$ to be C^1 in m and C^2 in p, the condition introduced by P.-L. Lions can be stated as the requirement that the following matrix be positive semi-definite:

$$
\begin{pmatrix}
m \, \partial_{pp}^2 H & \dfrac{1}{2} m \, \partial_{pm}^2 H \\[2ex]
\dfrac{1}{2} m \, (\partial_{pm}^2 H)^T & -\partial_m H
\end{pmatrix}
\geq 0 \qquad \forall (t, x, p, m) \,. \tag{1.87}
$$

Notice that condition (1.87) implies that H is convex with respect to p and nonincreasing with respect to m. In particular, when H has a separate form: $H = \tilde{H}(t, x, p) - f(x, m)$, condition (1.87) reduces to $\tilde{H}_{pp} \geq 0$ and $f_m \geq 0$. As usual, this condition needs to be taken in a strict form, so that Lions' result would state as follows in terms of smooth solutions.

Theorem 1.13 *Assume that $G(x, m)$ is nondecreasing in m and that $H = H(t, x, p, m)$ is a C^1 function satisfying (we omit the (t, x) dependence for simplicity)*

$$
\begin{aligned}
(H(p_2, m_2) - H(p_1, m_1))(m_2 - m_1) \\
-(m_2 H_p(p_2, m_2) - m_1 H_p(p_1, m_1)) \cdot (p_2 - p_1) \leq 0 \,,
\end{aligned} \tag{1.88}
$$

with equality if and only if $(m_1, p_1) = (m_2, p_2)$. Then system (1.86) has at most one classical solution.

Proof The proof is a straightforward extension of the usual monotonicity argument. Let (u_1, m_1) and (u_2, m_2) be solutions to (1.86). We set

$$
\tilde{m} = m_2 - m_1, \quad \tilde{u} = u_2 - u_1, \quad \tilde{H} = H(t, x, Du_2, m_2) - H(t, x, Du_1, m_1) \,.
$$

Then, subtracting the two equations we get

$$
\begin{aligned}
\frac{d}{dt} \int_{\mathbb{T}^d} (u_2(t) - u_1(t))(m_2(t) - m_1(t)) \\
= \int_{\mathbb{T}^d} (-\varepsilon \Delta \tilde{u} + \tilde{H}) \tilde{m} \\
+ \tilde{u}(\varepsilon \Delta \tilde{m} + \operatorname{div}(m_2 H_p(t, x, Du_2, m_2) - m_1 H_p(t, x, Du_1, m_1))) \\
= \int_{\mathbb{T}^d} \tilde{H} \tilde{m} - (m_2 H_p(t, x, Du_2, m_2) - m_1 H_p(t, x, Du_1, m_1)) \cdot D\tilde{u} \\
\leq 0
\end{aligned}
$$

by condition (1.88). Since $\int_{\mathbb{T}^d} (u_2(t) - u_1(t))(m_2(t) - m_1(t))$ vanishes at $t = 0$ and is nonnegative at $t = T$ (by monotonicity of $G(\cdot, \cdot)$), integrating the above equality between 0 and T gives

$$
\int_0^T \int_{\mathbb{T}^d} \tilde{H} \tilde{m} - (m_2 H_p(t, x, Du_2, m_2) - m_1 H_p(t, x, Du_1, m_1)) \cdot D\tilde{u} = 0 \,.
$$

Since (1.88) is assumed in strict form, this implies that $D\tilde{u} = 0$ and $\tilde{m} = 0$, so that $m_1 = m_2$ and $u_1 = u_2$. $\qquad\square$

Remark 1.14 It is immediate to check that if the matrix in (1.87) is positive definite, then (1.88) holds in the strict form. Indeed, set $\tilde{p} = p_2 - p_1$, $\tilde{m} = m_2 - m_1$ and, for $\theta \in [0, 1]$, $p_\theta = p_1 + \theta(p_2 - p_1)$, $m_\theta = m_1 + \theta(m_2 - m_1)$. Let

$$I(\theta) = (H(x, p_\theta, m_\theta) - H(x, p_1, m_1))\tilde{m} - \tilde{p} \cdot (m_\theta H_p(x, Du_\theta, m_\theta) - m_1 . H_p(x, Du_1, m_1))$$

Then

$$I'(\theta) = -\begin{pmatrix} \tilde{p}^T & \tilde{m} \end{pmatrix} \begin{pmatrix} m_\theta \, \partial_{pp}^2 H & \frac{1}{2} m_\theta \, \partial_{pm}^2 H \\ \frac{1}{2} m_\theta \, (\partial_{pm}^2 H)^T & -\partial_m H \end{pmatrix} \begin{pmatrix} \tilde{p} \\ \tilde{m} \end{pmatrix}.$$

If condition (1.87) holds with a strict sign, then the function $I(\theta)$ is decreasing and, for $(p_1, m_1) \neq (p_2, m_2)$, one has

$$I(0) = 0 > I(1) = (H(p_2, m_2) - H(p_1, m_1))(m_2 - m_1)$$
$$- (m_2 H_p(p_2, m_2) - m_1 H_p(p_1, m_1)) \cdot (p_2 - p_1).$$

We stress that another way of formulating (1.88) is exactly the requirement that $I(\theta)$ be decreasing for every $(p_1, m_1) \neq (p_2, m_2)$.

The main example of Hamiltonian satisfying (1.88) is given by the so-called congestion case.

Example 1.1 Assume that H is of the form: $H(x, p, m) = \frac{1}{2} \frac{|p|^2}{(\sigma + m)^\alpha}$, where $\sigma, \alpha > 0$. Then condition (1.87) holds if and only if $\alpha \in [0, 2]$. Notice that H is convex in p and nonincreasing in m if $\alpha \geq 0$. Checking condition (1.87) we find

$$\left(-\frac{\partial_m H}{m}\right) \partial_{pp}^2 H - \frac{1}{4} \partial_{pm}^2 H \otimes \partial_{pm}^2 H = \frac{\alpha |p|^2}{2m^{\alpha+2}} \frac{I_d}{m^\alpha} - \frac{\alpha^2}{4} \frac{p \otimes p}{m^{2\alpha+2}}$$
$$= \frac{2\alpha |p|^2 I_d}{4m^{2\alpha+2}} - \frac{\alpha^2}{4} \frac{p \otimes p}{m^{2\alpha+2}}$$

which is positive if and only if $\alpha \leq 2$.

This example (in the generalized version $H = \frac{|p|^q}{(\sigma+m)^\alpha}$ with $\alpha \leq \frac{4(q-1)}{q}$) was introduced by P.-L. Lions in [149] (Lesson 18/12 2009) as a possible mean field game model for crowd dynamics. In this case, the associated Lagrangian cost of the agents takes the form of $L(x, q) = \frac{1}{2}(\sigma + m)^\alpha |q|^2$, where q represents the velocity chosen by the controllers; the cost being higher in areas of higher density models the impact of the crowd in the individual motion. The case $\sigma = 0$ is also meaningful in this example and was treated by P.-L. Lions as well, even if it leads to a singular behavior of the Hamiltonian for $m = 0$.

As explained before, the existence of classical solutions with local couplings only holds in special cases, and this of course remains true for the general problem (1.86) (see e.g. [113, 116, 123] for a few results on smooth solutions of the congestion model). Therefore, the statement of Theorem 1.13 is of very little use. However, a satisfactory result of existence and uniqueness is proved in [5] for general Hamiltonians $H(t, x, Du, m)$ which include the congestion case (including the singular model with $\sigma = 0$). This is so far the unique general well-posedness result which exists for the local problem (1.86).

1.3.6   The Long Time Ergodic Behavior and the Turnpike Property of Solutions

It is a natural question to investigate the behavior of the MFG system (1.62) as the horizon T tends to infinity. Here we fix the diffusion constant $\varepsilon = 1$ and we consider nonlinearities F, H independent of t. As explained by Lions in [149] (see e.g. Lesson 20/11 2009), the limit of the MFG system, as the time horizon T tends to infinity, is given by the stationary ergodic problem

$$\begin{cases} \lambda - \Delta u + H(x, Du) = F(x, m) & \text{in } \mathbb{T}^d \\ -\Delta m - \text{div}\left(m\, H_p(x, Du)\right) = 0 & \text{in } \mathbb{T}^d \\ \displaystyle\int_{\mathbb{T}^d} m = 1\,, \quad \int_{\mathbb{T}^d} u = 0 \end{cases} \tag{1.89}$$

This system has also been introduced by Lasry and Lions in [143] as the limit, when the number of players tends to infinity, of Nash equilibria in ergodic differential games. Here the unknowns are (λ, u, m), where $\lambda \in \mathbb{R}$ is the so-called ergodic constant. The interpretation of the system is the following: each player wants to minimize his/her ergodic cost

$$\mathcal{J}(x, \alpha) := \inf_{\alpha} \limsup_{T \to +\infty} \mathbb{E}\left[\frac{1}{T} \int_0^T \left\{ H^*(X_t, -\alpha_t) + F(X_t, m(X_t)) \right\} dt \right]$$

where (X_t) is the solution to

$$\begin{cases} dX_t = \alpha_t dt + \sqrt{2}\, dB_t \\ X_0 = x \end{cases}$$

It turns out that, if (λ, u, m) is a classical solution to (1.89), then the optimal strategy of each player is given by the feedback $\alpha^*(x) := -H_p(x, Du(x))$ and, if X_t is the solution to

$$\begin{cases} dX_t = \alpha^*(X_t)dt + \sqrt{2}dB_t \\ X_0 = x \end{cases} \tag{1.90}$$

then $m(\cdot)$ is the invariant measure associated with (1.90) and, setting $\bar{\alpha}_t := \alpha^*(X_t)$, then $\mathcal{J}(x, \bar{\alpha}) = \lambda$ is independent of the initial position.

The "convergence" of the MFG system in $(0, T)$ towards the stationary ergodic system (1.89) was analyzed in [60, 61] when the Hamiltonian is purely quadratic (i.e. $H(x, p) = |p|^2$), in [58] where the long time behavior is completely described in case of smoothing coupling and uniformly convex Hamiltonian, and in [167] for the case of local couplings and globally Lipschitz Hamiltonian. The case of discrete time, finite states system is analyzed in [118].

The "long time stability" takes the form of a *turnpike pattern* for solutions (u^T, m^T) of system (1.62); namely, the solutions become nearly stationary for most of the time, which is related to the so-called turnpike property of optimality systems (see e.g. [170]). This pattern is clearly shown in numerical simulations as one can see in the contribution by Achdou & Lauriere in this volume. The strongest way to state this kind of behavior is through the proof of the exponential estimate

$$\|m^T(t) - \bar{m}\|_\infty + \|Du^T(t) - D\bar{u}\|_\infty \leq K(e^{-\omega t} + e^{-\omega(T-t)}) \qquad \forall t \in (0, T) \tag{1.91}$$

for some $K, \omega > 0$, where $(\bar{u}, \bar{m})$ is a solution to (1.89).

Notice that a weakest statement is also given by the time-average convergence (which is a consequence of (1.91), if this holds true)

$$\lim_{T \to +\infty} \frac{1}{T} \int_0^T \int_{\mathbb{T}^d} \left(|Du^T - D\bar{u}|^2 + |m^T - \bar{m}|^2 \right) dxdt = 0.$$

Of course this kind of convergence occurs provided the Lasry-Lions monotonicity condition holds true: in general, the behavior of the time-dependent problem can be much more complex. For instance it can exhibit periodic solutions. On that topic, see in particular [57, 88, 90, 153].

In this section we give a new proof of the turnpike property of solutions, by showing how it is possible to refine the usual fixed point argument in order to build directly the solution (u, m) embedded with the turnpike estimate (1.91). For simplicity, we develop this approach in the case of local couplings and globally Lipschitz Hamiltonian although, roughly speaking, a similar method would work for any case in which a global (in time) Lipschitz estimate is available for u.

Let us first remark that the stationary system (1.89) is well-defined.

Proposition 1.2 *Assume that (1.63) and (1.64) hold true and $F(x, m)$ is nonde-creasing in m. Then system (1.89) has a unique classical solution $(\bar{\lambda}, \bar{u}, \bar{m})$, and moreover $\bar{m} > 0$.*

The proof can be established by usual fixed point arguments, very similar as in Theorem 1.11, so we omit it.

Now we prove the exponential turnpike estimate for locally Lipschitz couplings F (without any growth restriction) and for globally Lipschitz, locally uniformly convex Hamiltonian.

Theorem 1.14 *Let $m_0 \in \mathcal{P}(\mathbb{T}^d)$. Assume that $F(x, m)$ is a Carathéodory function which is nondecreasing with respect to m and satisfies*

$$\forall K > 0, \ \exists \, c_K, \ \ell_K > 0 : \begin{cases} |F(x, m)| \leq c_K, \\ |F(x, m) - F(x, m')| \leq \ell_K |m - m'| \end{cases}$$

$$\forall x \in \mathbb{T}^d, m, m' \in \mathbb{R} : |m|, |m'| \leq K. \qquad (1.92)$$

Assume that $p \mapsto H(x, p)$ is a C^2 function which is globally Lipschitz (i.e. it satisfies (1.64)) and locally uniformly convex:

$$\forall K > 0, \ \exists \alpha_K, \beta_K > 0 : \quad \alpha_K I \leq H_{pp}(x, p) \leq \beta_K I \qquad \forall (x, p) \in \mathbb{T}^d \times \mathbb{R}^d : |p| \leq K. \qquad (1.93)$$

Then there exists $\omega, M > 0$ (independent of T) such that any solution (u^T, m^T) of problem (1.62) (with $\varepsilon = 1$) satisfies

$$\|m^T(t) - \bar{m}\|_\infty + \|Du^T(t) - D\bar{u}\|_\infty \leq M(e^{-\omega t} + e^{-\omega(T-t)}) \qquad \forall t \in (1, T-1). \qquad (1.94)$$

This kind of result is proved in [167] with a strategy based on the stabilization properties of the linearized system; an approach which explains the exact exponen-tial rate ω in (1.94) in terms of a Riccati feedback operator. Here we give a new direct proof of (1.94), mostly based on ideas in [58]. This approach is less precise in the rate ω but requires less demanding assumptions and avoids the formal use of the linearized system, though some form of linearization appears in obtaining the following a priori estimate.

Lemma 1.14 *Under the assumptions of Theorem 1.14, let $(\bar{\lambda}, \bar{u}, \bar{m})$ be the unique solution of (1.89).*

For $\sigma \in [0, 1]$, $m_0 \in L^\infty(\mathbb{T}^d) \cap \mathcal{P}(\mathbb{T}^d)$, $v_T \in C^{1,\alpha}(\mathbb{T}^d)$ for some $\alpha \in (0, 1)$, let (μ, v) be a solution of the system

$$\begin{cases} -\partial_t v - \Delta v + H(x, D\bar{u} + Dv) - H(x, D\bar{u}) = F(x, \bar{m} + \mu) - F(x, \bar{m}) \\ v(T) = v_T \\ \partial_t \mu - \Delta \mu - \mathrm{div}(\mu\, H_p(x, D\bar{u} + Dv)) = \sigma\, \mathrm{div}(\bar{m}\,[H_p(x, D\bar{u} + Dv) - H_p(x, D\bar{u})]) \\ \mu(0) = \sigma(m_0 - \bar{m}). \end{cases}$$

$$(1.95)$$

Then there exist constants $\omega, K > 0$ such that

$$\|\mu(t)\|_2 + \|Dv(t)\|_2 \le K(e^{-\omega t} + e^{-\omega(T-t)})[\|m_0 - \bar{m}\|_2 + \|Dv_T\|_2]. \quad (1.96)$$

Proof We first notice that, using the equation satisfied by $\bar{m}$, we can derive the equation satisfied by $\mu + \sigma\bar{m}$ and we deduce immediately that $\mu + \sigma\bar{m} \ge 0$. Since $\int_{\mathbb{T}^d} \mu(t) = 0$ for every t, this implies that $\|\mu(t)\|_{L^1(\mathbb{T}^d)} \le 2\sigma$ for every $t > 0$. Since H_p is globally bounded, and $\bar{m} \in L^\infty(\mathbb{T}^d)$, by standard (local in time) regularizing effect in the Fokker-Planck equation, we have $\|\mu(t)\|_\infty \le C\|\mu(t-1)\|_{L^1(\mathbb{T}^d)}$ for every $t > 1$ (see e.g. [142, Chapter V]). In addition, since $m_0 \in L^\infty(\mathbb{T}^d)$, we have that $\|\mu(t)\|_\infty$ is bounded for $t \in (0, 1)$ as well. From the global L^1 bound, we conclude therefore that $\|\mu(t)\|_\infty$ is bounded uniformly, for every $t \in (0, T)$, by a constant independent of the horizon T. Due to (1.92), this means that the function $F(x, \cdot)$ in the first equation can be treated as uniformly bounded and Lipschitz. The global bound on the right-hand side, together with the global Lipschitz character of the Hamiltonian, and the fact that $v_T \in C^{1,\alpha}(\mathbb{T}^d)$ for some $\alpha \in (0, 1)$, allow us to deduce the existence of a constant L, independent of T, such that $\|Dv(t)\|_\infty \le L$ for every $t \in (0, T)$. Due to (1.93), this means that $H(x, \cdot)$ can be treated as uniformly convex. Therefore, if we set

$$h(x, p) := H(x, D\bar{u}(x) + p) - H(x, D\bar{u}(x))$$
$$f(x, \mu) := F(x, \bar{m}(x) + \mu) - F(x, \bar{m}(x))$$
$$B(x, p) := \bar{m}(x)\,[H_p(x, D\bar{u}(x) + p) - H_p(x, D\bar{u}(x))],$$

we have that (v, μ) solves the system

$$\begin{cases} -\partial_t v - \Delta v + h(x, Dv) = f(x, \mu) \\ v(T) = v_T \\ \partial_t \mu - \Delta \mu - \mathrm{div}(\mu\, h_p(x, Dv)) = \sigma\, \mathrm{div}(B(x, Dv)) \\ \mu(0) = \sigma\, \mu_0 \end{cases}$$

$$(1.97)$$

where $\mu_0 = m_0 - \bar{m}$, and where $h(x, p)$, $f(x, s)$, $B(x, p)$ satisfy the following conditions for some constants c_0, C_0, C_1, C_2 and for every $s \in \mathbb{R}$, $p \in \mathbb{R}^d$, $x \in \mathbb{T}^d$:

$$h(x, 0) = 0, \quad |h_p(x, p)| \le c_0, \tag{1.98}$$

$$f(x, s)s \ge 0, \quad |f(x, s)| \le C_0, \quad |f(x, s)| \le C_1 |s| \tag{1.99}$$

$$B(x, p) \cdot p \ge C_2^{-1} |p|^2, \quad |B(x, p)| \le C_2 |p|. \tag{1.100}$$

In addition, since $\mu(t, x) \ge -\sigma \bar{m}(x)$, we also have, for some constant γ_0,

$$
\begin{aligned}
&\sigma B(x, p) \cdot p - \mu(t, x)(h(x, p) - h_p(x, p) \cdot p) \\
&\ge \sigma B(x, p) \cdot p - \sigma \bar{m}(x)(h_p(x, p) \cdot p - h(x, p)) \\
&= \sigma \bar{m}(x) \left[-H_p(x, D\bar{u}(x)) \cdot p + H(x, D\bar{u}(x) + p) - H(x, D\bar{u}(x)) \right] \\
&\ge \sigma \gamma_0 |p|^2 \quad \forall (t, x) \in Q_T, \forall p \in \mathbb{R}^d : |p| \le L,
\end{aligned}
\tag{1.101}
$$

where we used the local uniform convexity of H and that $\bar{m} > 0$.

We now derive the exponential estimate for system (1.97) under conditions (1.98)–(1.101).

Given $T > 0$, $\sigma \in [0, 1]$, $\mu_0 \in L^2(\mathbb{T}^d)$ with $\int_{\mathbb{T}^d} \mu_0 = 0$, $v_T \in H^1(\mathbb{T}^d)$, we denote by $S(T, \sigma, \mu_0, v_T)$ the solution (μ, v) of system (1.97). We will denote by $\langle v \rangle = \int_{\mathbb{T}^d} v$ and by $\tilde{v} = v - \langle v \rangle$. We first prove that there exists a constant C, independent of σ, T, μ_0, v_T, such that

$$\|\mu(t)\|_2 + \|Dv(t)\|_2 \le C(\|\mu_0\|_2 + \|Dv_T\|_2) \quad \forall (\mu, v) \in S(T, \sigma, \mu_0, v_T). \tag{1.102}$$

To prove (1.102), we observe that, due to (1.101),

$$
\begin{aligned}
-\frac{d}{dt} \int_{\mathbb{T}^d} \mu(t)v(t) &= \int_{\mathbb{T}^d} f(x, \mu)\mu + \sigma \int_{\mathbb{T}^d} B(x, Dv)Dv \\
&\quad - \int_{\mathbb{T}^d} \mu(h(x, Dv) - h_p(x, Dv) \cdot Dv) \\
&\ge \sigma \gamma_0 \int_{\mathbb{T}^d} |Dv|^2.
\end{aligned}
\tag{1.103}
$$

From the Fokker-Planck equation we also have (see e.g. [58, Lemma 1.1]) that there exists $\gamma, c > 0$:

$$\|\mu(t)\|_2^2 \le c\,e^{-\gamma t}\|\mu_0\|_2^2 + c\,\sigma^2 \int_0^t \int_{\mathbb{T}^d} |B(x, Dv)|^2$$

$$\le c\,e^{-\gamma t}\|\mu_0\|_2^2 + c\,\sigma \int_0^t \int_{\mathbb{T}^d} |Dv|^2$$

where we used (1.100). Here and after we denote by c possibly different constants independent of T, σ, μ_0, v_T. Putting together the above inequalities we get

$$\|\mu(t)\|_2^2 \le c\,e^{-\gamma t}\|\mu_0\|_2^2 + c \int_{\mathbb{T}^d} \mu_0 v(0) - c \int_{\mathbb{T}^d} \mu(T) v_T$$

which implies

$$\sup_{[0,T]} \|\mu(t)\|_2^2 \le c\,[\|\mu_0\|_2^2 + \|Dv_T\|_2^2] + c\,\|\mu_0\|_2\|\tilde{v}(0)\|_2. \tag{1.104}$$

Since the Hamilton-Jacobi equation implies (using $|f(x, \mu)| \le c\,|\mu|$ and [58, Lemma 1.2])

$$\|\tilde{v}(0)\|_2 \le c\,e^{-\gamma T}\|\tilde{v}_T\|_2 + c \int_0^T e^{-\gamma s}\|\mu(s)\|_2 ds \le c[\|\tilde{v}_T\|_2 + \sup_{[0,T]} \|\mu(t)\|_2]$$

coming back to (1.104) we deduce that

$$\sup_{[0,T]} \|\mu(t)\|_2^2 \le c\,[\|\mu_0\|_2^2 + \|\tilde{v}_T\|_2^2].$$

A similar estimate holds for $\sup_{[0,T]} \|\tilde{v}(t)\|_2$ as well. Finally, using e.g. [58, Lemma 1.2] we have

$$\|\nabla v(t)\|_2^2 \le c(\|\tilde{v}(t+1)\|_2^2 + c \int_t^{t+1} [\|\mu(s)\|_2^2 + \tilde{v}(s)^2])$$

hence

$$\|\nabla v(t)\|_2^2 \le c\,[\|\mu_0\|_2^2 + \|\tilde{v}_T\|_2^2] \quad \forall t < T - 1.$$

Standard parabolic estimates also imply that

$$\|\nabla v(t)\|_2^2 \le c\,[\sup_{[T-1,T]} \|\mu(t)\|_2^2 + \|Dv_T\|_2^2] \quad \forall t \in [T-1, T]$$

so that (1.102) is proved. Now we set

$$\rho(\tau) := \sup_{T \geq 2\tau} \sup_{\sigma, \mu_0, v_T} \left\{ \sup_{t \in [\tau, T-\tau]} \left| \frac{\int_{\mathbb{T}^d} v(t)\,\mu(t)}{[\|\mu_0\|_2 + \|Dv_T\|_2]^2} \right| , \quad (\mu, v) \in S(T, \sigma, \mu_0, v_T) \right\}.$$

We first remark that, by elementary inclusion property, one has $\rho(\tau + s) \leq \rho(\tau)$ for every $s > 0$. Hence $\rho(\cdot)$ is a non increasing function and we can define

$$\rho(\infty) := \lim_{\tau \to \infty} \rho(\tau).$$

As a first step, we shall prove that $\rho_\infty = 0$. To this purpose, we observe that by definition of ρ there exist sequences $\tau_n \to \infty$, $T_n \geq 2\tau_n$, $t_n \in [\tau_n, T_n - \tau_n]$, $\mu_0^n \in L^\infty(\mathbb{T}^d)$, $v_{T_n}^n \in W^{1,\infty}(\mathbb{T}^d)$ and $\sigma_n \in [0,1]$ such that

$$(\mu^n, v^n) \in S(T_n, \sigma_n, \mu_0^n, v_{T_n}^n), \qquad \left| \frac{\int_{\mathbb{T}^d} \mu^n(t_n) v^n(t_n)}{[\|\mu_0^n\|_2 + \|Dv_{T_n}^n\|_2]^2} \right| \geq \rho_\infty - 1/n.$$

We set, for $t \in [-t_n, T_n - t_n]$:

$$\tilde{\mu}^n(t, x) = \delta_n \mu^n(t_n + t, x), \quad \tilde{v}^n(t, x) = \delta_n(v^n(t_n + t, x) - \langle v^n(t_n) \rangle)$$

$$\delta_n := \frac{1}{\|\mu_0^n\|_2 + \|Dv_{T_n}^n\|_2}$$

and we notice that $(\tilde{\mu}^n, \tilde{v}^n)$ solve the system

$$\begin{cases} -\tilde{v}_t^n - \Delta \tilde{v}^n + \delta_n\, h(x, Dv^n) = \delta_n\, f(x, \mu^n) \\ \tilde{\mu}_t^n - \Delta \tilde{\mu}^n - \operatorname{div}(\tilde{\mu}^n\, h_p(x, Dv^n)) = \delta_n \sigma_n \operatorname{div}(B(x, Dv^n)) \end{cases}$$

where v^n, μ^n are computed at $t_n + t$. By estimate (1.102), $\|\tilde{\mu}^n(t)\|_2$ and $\|D\tilde{v}^n(t)\|_2$ are uniformly bounded. Hence, due to (1.98) and (1.99), $-\partial_t \tilde{v}^n - \Delta \tilde{v}^n$ is uniformly bounded in $L^2(\mathbb{T}^d)$, which implies that $\tilde{v}^n$ is relatively compact in $C^0([a,b]; L^2(\mathbb{T}^d))$, for every interval $[a,b]$. In particular, there exists $\tilde{v} \in L^2_{loc}(\mathbb{R}; L^2(\mathbb{T}^d))$ such that $\tilde{v}^n(t) \to \tilde{v}(t)$ in $L^2(\mathbb{T}^d)$ for every $t \in \mathbb{R}$, and $D\tilde{v}^n \to D\tilde{v}$ in $L^2((a,b) \times \mathbb{T}^d)$ for every bounded interval (a,b). Let us call respectively $\tilde{\mu}, \sigma$ a limit of (a subsequence of) $\tilde{\mu}^n, \sigma_n$; since $\tilde{\mu}^n(t)$ weakly converges to $\mu(t)$ in $L^2(\mathbb{T}^d)$, we have that the scalar product $\int_{\mathbb{T}^d} \tilde{\mu}^n(t) \tilde{v}^n(t)$ converges for every $t \in \mathbb{R}$. It follows from (1.103) (integrated between $t_n + t_1$ and $t_n + t_2$) and from (1.100), that

$$\sigma\, \gamma_0 \int_{t_1}^{t_2} \int_{\mathbb{T}^d} |D\tilde{v}|^2 \leq \liminf_{n \to \infty} \sigma_n\, \gamma_0\, \delta_n^2 \int_{t_n + t_1}^{t_n + t_2} \int_{\mathbb{T}^d} |Dv^n|^2$$

$$\leq \int_{\mathbb{T}^d} \tilde{\mu}(t_1) \tilde{v}(t_1) - \int_{\mathbb{T}^d} \tilde{\mu}(t_2) \tilde{v}(t_2) \qquad (1.105)$$

for every fixed $t_1, t_2 \in \mathbb{R}$. By construction, we also have

$$\rho_\infty - 1/n \leq \left| \frac{\int_{\mathbb{T}^d} \mu^n(t_n) v^n(t_n)}{[\|\mu_0^n\|_\infty + \|Dv_{T_n}^n\|_\infty]^2} \right| \leq \rho(\tau_n) \to \rho_\infty , \qquad (1.106)$$

hence

$$\rho_\infty = \lim_{n \to \infty} \left| \int_{\mathbb{T}^d} \tilde{\mu}^n(0)\tilde{v}^n(0) \right| = \left| \int_{\mathbb{T}^d} \tilde{\mu}(0)\tilde{v}(0) \right| .$$

On another hand, for any $t \in \mathbb{R}$ and for n large enough, we have that $t_n + t \in [\tau_n - |t|, T_n - (\tau_n - |t|)]$, so that

$$\left| \int_{\mathbb{T}^d} \tilde{\mu}(t)\tilde{v}(t) \right| = \lim_n \left| \frac{\int_{\mathbb{T}^d} \mu^n(t_n + t) v^n(t_n + t)}{[\|\mu_0^n\|_\infty + \|Dv_{T_n}^n\|_\infty]^2} \right| \leq \lim_n \rho(\tau_n - |t|) = \rho_\infty.$$
$$(1.107)$$

Now suppose that $\rho_\infty > 0$ and $\sigma > 0$. If $\rho_\infty = \int_{\mathbb{T}^d} \tilde{\mu}(0)\tilde{v}(0) > 0$; using (1.105) with $t_2 = 0$ we deduce, due to (1.107), that $\int_{t_1}^0 \int_{\mathbb{T}^d} |D\tilde{v}|^2 \leq 0$. This implies that $|\tilde{v}(0)| = 0$. If $\rho_\infty = -\int_{\mathbb{T}^d} \tilde{\mu}(0)\tilde{v}(0)$, we get at the same conclusion by choosing now $t_1 = 0$ in (1.105). But $\tilde{v}(0) = 0$ is impossible unless $\rho_\infty = 0$. It remains the case that $\sigma = 0$; this means that $\tilde{\mu}$ satisfies

$$\partial_t \tilde{\mu} - \Delta \tilde{\mu} - \mathrm{div}(\tilde{\mu} \, b) = 0$$

for a bounded drift $b(t, x)$. But this readily leads to $\tilde{\mu} = 0$ (because $\|\tilde{\mu}(t)\| \leq e^{-\omega(t-t_0)}\|\tilde{\mu}(t_0)\|$ for all t_0, t and $\tilde{\mu}$ is uniformly bounded), and again this implies $\rho_\infty = 0$.

So we proved that $\rho_\infty = 0$. We claim now that this implies the existence of t_0 such that

$$\|\mu(t)\|_2 + \|Dv(t)\|_2 \leq \frac{1}{2}[\|\mu_0\|_2 + \|Dv_T\|_2] \qquad \forall t \in [t_0, T - t_0]. \qquad (1.108)$$

In fact, using the Fokker-Planck equation and (1.103), for every $t \in [\tau, T - \tau]$ we have

$$\|\mu(t)\|_2^2 \leq c \, e^{-\gamma(t-\tau)}\|\mu(\tau)\|_2^2 + c\sigma^2 \int_\tau^{T-\tau} \int_{\mathbb{T}^d} |B(x, Dv)|^2$$

$$\leq c \, e^{-\gamma(t-\tau)}[\|\mu_0\|_2^2 + \|Dv_T\|_2]^2$$

$$+ c \left\{ \left| \int_{\mathbb{T}^d} \mu(\tau)v(\tau) \right| + \left| \int_{\mathbb{T}^d} \mu(T-\tau)v(T-\tau) \right| \right\},$$

hence

$$\|\mu(t)\|_2^2 \le c[\|\mu_0\|_2^2 + \|Dv_T\|_2]^2 \left(e^{-\gamma(t-\tau)} + \rho(\tau)\right). \tag{1.109}$$

Similarly we have, using now the estimate for μ,

$$\|\tilde{v}(t)\|_2^2 \le c\, e^{-\gamma(T-\tau-t)}\|\tilde{v}(T-\tau)\|_2^2 + c \int_t^{T-\tau} e^{-\gamma(s-t)}\|\mu(s)\|_2^2 ds$$

$$\le c\,[\|\mu_0\|_2 + \|Dv_T\|_2]^2 \left(e^{-\gamma(T-\tau-t)} + e^{-\gamma(t-\tau)} + \rho(\tau)\right)$$

which implies, for every $t \in (\tau, T - \tau - 1)$:

$$\|\nabla v(t)\|_2^2 \le c(\|\tilde{v}(t+1)\|_2^2 + c \int_t^{t+1} [\|\mu(s)\|_2^2 + \tilde{v}(s)^2])$$

$$\le c\,[\|\mu_0\|_2 + \|Dv_T\|_2]^2 \left(e^{-\gamma(T-\tau-t)} + e^{-\gamma(t-\tau)} + \rho(\tau)\right). \tag{1.110}$$

Since $\rho(\tau) \xrightarrow{\tau \to \infty} 0$, from (1.109)–(1.110) we obtain (1.108) by choosing τ and t conveniently. Finally, by iteration of (1.108), we deduce the exponential estimate (1.96). $\qquad\square$

Proof of Theorem 1.14 Let us first assume that $m_0 \in C^\alpha(\mathbb{T}^d)$. We set $X = C^0([0, T]; L^2(\mathbb{T}^d))$ and we introduce the following norm in X:

$$|||u|||_X := \sup_{[0,T]} \left(\frac{\|u(t)\|_{L^2(\mathbb{T}^d)}}{e^{-\omega t} + e^{-\omega(T-t)}}\right)$$

where $\omega > 0$ is given by Lemma 1.14. It is easy to verify that $(X, |||u|||_X)$ is a Banach space and $||| \cdot |||$ is equivalent to the standard norm $\|u\| = \sup_{[0,T]} \|u(t)\|_{L^2(\mathbb{T}^d)}$.

We define the operator T on X as follows: given $\mu \in X$, let (v, ρ) be the solution to the system

$$\begin{cases} -v_t - \Delta v + H(x, D\bar{u} + Dv) - H(x, D\bar{u}) = F(x, \bar{m} + \mu) - F(x, \bar{m}) \\ v(T) = G(x, \bar{m} + \mu(T)) - \bar{u} \\ \rho_t - \Delta\rho - \text{div}(\rho\, H_p(x, D\bar{u} + Dv)) = -\text{div}(\bar{m}\,[H_p(x, D\bar{u} + Dv) - H_p(x, D\bar{u})]) \\ \rho(0) = m_0 - \bar{m} \end{cases}$$
$$\tag{1.111}$$

then we set $\rho = T\mu$. Since H_p is globally bounded, and $\bar{m}, m_0 \in C^{0,\alpha}$, by standard regularity results (see [142, Chapter V, Thms 1.1 and 2.1]) we notice that the range of T is bounded in $C^{\alpha/2,\alpha}(Q_T)$, in particular the range of T lies in a bounded set of L^∞ and its closure is compact in X. As a consequence, due to (1.92), there is no loss of generality if we consider $F(x, \cdot)$ to be globally bounded and Lipschitz. Using now the global bound on μ and proceeding as in Lemma 1.14, a global bound for $\|Dv(t)\|_2$ follows, and then, by (local) regularizing effect of parabolic equations, we deduce that there exists a constant $L > 0$ such that

$$\|Dv(t)\|_\infty \le L \qquad \forall t \le T - 1, \qquad \forall \mu \in X. \qquad (1.112)$$

We now check that the operator T is continuous: if $\mu_n \to \mu$ in X, then $\mu_n(T)$ is strongly convergent in $L^2(\mathbb{T}^d)$, and $F(x, \bar{m} + \mu_n) - F(x, \bar{m})$ converges in $L^2(Q_T)$ as well. By standard parabolic theory, we have that Dv_n converges in $L^2(Q_T)$ to Dv where v is a solution corresponding to μ. The convergence of Dv_n in L^2 and the boundedness of H_p imply that the drift and source terms in the equation of ρ_n converge in $L^p(Q_T)$ for every $p < \infty$. This immediately implies the convergence of ρ_n in $L^2(0, T; H^1(\mathbb{T}^d))$ and then in $C^0([0, T]; L^2(\mathbb{T}^d))$ as well. By uniqueness, we deduce that ρ_n converges to $T\mu$. This concludes the continuity of T. Thus, T is a compact and continuous operator. We are left with the following claim: there exists a constant $M > 0$ such that

$$|||\mu||| \le M \quad \text{for every } \mu \in X \text{ and every } \sigma \in [0, 1] \text{ such that } \mu = \sigma T(\mu). \qquad (1.113)$$

In order to prove (1.113), we use Lemma 1.14 in the interval $(0, T - 1)$; indeed if $\mu = \sigma T(\mu)$, then (μ, v) is a solution to (1.95) with $v_{T-1} = v(T - 1)$. Therefore, there exists $K > 0$ (only depending on $\|m_0\|_\infty$, and $F, H, \bar{u}, \bar{m}$) such that

$$\|\mu(t)\|_2 + \|Dv(t)\|_2 \le K(e^{-\omega t} + e^{-\omega(T-t)}) \qquad \forall t \in (0, T - 1).$$

Since $\|\mu(t)\|_2$ is uniformly bounded for $t \in (T - 1, T)$, we conclude that (1.113) holds true for some $M > 0$. By the Schaefer's fixed point theorem [110, Thm 11.3], we conclude that there exists $\mu \in X$ such that $\mu = T\mu$. Setting $m = \bar{m} + \mu$, $u = \bar{u} + \bar{\lambda}(T - t) + v$, we have found a solution of the MFG system (1.62) which satisfies the estimate

$$\|m(t) - \bar{m}\|_2 + \|Du(t) - D\bar{u}\|_2 \le C(e^{-\omega t} + e^{-\omega(T-t)}) \qquad \forall t \in (0, T - 1).$$

To conclude with the general case, let $m_0 \in \mathcal{P}(\mathbb{T}^d)$ and let (u, m) be any solution to system (1.62). By mass conservation and the global Lipschitz bound (1.64), there exists $\alpha \in (0, 1)$ and a constant C, only depending on β, such that

$$\|m(t)\|_{C^\alpha(\mathbb{T}^d)} \le C \qquad \forall t \ge \frac{1}{2}.$$

In turn, this implies that

$$\|Du(t)\|_\infty \le C \qquad \forall t \le T - \frac{1}{2}.$$

for a possibly different constant only depending on β, F, G. By monotonicity of $F(x, \cdot)$, (u, m) is the unique solution of the MFG system in $(\frac{1}{2}, T - \frac{1}{2})$ with initial-terminal conditions given by $m(\frac{1}{2})$ and $u(T - \frac{1}{2})$ respectively. By the first part of the proof, we know that this unique MFG solution satisfies the exponential turnpike estimate. Hence there exists $M > 0$ such that

$$\|m^T(t) - \bar{m}\|_2 + \|Du^T(t) - D\bar{u}\|_2 \le M(e^{-\omega t} + e^{-\omega(T-t)}) \qquad \forall t \in (1/2, T-1/2).$$

Using the regularizing effect of the two equations, this estimate is upgraded to L^∞-norms and yields (1.94).   □

Let us stress that the turnpike estimate (1.94) gives an information in a long intermediate time of the interval $(0, T)$. A stronger result can also be obtained, by showing the convergence of $(u^T(t), m^T(t))$ at any time scale, i.e. for every $t \in (0, T)$. More precisely, there exists (u, m) solution of the problem in $(0, \infty)$:

$$\begin{cases} -\partial_t u + \bar{\lambda} - \Delta u + H(x, Du) = F(x, m) & \text{in } (0, \infty) \times \mathbb{T}^d, \\ \partial_t m - \Delta m - \text{div}(m\, H_p(x, Du)) = 0 & \text{in } (0, \infty) \times \mathbb{T}^d, \\ m(0) = m_0, \qquad Du \in D\bar{u} + L^2((0, \infty) \times \mathbb{T}^d), \ u \text{ bounded} \end{cases}$$

$$(1.114)$$

such that

$$u^T(t) + \bar{\lambda}(T - t) \overset{T\to\infty}{\to} u(t) \qquad ; \qquad m^T(t) \overset{T\to\infty}{\to} m(t)$$

where the convergence is uniform (locally in time). We notice that, since $F(x, \cdot)$ is nondecreasing, there is a unique m which solves problem (1.114), while u is unique up to addition of a constant. Nevertheless the above convergence holds for the whole sequence $T \to \infty$. i.e. there is a unique solution u of the infinite horizon problem which is selected in the limit of $u^T(t) + \bar{\lambda}(T - t)$. We also point out that $\bar{m}$ is the unique invariant measure of the Fokker-Planck equation (hence $m(t) \to \bar{m}$ as $t \to \infty$). Finally, the same problem (1.114) is obtained as the limit of the discounted MFG problem when the discount factor vanishes. The discounted (infinite horizon) problem is described by the system

$$\begin{cases} -\partial_t u + \delta u - \Delta u + H(x, Du) = F(x, m) & \text{in } (0, \infty) \times \mathbb{T}^d, \\ \partial_t m - \Delta m - \text{div}(m\, H_p(x, Du)) = 0 & \text{in } (0, \infty) \times \mathbb{T}^d, \\ m(0) = m_0, \ u \text{ bounded} \end{cases}$$

and corresponds to the following minimization problem for each agent:

$$\mathcal{J}(x, \alpha) = \inf_{\alpha} \mathbb{E}\left[\int_0^{+\infty} e^{-\delta t}\left(H^*(X_t, -\alpha_t) + F(X_t, m(t))\right) dt\right]$$

where $\delta > 0$ is a fixed discount rate. In case of monotone couplings, the limit as $\delta \to 0$ produces a unique solution of (1.114) and is, once more, related to the ergodic behavior of the controlled system. We refer to [58] where all the above mentioned results are proved for smoothing couplings in connection with the long time behavior and the ergodic properties of the master equation. A different proof is also given in [91] for local couplings and Lipschitz Hamiltonians.

We conclude by mentioning that the aforementioned results, and specifically the exponential convergence, mostly rely on the presence of the diffusion term in the equations (the individual noise of the agents). Indeed, in case of first order MFG systems, only partial results are known, even in case of monotone couplings. The typical result proved so far consists in the long time average convergence towards the ergodic first order system, see [52, 53] for the case of, respectively, smoothing and local couplings.

Remark 1.15 It is well-known that the ergodic behavior of Hamilton-Jacobi equations has strict connections with the study of homogenization problems. To this respect, the study of MFG systems is still largely open. MFG problems with fast oscillation in the space variable (homogenization) are studied in [82, 151]. Interestingly, the monotonicity structure of MFG might be lost after homogenization (although very recent results by Lions show that some structure is preserved).

1.3.7 The Vanishing Viscosity Limit and the First Order System with Local Couplings

1.3.7.1 Existence and Uniqueness of Solutions

We now analyze the vanishing viscosity limit of weak solutions. Compared to the case of smoothing couplings, now we cannot rely anymore on the semi concavity estimates for u, and the relaxed solutions obtained for the deterministic problem fall outside the viscosity solutions setting. However, the monotonicity of the couplings, the coercivity of the Hamiltonian and, eventually, the stability properties of the system, will allow us to handle the two equations in a purely distributional sense.

To fix the ideas, we still assume that the Hamiltonian $H(x, p)$ is convex and C^1 with respect to p and satisfies assumptions (1.66)–(1.68). We also assume that

$F, G \in C(\mathbb{T}^d \times \mathbb{R}_+)$ are nondecreasing functions of m which verify, for some constants $C_i > 0$,

$\exists f \in C(\mathbb{R}_+, \mathbb{R}_+)$ nondecreasing, with $\lim_{s \to +\infty} f(s) = +\infty$

and $f(s)s$ convex, such that

$$C_0 f(m) - C_1 \le F(x, m) \le f(m) + C_1, \qquad \forall(x, m) \in \mathbb{T}^d \times \mathbb{R}_+$$

$$(1.115)$$

$\exists g \in C(\mathbb{R}_+, \mathbb{R}_+)$ nondecreasing, with $\lim_{s \to +\infty} g(s) = +\infty$

and $g(s)s$ convex, such that

$$C_2 g(m) - C_3 \le G(x, m) \le g(m) + C_3, \qquad \forall(x, m) \in \mathbb{T}^d \times \mathbb{R}_+.$$

$$(1.116)$$

Of course the simplest example occurs when $f(s)$ and $g(s)$ are power-type functions, as considered e.g. in [62]. Both nonlinearities F and H could also depend (in a measurable way) on t, but this would not add any change in the following, so we omit this dependence to shorten notations.

The key point here is to consider the duality between weak *sub solutions* of Hamilton-Jacobi equation and weak solutions of the continuity equation. This topic has an independent interest for PDEs especially in connection with the theory of optimal transport.

Definition 1.6 Given $f \in L^1(Q_T)$, $g \in L^1(\mathbb{T}^d)$, a function $u \in L^2(0, T; H^1(\mathbb{T}^d))$ is a weak sub solution of

$$\begin{cases} -\partial_t u + H(x, Du) = f(t, x) \\ u(T, x) = g(x) \end{cases} \qquad (1.117)$$

if it satisfies

$$\int_0^T \int_{\mathbb{T}^d} u\, \partial_t \varphi + \int_0^T \int_{\mathbb{T}^d} H(x, Du)\, \varphi \le \int_0^T \int_{\mathbb{T}^d} f\, \varphi + \int_{\mathbb{T}^d} g\varphi(T)$$

$$\forall \varphi \in C_c^1((0, T] \times \mathbb{T}^d),\ \varphi \ge 0. \qquad (1.118)$$

Hereafter, we will shortly write $-\partial_t u + H(x, Du) \le f$ and $u(T) \le g$ to denote the previous inequality.

Let us point out that, since H is bounded below thanks to condition (1.66), any sub solution u according to the above definition is time-increasing up to an absolutely continuous function; in particular, u admits a right-continuous Borel representative and admits one-sided limits at any $t \in [0, T]$. We refer the reader to

[158, Section 4.2] for the analysis of trace properties of u. We will use in particular the existence of a trace at time $t = 0$ for u; this trace should be understood in the sense of limits of measurable functions (convergence in measure of $u(t, x)$ as $t \to 0^+$).

Definition 1.7 Let $m_0 \in \mathcal{P}(\mathbb{T}^d)$. Given a measurable vector field $b : Q_T \to \mathbb{R}$, a function $m \in L^1(Q_T)$ is a weak solution of the continuity equation

$$\begin{cases} \partial_t m - \mathrm{div}(m\,b) = 0 \\ m(0) = m_0 \end{cases} \tag{1.119}$$

if $m \in C^0([0, T]; \mathcal{P}(\mathbb{T}^d))$, $\int_0^T \int_{\mathbb{T}^d} m\,|b|^2 < \infty$ and the distributional equality holds:

$$-\int_0^T \int_{\mathbb{T}^d} m\,\partial_t \varphi + \int_0^T \int_{\mathbb{T}^d} m\,b \cdot D\varphi = \int_{\mathbb{T}^d} m_0 \varphi(0) \qquad \forall \varphi \in C_c^1([0, T) \times \mathbb{T}^d).$$

$$\tag{1.120}$$

Let us recall that the requirement that $\int_0^T \int_{\mathbb{T}^d} m\,|b|^2 < \infty$ is very natural in the framework of weak solutions to the continuity equation, and this is related with the fact that $m(t)$ is an absolutely continuous curve in $\mathcal{P}(\mathbb{T}^d)$ with L^2 metric velocity, see [18].

Standing on the above two definitions, we have a weak setting for the deterministic MFG system. For simplicity, we restrict hereafter to the case that $m_0 \in L^1(\mathbb{T}^d)$.

Definition 1.8 A pair $(u, m) \in L^2(0, T; H^1(\mathbb{T}^d)) \times L^1(Q_T)_+$ is a weak solution to the first order MFG system

$$\begin{cases} -\partial_t u + H(x, Du) = F(x, m) \\ u(T) = G(x, m(T)) \end{cases} \tag{1.121}$$

$$\begin{cases} \partial_t m - \mathrm{div}(m\,H_p(x, Du)) = 0 \\ m(0) = m_0 \end{cases} \tag{1.122}$$

if

(i) $F(x, m)m \in L^1(Q_T)$, $G(x, m(T))m(T) \in L^1(\mathbb{T}^d)$, $m|Du|^2 \in L^1(Q_T)$, and u is bounded below
(ii) u is a weak sub solution of (1.121), $m \in C^0([0, T]; \mathcal{P}(\mathbb{T}^d))$ is a weak solution of (1.122)

(iii) u and m satisfy the following identity:

$$\int_{\mathbb{T}^d} m_0\, u(0)\, dx = \int_{\mathbb{T}^d} G(x, m(T))\, m(T)\, dx + \int_0^T \int_{\mathbb{T}^d} F(t, x, m)m\, dx dt$$

$$+ \int_0^T \int_{\mathbb{T}^d} m\, \big[H_p(x, Du) \cdot Du - H(x, Du)\big] dx dt$$

$$\tag{1.123}$$

A key point is played by the following lemma, which justifies the duality between weak sub solutions of the Hamilton-Jacobi equation and weak solutions of the continuity equation. This also gives sense to the first term in (1.123), where we recall that the value $u(0)$ is the trace of $u(t)$ as explained before.

In the following, for a convex super linear function $\phi : \mathbb{R}^d \to \mathbb{R}$, we denote by ϕ^* its Legendre transform defined as $\phi^*(q) = \sup_{p \in \mathbb{R}^d}[q \cdot p - \phi(p)]$.

Lemma 1.15 *Let u be a weak sub solution of (1.117) and m be a weak solution of (1.119). Assume that f, g, u are bounded below and there exist convex increasing and superlinear functions ϕ_1, ϕ_2 such that $\phi_1(m), \phi_1^*(f) \in L^1(Q_T)$ and $\phi_2(m(T)), \phi_2^*(g) \in L^1(\mathbb{T}^d)$.*

Then we have $m|Du|^2 \in L^1(Q_T)$, $u(0)m_0 \in L^1(\mathbb{T}^d)$ and

$$\int_{\mathbb{T}^d} m_0\, u(0)\, dx \; \le \; \int_{\mathbb{T}^d} g\, m(T)\, dx + \int_0^T \int_{\mathbb{T}^d} f\, m\, dx dt$$

$$+ \int_0^T \int_{\mathbb{T}^d} m\, [b \cdot Du - H(x, Du)]\, dx dt \qquad (1.124)$$

Proof Let $\rho_\delta(\cdot)$ be a sequence of standard symmetric mollifiers in $\mathbb{R}^d$. We set $m_\delta(t, x) = m(t) \star \rho_\delta$. We also take a sequence of 1-d mollifiers $\xi_\varepsilon(t)$ such that $\text{supp}(\xi_\varepsilon) \subset (-\varepsilon, 0)$, and we set

$$m_{\delta,\varepsilon} := \int_0^T \xi_\varepsilon(s - t)\, m_\delta(s)ds = \int_0^T \int_{\mathbb{R}^N} m(s, y)\xi_\varepsilon(s - t)\rho_\delta(x - y)\, dy ds\,.$$

Notice that this function vanishes near $t = 0$, so we can take it as test function in (1.118). We get

$$\int_0^T \int_{\mathbb{T}^d} u\, \partial_t m_{\delta,\varepsilon} + \int_0^T \int_{\mathbb{T}^d} H(x, Du)m_{\delta,\varepsilon} \le \int_0^T \int_{\mathbb{T}^d} f\, m_{\delta,\varepsilon} + \int_{\mathbb{T}^d} g\, m_{\delta,\varepsilon}(T)\,.$$

$$\tag{1.125}$$

The first integral is equal to $\int_0^T \int_{\mathbb{T}^d} -\partial_s u_{\delta,\varepsilon} m(s, y) \, ds dy$, where $u_{\delta,\varepsilon}(s, y) = \int_0^T \int_{\mathbb{T}^d} u(t, x) \xi_\varepsilon(s - t) \rho_\delta(x - y) \, dt dx$. Notice that this function vanishes near $s = T$ so it can be used as test function in (1.120). Therefore we have

$$\int_0^T \int_{\mathbb{T}^d} u(t, x) \, \partial_t m_{\delta,\varepsilon}(t, x) \, dx dt$$

$$= -\int_0^T \int_{\mathbb{T}^d} m_{\delta,\varepsilon}(s, y) \, \partial_s u_{\delta,\varepsilon}(s, y) \, ds dy$$

$$= -\int_0^T \int_{\mathbb{T}^d} m(s, y) b(s, y) \cdot D_y u_{\delta,\varepsilon} \, ds dy + \int_{\mathbb{T}^d} m_0(y) u_{\delta,\varepsilon}(0, y) dy \,.$$

We shift the convolution kernels from u to m in the right-hand side and we use this equality in (1.125). We get

$$-\int_0^T \int_{\mathbb{T}^d} Du \cdot w_{\delta,\varepsilon} + \int_0^T \int_{\mathbb{T}^d} H(x, Du) m_{\delta,\varepsilon} + \int_{\mathbb{T}^d} (m_0 \star \rho_\delta) \left(\int_0^T u(t) \xi_\varepsilon(-t) \, dt \right)$$

$$\leq \int_0^T \int_{\mathbb{T}^d} f \, m_{\delta,\varepsilon} + \int_{\mathbb{T}^d} g \, m_{\delta,\varepsilon}(T)$$

$$(1.126)$$

where we denote $w_\delta = [(b \, m) \star \rho_\delta]$ and $w_{\delta,\varepsilon}(t, x) = \int_0^T w_\delta(s, x) \xi_\varepsilon(s - t) \, ds$.

Now we let first $\varepsilon \to 0$, and then $\delta \to 0$. Since u is time increasing (up to an absolutely continuous function), we have

$$\liminf_{\varepsilon \to 0} \int_{\mathbb{T}^d} (m_0 \star \rho_\delta) \left(\int_0^T u(t) \xi_\varepsilon(-t) \, dt \right) \geq \int_{\mathbb{T}^d} (m_0 \star \rho_\delta) \, u(0)$$

and since u is bounded below once we let $\delta \to 0$ we have $m_0 \, u(0) \in L^1(\mathbb{T}^d)$ and

$$\liminf_{\delta \to 0} \liminf_{\varepsilon \to 0} \int_{\mathbb{T}^d} (m_0 \star \rho_\delta) \left(\int_0^T u(t) \xi_\varepsilon(-t) \, dt \right) \geq \int_{\mathbb{T}^d} m_0 \, u(0) \,. \qquad (1.127)$$

Using the time continuity of m into $\mathcal{P}(\mathbb{T}^d)$, we have

$$\|m_{\delta,\varepsilon}(T) - (m(T) \star \rho_\delta)\|_\infty \leq \|D\rho_\delta\|_\infty \int_0^T \xi_\varepsilon(s - T) \mathbf{d}_1(m(s), m(T)) ds \overset{\varepsilon \to 0}{\to} 0 \,,$$

so we handle the term at $t = T$:

$$\lim_{\varepsilon \to 0} \int_{\mathbb{T}^d} g \, m_{\delta,\varepsilon}(T) = \int_{\mathbb{T}^d} g(m(T) \star \rho_\delta) \,.$$

Now we can pass to the limit in the last term due to Lebesgue's theorem, since by the assumptions we dominate

$$g(m(T) \star \rho_\delta) \le \phi_2^*(g) + \phi_2(m(T) \star \rho_\delta) \le \phi_2^*(g) + \phi_2(m(T)) \star \rho_\delta$$

where we used Jensen's inequality in the last step. Since $\phi_2^*(g), \phi_2(m(T)) \in L^1(\mathbb{T}^d)$, last term strongly converges in $L^1(\mathbb{T}^d)$, and since g is also bounded below we deduce that $|g(m(T) \star \rho_\delta)|$ is dominated in $L^1(\mathbb{T}^d)$. Therefore

$$\lim_{\delta \to 0} \lim_{\varepsilon \to 0} \int_{\mathbb{T}^d} g \, m_{\delta,\varepsilon}(T) = \int_{\mathbb{T}^d} g \, m(T) . \tag{1.128}$$

We reason in a similar way for the term with f, which satisfies, for some constant c_0,

$$c_0 \, m_{\delta,\varepsilon} \le f \, m_{\delta,\varepsilon} \le \phi_1^*(f) + (\phi_1(m) \star \rho_{\delta,\varepsilon}) .$$

By dominated convergence again we deduce

$$\lim_{\delta \to 0} \lim_{\varepsilon \to 0} \int_0^T \int_{\mathbb{T}^d} f \, m_{\delta,\varepsilon} = \int_0^T \int_{\mathbb{T}^d} f \, m . \tag{1.129}$$

Finally, using (1.66) we have

$$-Du \cdot w_{\delta,\varepsilon} + H(x, Du)m_{\delta,\varepsilon} \ge \frac{\alpha}{2} m_{\delta,\varepsilon}|Du|^2 - C \left[m_{\delta,\varepsilon} + \frac{|w_{\delta,\varepsilon}|^2}{m_{\delta,\varepsilon}} \right]$$

Now we define the lower semi-continuous function Ψ on $\mathbb{R}^N \times \mathbb{R}$ by

$$\Psi(w, m) = \begin{cases} \frac{|w|^2}{m} & \text{if } m > 0, \\ 0 & \text{if } m = 0 \text{ and } w = 0, \\ +\infty & \text{otherwise,} \end{cases} \tag{1.130}$$

and we observe that Ψ is convex in the couple (w, m). So by Jensen's inequality we have $\frac{|w_{\delta,\varepsilon}|^2}{m_{\delta,\varepsilon}} \le (\frac{|w|^2}{m}) \star \rho_{\delta,\varepsilon}$. Recalling that $w = bm$ in our setting (hence $\frac{|w|^2}{m} = m|b|^2$), we deduce that

$$-Du \cdot w_{\delta,\varepsilon} + H(x, Du)m_{\delta,\varepsilon} \ge \frac{\alpha}{2} m_{\delta,\varepsilon}|Du|^2 - C \left[m_{\delta,\varepsilon} + (|b|^2 m) \star \rho_{\delta,\varepsilon} \right] .$$

From the previous inequality we are allowed to use Fatou's lemma as $\varepsilon, \delta \to 0$, obtaining

$$\liminf_{\delta \to 0} \liminf_{\varepsilon \to 0} \int_0^T \int_{\mathbb{T}^d} [-Du \cdot w_{\delta,\varepsilon} + H(x, Du) m_{\delta,\varepsilon}]$$

$$\geq \int_0^T \int_{\mathbb{T}^d} m[-Du \cdot b + H(x, Du)] \qquad (1.131)$$

and we also deduce in between that $m|Du|^2 \in L^1(Q_T)$. Finally, collecting (1.127), (1.128), (1.129) and (1.131), we obtain (1.124). □

We are now able to discuss the vanishing viscosity limit of the MFG system. In the following, we will make use of the family of Young measures generated by the sequence $\{m^\varepsilon\}$. To this purpose, we recall the fundamental result concerning Young measures [181], see e.g. [22, 161]. Here $\mathcal{P}(\mathbb{R})$ denotes the space of probability measures on $\mathbb{R}$.

Proposition 1.3 *Let Q be a bounded subset in $\mathbb{R}^N$, and let $\{w_n\}$ be a sequence which is weakly converging in $L^1(Q)$. Then there exists a subsequence $\{w_{n_k}\}$ and a weakly* measurable function $v : Q \mapsto \mathcal{P}(\mathbb{R})$ such that if $f(y, s)$ is a Carathéodory function and $\{f(y, w_{n_k}(y))\}$ is an equi-integrable sequence in $L^1(Q)$, then*

$$f(y, w_{n_k}(y)) \rightharpoonup \bar{f}(y) \quad \text{weakly in } L^1(Q), \quad \text{where} \quad \bar{f}(y) = \int_{\mathbb{R}} f(y, \lambda) dv_y(\lambda).$$

□

Theorem 1.15 *Assume that F, G satisfy (1.115)–(1.116), and that $H(x, p)$ satisfies (1.66)–(1.68). Let $m_0 \in L^\infty(\mathbb{T}^d)$ and let $(u^\varepsilon, m^\varepsilon)$ be a solution of (1.62). Then there exists a subsequence, not relabeled, and a couple $(u, m) \in L^2(0, T; H^1(\mathbb{T}^d)) \times L^1(Q_T)$ such that $(u^\varepsilon, m^\varepsilon) \to (u, m)$ in $L^1(Q_T)$, and (u, m) is a weak solution to (1.121)–(1.122) in the sense of Definition 1.8.*

Proof By Lemma 1.13, $(u^\varepsilon, m^\varepsilon)$ satisfy the a priori estimates (1.69). On account of conditions (1.66)–(1.68), this implies that there exists a constant C, independent of ε, such that

$$\int_0^T \int_{\mathbb{T}^d} F(x, m^\varepsilon) m^\varepsilon + \int_{\mathbb{T}^d} G(x, m^\varepsilon(T)) m^\varepsilon(T) + \int_0^T \int_{\mathbb{T}^d} m^\varepsilon |Du^\varepsilon|^2 + \int_0^T \int_{\mathbb{T}^d} |Du^\varepsilon|^2 \leq C.$$

Hence there exists a subsequence, not relabeled, and a function $u \in L^2(0, T; H^1(\mathbb{T}^d))$ such that $u^\varepsilon \to u$ weakly in $L^2(0, T; H^1(\mathbb{T}^d))$. Notice that, since $\partial_t u^\varepsilon$ is bounded in $L^2(0, T; (H^1(\mathbb{T}^d))') + L^1(Q_T)$, by standard compactness results the convergence of u^ε to u is strong in $L^2(Q_T)$. Moreover, since F, G are bounded below, by maximum principle we also have that u^ε is bounded below.

As for m^ε, from (1.115) we have that $f(m^\varepsilon)m^\varepsilon$ is bounded in $L^1(Q_T)$. This implies that m^ε is equi-integrable and so, by Dunford-Pettis theorem, it is relatively compact in the weak topology of $L^1(Q_T)$; there exists a subsequence, not relabeled, and a function $m \in L^1(Q_T)$ such that $m^\varepsilon \to m$ weakly in $L^1(Q_T)$. Let us denote by $v_{(t,x)}(\cdot)$ the family of Young measures generated by m^ε, according to Proposition 1.3. Since $F(x, m^\varepsilon)m^\varepsilon$ is bounded in $L^1(Q_T)$, then $F(x, m^\varepsilon)$ is equi-integrable and then we have

$$F(x, m^\varepsilon) \to \bar{f} \quad \text{weakly in } L^1(Q_T), \quad \text{where} \quad \bar{f} = \int_{\mathbb{R}} F(x, \lambda) dv_{(t,x)}(\lambda).$$

We notice that the bound on $F(x, m^\varepsilon)m^\varepsilon$ implies that $\int_{\mathbb{R}} F(x, \lambda)\lambda dv_{(t,x)}(\lambda) \in L^1(Q_T)$. Indeed, applying Proposition 1.3 to the function $F(x, m)T_k(m)$, where $T_k(m) = \min(m, k)$, implies

$$\int_0^T \int_{\mathbb{T}^d} \int_{\mathbb{R}} F(x, \lambda)T_k(\lambda) \, dv_{(t,x)}(\lambda) = \lim_{\varepsilon \to 0} \int_0^T \int_{\mathbb{T}^d} F(x, m^\varepsilon)T_k(m^\varepsilon) \leq C,$$

and then, by letting $k \to \infty$, by monotone convergence we get

$$\int_0^T \int_{\mathbb{T}^d} \int_{\mathbb{R}} F(x, \lambda)\lambda \, dv_{(t,x)}(\lambda) < \infty. \tag{1.132}$$

Similarly we reason for the sequence $m^\varepsilon(T)$. This is equi-integrable in $L^1(\mathbb{T}^d)$ and then, up to subsequences, it converges weakly in $L^1(\mathbb{T}^d)$; in addition, denoting $\{\gamma_x(\cdot)\}$ the sequence of Young measures generated by $m^\varepsilon(T)$, we have that $G(x, m^\varepsilon(T))$ is weakly relatively compact in $L^1(\mathbb{T}^d)$ and

$$G(x, m^\varepsilon) \to \bar{g} \quad \text{weakly in } L^1(Q_T), \quad \text{where} \quad \bar{g} = \int_{\mathbb{R}} G(x, \lambda) d\gamma_x(\lambda).$$

In addition, as before, we deduce that $\int_{\mathbb{R}} G(x, \lambda)\lambda d\gamma_x(\lambda) \in L^1(\mathbb{T}^d)$.

We can now pass to the limit in the two equations. As for the HJ equation, since $p \mapsto H(x, p)$ is convex, then by weak lower semi-continuity we deduce that u satisfies

$$\begin{cases} -\partial_t u + H(x, Du) \leq \bar{f} \\ u(T) \leq \bar{g} \end{cases} \tag{1.133}$$

in the sense of Definition 1.6. As for m^ε, we observe that (1.69) and (1.67) imply that $m^\varepsilon |H_p(x, Du^\varepsilon)|^2$ is bounded in $L^1(Q_T)$. It follows (see also Remark 1.6) that $d_2(m^\varepsilon(t), m^\varepsilon(s)) \leq C|t - s|^{\frac{1}{2}}$, where d_2 is the Wasserstein distance in $\mathcal{P}(\mathbb{T}^d)$. Therefore, $m^\varepsilon(t)$ is equi-continuous and converges uniformly in the weak* topology. This implies that the L^1-weak limit m belongs to $C^0([0, T]; \mathcal{P}(\mathbb{T}^d))$, and $m(0) =$

m_0. Finally, $m^\varepsilon H_p(x, Du^\varepsilon)$ is equi-integrable and therefore weakly converges (up to subsequences) in $L^1(Q_T)$ to some vector field w. If Ψ is defined in (1.130), we deduce

$$\int_0^T \int_{\mathbb{T}^d} \Psi(w, m) \le \liminf_{\varepsilon \to 0} \int_0^T \int_{\mathbb{T}^d} \Psi(m^\varepsilon, w^\varepsilon) = \int_0^T \int_{\mathbb{T}^d} m^\varepsilon |H_p(x, Du^\varepsilon)|^2 \le C$$

hence $\Psi(w, m) \in L^1(Q_T)$. In particular we can set $b := \frac{w}{m} 1_{\{m > 0\}}$, then m is a weak solution of (1.119), with $m |b|^2 \in L^1(Q_T)$. Eventually, since m^ε weakly converges to m and $f(s)s$ is convex, by lower semicontinuity we deduce that

$$\int_0^T \int_{\mathbb{T}^d} f(m)m \le \liminf_{\varepsilon \to 0} \int_0^T \int_{\mathbb{T}^d} f(m^\varepsilon)m^\varepsilon \le \int_0^T \int_{\mathbb{T}^d} F(x, m^\varepsilon)m^\varepsilon + C_0 \le C.$$

Similarly we have for $m^\varepsilon(T)$, hence we conclude that

$$f(m)m \in L^1(Q_T), \qquad g(m(T))m(T) \in L^1(\mathbb{T}^d).$$

Now we observe that, using the monotonicity of $F(x, \cdot)$ and condition (1.115) we can estimate

$$\bar{f} = \int_{\mathbb{R}} F(x, \lambda) d\nu_{(t,x)}(\lambda) \le F(x, s) + \frac{1}{s} \int_{\mathbb{R}} F(x, \lambda) \lambda \, d\nu_{(t,x)}(\lambda)$$

$$\le f(s) + C_1 + \frac{1}{s} \int_{\mathbb{R}} F(x, \lambda) \lambda \, d\nu_{(t,x)}(\lambda).$$

hence

$$[\bar{f} - C_1]s - f(s)s \le \int_{\mathbb{R}} F(x, \lambda) \lambda \, d\nu_{(t,x)}(\lambda) \qquad \forall s \ge 0.$$

Recall that $f(s)s$ is convex and the right-hand side belongs to $L^1(Q_T)$; we deduce from the above inequality that $\phi_1^*(\bar{f} - C_1) \in L^1(Q_T)$, where ϕ_1^* is the convex conjugate of $\phi_1(s) := f(s)s$. Similarly we reason for $\bar{g}$, obtaining that $\phi_2^*(\bar{g} - C_3) \in L^1(Q_T)$ where $\phi_2(s) = g(s)s$. Notice that the addition of constants to $\bar{f}, \bar{g}$ in (1.133) is totally innocent up to replacing u with $u + a(T - t) + b$. Collecting all the above properties, we can apply Lemma 1.15 to u and m and we obtain that the following inequality holds:

$$\int_{\mathbb{T}^d} m_0 u(0) \le \int_{\mathbb{T}^d} \bar{g} m(T) + \int_0^T \int_{\mathbb{T}^d} \bar{f} m + \int_0^T \int_{\mathbb{T}^d} m [b \cdot Du - H(x, Du)].$$

$$(1.134)$$

Now we conclude by identifying the weak limits $\bar{f}, \bar{g}$ and b. We start from the equality (1.70)

$$\int_{\mathbb{T}^d} G(x, m^{\varepsilon}(T)) m^{\varepsilon}(T) + \int_0^T \int_{\mathbb{T}^d} F(x, m^{\varepsilon}) m^{\varepsilon}$$

$$+ \int_0^T \int_{\mathbb{T}^d} m^{\varepsilon} [H_p(x, Du^{\varepsilon}) \cdot Du^{\varepsilon} - H(x, Du^{\varepsilon})] = \int_{\mathbb{T}^d} m_0 u^{\varepsilon}(0).$$

We observe that $u^{\varepsilon}(0)$ is equi-integrable: indeed, $(u^{\varepsilon} - k)_+$ is a sub solution of the Bellman equation, so that

$$\int_{\mathbb{T}^d} (u^{\varepsilon}(0) - k)_+ + \int_0^T \int_{\mathbb{T}^d} H(x, Du^{\varepsilon}) 1_{\{u^{\varepsilon} > k\}}$$

$$\leq \int_0^T \int_{\mathbb{T}^d} F(x, m^{\varepsilon}) 1_{\{u^{\varepsilon} > k\}} + \int_{\mathbb{T}^d} (G(x, m^{\varepsilon}(T)) - k)_+ .$$

Hence the bound from below of H (see (1.66)) and the equi-integrability of $F(x, m^{\varepsilon})$, $G(x, m^{\varepsilon}(T))$ imply that $\int_{\mathbb{T}^d} (u^{\varepsilon}(0) - k)_+ \to 0$ as $k \to \infty$ uniformly with respect to ε. This implies that $u^{\varepsilon}(0)$ is equi-integrable, and then it weakly converges in $L^1(\mathbb{T}^d)$ to some function χ. In particular, when we pass to the limit in (1.62), we have

$$\int_{\mathbb{T}^d} \varphi(0) \chi + \int_0^T \int_{\mathbb{T}^d} u \varphi_t + \int_0^T \int_{\mathbb{T}^d} H(x, Du) \varphi \leq \int_0^T \int_{\mathbb{T}^d} \bar{f} \varphi + \int_{\mathbb{T}^d} \bar{g} \varphi(T)$$

$$\forall \varphi \in C^1(\overline{Q_T}), \ \varphi \geq 0.$$

By choosing a sequence φ_j such that $\varphi_j(0) = 1$ and φ_j approximates the Dirac mass at $t = 0$, we conclude that $\chi \leq u(0)$, where $u(0)$ is the trace of u at time $t = 0$ in the sense explained above. Finally, we have

$$\int_{\mathbb{T}^d} G(x, m^{\varepsilon}(T)) m^{\varepsilon}(T) + \int_0^T \int_{\mathbb{T}^d} F(x, m^{\varepsilon}) m^{\varepsilon}$$

$$+ \int_0^T \int_{\mathbb{T}^d} m^{\varepsilon} [H_p(x, Du^{\varepsilon}) \cdot Du^{\varepsilon} - H(x, Du^{\varepsilon})] \qquad (1.135)$$

$$\xrightarrow{\varepsilon \to 0} \int_{\mathbb{T}^d} m_0 \chi \leq \int_{\mathbb{T}^d} m_0 u(0)$$

and using (1.134) we get

$$\limsup_{\varepsilon \to 0} \left\{ \int_{\mathbb{T}^d} G(x, m^\varepsilon(T)) m^\varepsilon(T) + \int_0^T \int_{\mathbb{T}^d} F(x, m^\varepsilon) m^\varepsilon \right.$$

$$\left. + \int_0^T \int_{\mathbb{T}^d} m^\varepsilon \left[H_p(x, Du^\varepsilon) \cdot Du^\varepsilon - H(x, Du^\varepsilon) \right] \right\}$$

$$\leq \int_{\mathbb{T}^d} \bar{g} \, m(T) + \int_0^T \int_{\mathbb{T}^d} \bar{f} \, m + \int_0^T \int_{\mathbb{T}^d} m \left[b \cdot Du - H(x, Du) \right] .$$

$$(1.136)$$

Let us denote $w^\varepsilon := m^\varepsilon H_p(x, Du^\varepsilon)$. We have called w its weak limit in $L^1(Q_T)$; since we have

$$\int_0^T \int_{\mathbb{T}^d} m^\varepsilon \left[H_p(x, Du^\varepsilon) \cdot Du^\varepsilon - H(x, Du^\varepsilon) \right]$$

$$= \int_0^T \int_{\mathbb{T}^d} m^\varepsilon H^*(x, H_p(x, Du^\varepsilon)) = \int_0^T \int_{\mathbb{T}^d} m^\varepsilon H^* \left(x, \frac{w^\varepsilon}{m^\varepsilon} \right)$$

where H^* is the convex conjugate of H, and since $m H^* \left(x, \frac{w}{m} \right)$ is a convex function of (m, w), by weak lower semicontinuity we have

$$\liminf_{\varepsilon \to 0} \int_0^T \int_{\mathbb{T}^d} m^\varepsilon \left[H_p(x, Du^\varepsilon) \cdot Du^\varepsilon - H(x, Du^\varepsilon) \right] \geq \int_0^T \int_{\mathbb{T}^d} m H^* \left(x, \frac{w}{m} \right) .$$

$$(1.137)$$

Therefore we deduce from (1.136)

$$\limsup_{\varepsilon \to 0} \left\{ \int_{\mathbb{T}^d} G(x, m^\varepsilon(T)) m^\varepsilon(T) + \int_0^T \int_{\mathbb{T}^d} F(x, m^\varepsilon) m^\varepsilon \right\} \leq \int_{\mathbb{T}^d} \bar{g} \, m(T) + \int_0^T \int_{\mathbb{T}^d} \bar{f} \, m$$

$$+ \int_0^T \int_{\mathbb{T}^d} m \left[b \cdot Du - H(x, Du) \right] - \int_0^T \int_{\mathbb{T}^d} m H^* \left(x, \frac{w}{m} \right)$$

$$\leq \int_{\mathbb{T}^d} \bar{g} \, m(T) + \int_0^T \int_{\mathbb{T}^d} \bar{f} \, m$$

$$(1.138)$$

where we have used that $w = bm$. We use now the monotonicity of F, G to identify their limits. Indeed, denoting $T_k(s) = \min(s, k)$, we have (there is no loss of generality here in assuming F, G positive, which is true up to addition of constants):

$$\int_0^T \int_{\mathbb{T}^d} [F(x, m^\varepsilon) - F(x, m)] (m^\varepsilon - m)$$

$$+ \int_{\mathbb{T}^d} [G(x, m^\varepsilon(T)) - G(x, m(T))] (m^\varepsilon(T) - m(T))$$

$$\leq \int_0^T \int_{\mathbb{T}^d} F(x, m^\varepsilon) m^\varepsilon + \int_{\mathbb{T}^d} G(x, m^\varepsilon(T)) m^\varepsilon(T)$$

$$- \int_0^T \int_{\mathbb{T}^d} F(x, m^\varepsilon) T_k(m) - \int_0^T \int_{\mathbb{T}^d} T_k(F(x, m)) m^\varepsilon + \int_0^T \int_{\mathbb{T}^d} F(x, m) m$$

$$- \int_{\mathbb{T}^d} G(x, m^\varepsilon(T)) T_k(m(T))$$

$$- \int_{\mathbb{T}^d} T_k(G(x, m(T))) m^\varepsilon(T) + \int_{\mathbb{T}^d} G(x, m(T)) m(T) .$$

Hence, using (1.138) and the weak convergences of m^ε, $F(x, m^\varepsilon)$, $G(x, m^\varepsilon(T))$ we get

$$\limsup_{\varepsilon \to 0} \left\{ \int_0^T \int_{\mathbb{T}^d} [F(x, m^\varepsilon) - F(x, m)] (m^\varepsilon - m) \right.$$

$$\left. + \int_{\mathbb{T}^d} [G(x, m^\varepsilon(T)) - G(x, m(T))] (m^\varepsilon(T) - m(T)) \right\}$$

$$\leq \int_0^T \int_{\mathbb{T}^d} \bar{f} [m - T_k(m)] + \int_0^T \int_{\mathbb{T}^d} [F(x, m) - T_k(F(x, m))] m$$

$$+ \int_{\mathbb{T}^d} \bar{g} [m(T) - T_k(m(T))] + \int_{\mathbb{T}^d} [G(x, m(T)) - T_k(G(x, m(T)))] m(T) .$$

Letting $k \to \infty$ the right-hand side vanishes due to Lebesgue 's theorem, so we conclude that

$$\limsup_{\varepsilon \to 0} \left\{ \int_0^T \int_{\mathbb{T}^d} [F(x, m^\varepsilon) - F(x, m)] (m^\varepsilon - m) \right.$$

$$\left. + \int_{\mathbb{T}^d} [G(x, m^\varepsilon(T)) - G(x, m(T))] (m^\varepsilon(T) - m(T)) \right\} = 0 .$$

This means that $[F(x, m^\varepsilon) - F(x, m)] (m^\varepsilon - m) \to 0$ in $L^1(Q_T)$, and almost everywhere in Q_T up to subsequences. In particular, we deduce that $F(x, m^\varepsilon) \to F(x, m)$ a.e. in Q_T up to subsequences, hence $\bar{f} = F(x, m)$ and the convergence (of both $F(x, m^\varepsilon)$ and $F(x, m^\varepsilon)m^\varepsilon$) is actually strong in $L^1(Q_T)$. Similarly we reason for $G(x, m^\varepsilon(T))$, which implies that $\bar{g} = G(x, m(T))$. If we come back to (1.138), now the limit of the left-hand side coincides with the right-hand side, and trapped in between we deduce that

$$\int_0^T \int_{\mathbb{T}^d} m \left[\frac{w}{m} \cdot Du - H(x, Du) - H^*\left(x, \frac{w}{m}\right) \right] = 0$$

which yields that $\frac{w}{m} = H_p(x, Du)$ m-a.e. in Q_T. This implies that $w = m\, H_p(x, Du)$. Finally, all the weak limits are identified. Coming back to (1.135), now we know that $F(x, m^\varepsilon)m^\varepsilon \to F(x, m)m$ and $G(x, m^\varepsilon(T))m^\varepsilon(T) \to G(x, m(T))m(T)$, and in addition (1.137) holds with $w = m\, H_p(x, Du)$. Therefore, we have

$$\int_{\mathbb{T}^d} G(x, m(T))m(T) + \int_0^T \int_{\mathbb{T}^d} F(x, m)m$$

$$+ \int_0^T \int_{\mathbb{T}^d} m\, [H_p(x, Du) \cdot Du - H(x, Du]$$

$$\leq \liminf_{\varepsilon \to 0} \left\{ \int_{\mathbb{T}^d} G(x, m^\varepsilon(T))m^\varepsilon(T) + \int_0^T \int_{\mathbb{T}^d} F(x, m^\varepsilon)m^\varepsilon \right.$$

$$\left. + \int_0^T \int_{\mathbb{T}^d} m^\varepsilon\, [H_p(x, Du^\varepsilon) \cdot Du^\varepsilon - H(x, Du^\varepsilon]\right\}$$

$$\leq \int_{\mathbb{T}^d} u(0)\, m_0 .$$

Combining this information with (1.134), where $\bar{f}, \bar{g}, b$ are now identified, yields the energy equality (1.123). Thus, we conclude that (u, m) is actually a weak solution of the MFG system in the sense of Definition 1.8. $\qquad\square$

Now we conclude the analysis with a uniqueness result. To this purpose, we need a refined version of Lemma 1.15, as follows.

Lemma 1.16 *Assume that F, G satisfy (1.115)–(1.116), and that $H(x, p)$ satisfies (1.66)–(1.68). Let (u, m) be a weak solution to (1.121)–(1.122). Then $u(t)m(t) \in L^1(\mathbb{T}^d)$ for a.e. $t \in (0, T)$, and the following equality holds:*

$$\int_{\mathbb{T}^d} m(t)\, u(t)\, dx = \int_{\mathbb{T}^d} G(x, m(T))\, m(T)\, dx + \int_t^T \int_{\mathbb{T}^d} F(t, x, m)m\, dx dt$$

$$+ \int_t^T \int_{\mathbb{T}^d} m\, \left[H_p(x, Du) \cdot Du - H(x, Du)\right] dx dt$$

for a.e. $t \in (0, T)$.

Proof Let us take t such that $m(t) \in L^1(\mathbb{T}^d)$ (this is true for a.e. $t \in (0, T)$). First we apply Lemma 1.15 in the time interval (t, T). Notice that, since $F(x, m)m \in L^1(Q_T)$, $G(x, m(T))m(T) \in L^1(\mathbb{T}^d)$, then the requirements of the Lemma hold with $\phi_1(s) = f(s)s$ and $\phi_2(s) = g(s)s$, where f, g are given by (1.115)–(1.116). We obtain that $u(t)m(t) \in L^1(\mathbb{T}^d)$ (where $u(t)$ is the right-continuous Borel representative of u) and

$$
\int_{\mathbb{T}^d} m(t)\, u(t)\, dx \leq \int_{\mathbb{T}^d} G(x, m(T))\, m(T)\, dx + \int_t^T \int_{\mathbb{T}^d} F(x, m)m\, dx dt
$$

$$
+ \int_t^T \int_{\mathbb{T}^d} m \left[H_p(x, Du) \cdot Du - H(x, Du) \right] dx dt
$$

$$
= \int_{\mathbb{T}^d} u(0)m_0 - \int_0^t \int_{\mathbb{T}^d} F(x, m)m\, dx dt
$$

$$
- \int_0^t \int_{\mathbb{T}^d} m \left[H_p(x, Du) \cdot Du - H(x, Du) \right] dx dt
$$

$$
\tag{1.139}
$$

where we used (1.123) in the last equality. Now we wish to apply once more Lemma 1.15 in the interval $(0, t)$; but this needs to be done in two steps. First of all, we replace u with $u_k = \min(u, k)$; u_k is itself a sub solution and satisfies (see e.g. [158, Lemma 5.3])

$$
-\partial_t u_k + H(x, Du_k) \leq F(x, m)\, 1_{\{u < k\}} + c\, 1_{\{u > k\}}
$$

for some constant $c > 0$. Since $u_k(t) \in L^\infty(\mathbb{T}^d)$ and $m(t) \in L^1(\mathbb{T}^d)$, we can apply Lemma 1.15 in $(0, t)$ to get

$$
\int_{\mathbb{T}^d} u_k(0)m_0 \leq \int_{\mathbb{T}^d} u_k(t)\, m(t) + \int_0^t \int_{\mathbb{T}^d} [F(x, m)\, 1_{\{u < k\}} + c\, 1_{\{u > k\}}]m
$$

$$
+ \int_0^t \int_{\mathbb{T}^d} m \left[H_p(x, Du) \cdot Du_k - H(x, Du_k) \right].
$$

Letting $k \to \infty$ is allowed since $u(t)m(t) \in L^1(\mathbb{T}^d)$, and we deduce that

$$
\int_{\mathbb{T}^d} u(0)m_0 \leq \int_{\mathbb{T}^d} u(t)\, m(t) + \int_0^t \int_{\mathbb{T}^d} F(x, m)m
$$

$$
+ \int_0^t \int_{\mathbb{T}^d} m \left[H_p(x, Du) \cdot Du - H(x, Du) \right].
$$

Using this information in (1.139) we conclude the proof of the desired equality. □

We are ready for the uniqueness result, where we further invoke the following lemma. This is a particular case of what proved in [158, Lemma 5.3] for solutions in the whole space; the proof follows the same steps in the setting of $x \in \mathbb{T}^d$. A similar statement is also contained in [62, Thm 6.2].

Lemma 1.17 ([158]) *Let u_1, u_2 be two weak sub solutions of (1.117). Then $v :=$ $\max(u_1, u_2)$ is also a sub solution of the same problem.*

We have all ingredients for the uniqueness result.

Theorem 1.16 *Assume that F, G satisfy (1.115)–(1.116), and that $H(x, p)$ satisfies (1.66)–(1.68). Let $m_0 \in L^\infty(\mathbb{T}^d)$, and let (u, m), $(\tilde{u}, \tilde{m})$ be two weak solutions of (1.121)–(1.122), in the sense of Definition 1.8. Then we have $F(x, m) = F(x, \tilde{m})$ and, if $F(x, \cdot)$ is an increasing function, then $m = \tilde{m}$ and $u = \tilde{u}$ m-almost everywhere.*

Proof After condition (1.115), there is no loss of generality in assuming that $F(x, s) \le f(s)$ (which is the case up to addition of a (same) constant to both H ad F). Therefore, using the monotonicty of $F(x, \cdot)$, we have

$$F(x, m)s \le F(x, m)m + F(x, s)s \le F(x, m)m + f(s)s$$

Hence, if we denote by $\phi_1(s) = f(s)s$, we have that $\phi_1^*(F(x, m)) \in L^1(Q_T)$, while $\phi_1(m) \in L^1(Q_T)$. This is of course true for $\tilde{m}$ as well. Similarly we reason for $G(x, m(T))$ with $\phi_2(s) = g(s)s$ given by (1.116). Therefore, we can apply Lemma 1.15 to u and to $\tilde{m}$ as well as to $\tilde{u}$ and m. We obtain

$$\int_{\mathbb{T}^d} u(0)m_0 \le \int_0^T \int_{\mathbb{T}^d} F(x, m)\tilde{m} + \int_{\mathbb{T}^d} G(x, m(T))\tilde{m}(T)$$
$$+ \int_0^T \int_{\mathbb{T}^d} \tilde{m}[H_p(x, D\tilde{u})Du - H(x, Du)],$$

$$\int_{\mathbb{T}^d} \tilde{u}(0)m_0 \le \int_0^T \int_{\mathbb{T}^d} F(x, \tilde{m})m + \int_{\mathbb{T}^d} G(x, \tilde{m}(T))m(T)$$
$$+ \int_0^T \int_{\mathbb{T}^d} m[H_p(x, Du)D\tilde{u} - H(x, D\tilde{u})].$$

We use (1.123) in the first inequality, and similarly we use (1.123) written for $(\tilde{u}, \tilde{m})$ in the second one. When we add the two contributions we deduce the usual inequality

$$\int_0^T \int_{\mathbb{T}^d} (F(x, m) - F(x, \tilde{m}))(m - \tilde{m})$$
$$+ \int_{\mathbb{T}^d} [G(x, m(T)) - G(x, \tilde{m}(T))][m(T) - \tilde{m}(T)]$$

$$\int_0^T \int_{\mathbb{T}^d} m \left[H(x, D\tilde{u}) - H(x, Du) - H_p(x, Du)(D\tilde{u} - Du) \right]$$

$$+ \int_0^T \int_{\mathbb{T}^d} \tilde{m} \left[H(x, Du) - H(x, D\tilde{u}) - H_p(x, D\tilde{u})(Du - D\tilde{u}) \right] \le 0 \,.$$

This implies that $F(x, m) = F(x, \tilde{m})$ and $G(x, m(T)) = G(x, \tilde{m}(T))$, and we have

$$H(x, D\tilde{u}) - H(x, Du) = H_p(x, Du)(D\tilde{u} - Du) \qquad \text{in } \{(t, x) : \ m(t, x) > 0\}$$

$$H(x, Du) - H(x, D\tilde{u}) = H_p(x, D\tilde{u})(Du - D\tilde{u}) \qquad \text{in } \{(t, x) : \ \tilde{m}(t, x) > 0\}.$$

$$(1.140)$$

Of course, if $F(x, \cdot)$ is increasing, we deduce that $m = \tilde{m}$ almost everywhere.

Now we use Lemma 1.17, which says that $z := \max(u, \tilde{u})$ is a sub solution of the HJ equation. Then we can apply Lemma 1.15 and we obtain, for a.e. $t \in (0, T)$:

$$\int_{\mathbb{T}^d} m(t) z(t) \ \le \ \int_{\mathbb{T}^d} G(x, m(T)) m(T) + \int_t^T \int_{\mathbb{T}^d} F(x, m) m$$

$$+ \int_t^T \int_{\mathbb{T}^d} m \left[H_p(x, Du) \cdot Dz - H(x, Dz) \right]. \quad (1.141)$$

Now we have

$$\int_t^T \int_{\mathbb{T}^d} m \left[H_p(x, Du) \cdot Dz - H(x, Dz) \right]$$

$$= \int_t^T \int_{\mathbb{T}^d} m \left[H_p(x, Du) \cdot D\tilde{u} - H(x, D\tilde{u}) \right] 1_{\{u \le \tilde{u}\}}$$

$$+ \int_t^T \int_{\mathbb{T}^d} m \left[H_p(x, Du) \cdot Du - H(x, Du) \right] 1_{\{u > \tilde{u}\}}$$

$$= \int_t^T \int_{\mathbb{T}^d} m \left[H_p(x, Du) \cdot Du - H(x, Du) \right]$$

thanks to (1.140); thus we deduce from (1.141)

$$\int_{\mathbb{T}^d} m(t) z(t) \ \le \ \int_{\mathbb{T}^d} G(x, m(T)) m(T) + \int_t^T \int_{\mathbb{T}^d} F(x, m) m$$

$$+ \int_t^T \int_{\mathbb{T}^d} m \left[H_p(x, Du) \cdot Du - H(x, Du) \right]$$

$$= \int_{\mathbb{T}^d} m(t) u(t)$$

where we used Lemma 1.16. We conclude that

$$\int_{\mathbb{T}^d} m(t)\,[z(t) - u(t)] \leq 0$$

which implies that $u(t) = z(t)$ (i.e. $u(t) \geq \tilde{u}(t)$) m-almost everywhere. Reversing the roles of the two functions we conclude that $u = \tilde{u}$ m-almost everywhere.   □

Remark 1.16 There are other approaches to study the first order MFG system (1.121)–(1.122), especially if model cases are considered. One possible strategy, introduced in [145], consists in transforming the system into a second order elliptic equation for u in time space. More precisely, using that F is one-to-one and replacing m in the continuity equation by

$$m(t, x) = F^{-1}(x, -\partial_t u + H(x, Du)),$$

one finds an elliptic equation in (t, x) for u. This elliptic equation is fully nonlinear and degenerate (at least on the points (t, x) where $m(t, x) = 0$). This strategy is the starting point of regularity results proved by P.-L. Lions in [149] (Lessons 6-27/11 2009), which in particular lead to uniform bounds for the density m.

Other regularity results, including L^∞- bounds or Sobolev regularity for the density, were obtained by F. Santambrogio in [147, 148, 175] using completely different techniques inspired by optimal transport theory. Those results are just one by-product of the Lagrangian approach developed by F. Santambrogio, for which we refer to his presentation in this same volume.

1.3.7.2 Variational Approach and Optimality Conditions

Following [144] the MFG system (1.62) can be viewed as an optimality condition for two optimal control problems: the first one is an optimal control of Hamilton-Jacobi equations and the second one an optimal control of the Fokker-Planck equation.

In order to be more precise, let us first introduce some assumptions and notations: without loss of generality, we suppose that $F(x, 0) = 0$ (indeed we can always subtract $F(x, 0)$ to both sides of (1.62)). Then we set

$$\Phi(x, m) = \begin{cases} \int_0^m F(x, \rho)d\rho & \text{if } m \geq 0 \\ 0 & \text{otherwise}. \end{cases}$$

As F is nondecreasing with respect to the second variable, $\Phi(x, m)$ is convex with respect to m, so we denote by $\Phi^* = \Phi^*(x, \alpha) = \sup_{m \in \mathbb{R}} (\alpha m - \Phi(x, m))$ its convex conjugate. Note that Φ^* is convex and nondecreasing with respect to the second variable. We also recall the convex conjugate $H^*(x, q) =$

$\sup_{p \in \mathbb{R}^d} (q \cdot p - H(x, p))$ already used before. For simplicity, we neglect here
the coupling at $t = T$ and we let $G = G(x)$.

The first optimal control problem is the following: the distributed control is α :
$\mathbb{T}^d \times [0, T] \to \mathbb{R}$ and the state function is u. The goal is to minimize the functional

$$J^{HJ}(\alpha) = \int_0^T \int_{\mathbb{T}^d} \Phi^*(x, \alpha(t, x))\ dxdt - \int_{\mathbb{T}^d} u(0, x)dm_0(x)$$

over Lipschitz continuous maps α : $(0, T) \times \mathbb{T}^d \to \mathbb{R}^d$, where u is the unique
classical solution to the backward Hamilton-Jacobi equation

$$\begin{cases} -\partial_t u - \varepsilon \Delta u + H(x, Du) = \alpha(t, x) & \text{in } (0, T) \times \mathbb{T}^d \\ u(T, x) = G(x) & \text{in } \mathbb{T}^d. \end{cases} \tag{1.142}$$

The second is an optimal control problem where the state is the solution m of the
Fokker-Planck equation: the (distributed and vector valued) control is now the drift
term $v : [0, T] \times \mathbb{T}^d \to \mathbb{R}^d$. The goal here is to minimize the functional

$$J^{FP}(v) = \int_0^T \int_{\mathbb{T}^d} [m\, H^*(x, -v) + F(x, m)]\ dxdt + \int_{\mathbb{T}^d} G(x)m(T)dx,$$

where the pair (m, v) solves the Fokker-Planck equation

$$\partial_t m - \varepsilon \Delta m + \text{div}(mv) = 0 \text{ in } (0, T) \times \mathbb{T}^d, \qquad m(0) = m_0. \tag{1.143}$$

Assuming that F^* and H^* are smooth enough, the equivalence between the MFG
system and the optimality conditions of the previous two problems can be checked
with a direct verification.

Theorem 1.17 ([149]) *Assume that $(\bar{m}, \bar{u})$ is of class $C^2((0, T) \times \mathbb{T}^d)$, with $\bar{m}(0) = m_0$ and $\bar{u}(T, x) = G(x)$. Suppose furthermore that $\bar{m}(t, x) > 0$ for any $(t, x) \in (0, T) \times \mathbb{T}^d$. Then the following statements are equivalent:*

- (i) *$(\bar{u}, \bar{m})$ is a solution of the MFG system (1.62).*
- (ii) *The control $\bar{\alpha}(t, x) := F(x, \bar{m}(t, x))$ is optimal for J^{HJ} and the solution to (1.142) is given by $\bar{u}$.*
- (iii) *The control $\bar{v}(t, x) := -H_p(x, D\bar{u}(t, x))$ is optimal for J^{FP}, $\bar{m}$ being the solution of (1.143).*

Let us stress that the above equivalence holds even for $\varepsilon = 0$, say for the
deterministic problem. But of course a formal equivalence for smooth solutions is
of very little help. However, it is possible to exploit the convexity of the optimal
control problems in order to export the equivalence principle to suitably relaxed

optimization problems. To this purpose, observe that the optimal control problem of Hamilton-Jacobi equation can be rewritten as

$$\inf_u \int_0^T \int_{\mathbb{T}^d} F^*\left(x, -\partial_t u(x, t) - \varepsilon \Delta u(x, t) + H(x, Du(x, t))\right) dxdt - \int_{\mathbb{T}^d} u(0, x) dm_0(x)$$

under the constraint that u is sufficiently smooth, with $u(\cdot, T) = G(\cdot)$. Remembering that H is convex with respect to the last variable and that F is convex and increasing with respect to the last variable, it is clear that the above problem is convex.

The optimal control problem of the Fokker-Planck equation is also a convex problem, up to a change of variables which appears frequently in optimal transportation theory since the pioneering paper [28]. In fact, if we set $w = mv$, then the problem can be rewritten as

$$\inf_{(m,w)} \int_0^T \int_{\mathbb{T}^d} [m(t, x) H^*\left(x, -\frac{w(t, x)}{m(t, x)}\right) + F(x, m(t, x))] \, dxdt + \int_{\mathbb{T}^d} G(x) m(T, x) dx,$$

where the pair (m, w) solves the Fokker-Planck equation

$$\partial_t m - \varepsilon \Delta m + \operatorname{div}(w) = 0 \text{ in } (0, T) \times \mathbb{T}^d, \qquad m(0) = m_0. \tag{1.144}$$

This problem is convex because the constraint (1.144) is linear and the map $(m, w) \to m H^*\left(x, -\frac{w}{m}\right)$ is convex.

It turns out that the two optimal control problems just defined are conjugate in the Fenchel-Rockafellar sense (see, for instance, [100]) and they share the same optimality condition. Minimizers of such problems are expected to provide with weak solutions for (1.62). This approach has been extensively used for first order problems since [53, 54], leading to weak solutions in the sense of Definition 1.8. A similar analysis was later extended to second order degenerate MFG problems in [62], as well as to problems with density constraints, in which case one enforces the constraint that the density $m = m(t, x)$ is below a certain threshold. In this case, a penalization term appears in the HJ equation as an extra price to go through the zones where the density saturates the constraint (see [63, 124, 155, 173]).

A similar variational approach was also specially developed for the planning problem in connection with optimal transportation [126, 158].

We do not comment more on the optimal control approach because this is also extensively recalled in the contributions by Y. Achdou & M. Lauriere, and in the one by F. Santambrogio, in this volume.

1.3.8   Further Comments, Related Topics and References

1.3.8.1   Boundary Conditions, Exit Time Problems, State Constraints, Planning Problem

The existence and uniqueness results presented here for second order problems remain valid, with no additional difficulty, for the case of Neumann boundary conditions, i.e. when the controlled process lives in a bounded domain $\Omega \subset \mathbb{R}^d$ with reflection on the boundary. Results in this setting can be found e.g. in [165], or in [86]. A similar situation occurs when the domain happens to be invariant for the controlled process, and the trajectory cannot reach the boundary because of the degeneracy of the diffusion or due to the direction of the controlled drift. The study of the MFG system in this situation appears in [168].

By contrast, in many models, players can leave the game before the terminal time and the population is not constant: this is for instance the case of MFG with exit time, which lead to Dirichlet boundary conditions for the two unknowns (u, m) in the system. An interesting problem arises when the agents also control the time they stay in the game. The optimal control problem then becomes

$$
u(t, x) = \inf_{\tau, \alpha} \mathbb{E}\left[\int_0^\tau f(X_t, \alpha_t, m(t))dt + g(X_T, m(T)) \right]
$$

where (X_t) is driven by the usual controlled diffusion process. Here the controls are α and the stopping time τ. The measure $m(t)$ can be (depending on the model) either the law of X_t given $\{\tau \geq t\}$ (in which case the mass of $m(t)$ is constant, but the equation for m is no longer a simple Fokker-Planck equation) or simply the measure $m(t)$ defined by

$$
\int_{\mathbb{R}^d} \phi(x)m(t, dx) = \mathbb{E}\left[\phi(X_t)\mathbf{1}_{\tau \geq t}\right] \qquad \forall \phi \in C_c^\infty(\mathbb{R}^d).
$$

In this case the mass of $m(t)$ is non increasing in time.

Such models have been studied in the framework of bank run in [74] or in exhaustible commodities trade [84, 125]. In [32], the author provides a general PDE framework to study the problem and shows that the players might be led to use random strategies (see also [38]). An early work on the topic is [112] while surprising phenomena in the mean field analysis are discussed in [46, 112, 156, 157]. Minimal exit time problems for first order MFG problems are also studied in [154].

In many applications, the MFG system also involves state constraints. Namely, the optimal control problem for a small player takes the form

$$
u(t, x) = \inf_\gamma \int_t^T L(\gamma(s), \alpha(s), m(s))ds + G(\gamma(T), m(T)),
$$

where the infimum is taken over solution of

$$\begin{cases} \dot{\gamma}(s) = b(\gamma(s), m(s))ds, & \gamma(s) \in \overline{\Omega} \quad \forall s \in [t, T], \\ \gamma(t) = x \end{cases}$$

where Ω is an open subset of $\mathbb{R}^d$ (in general with a smooth boundary). This is the case of the Aiyagari problem [13] in economy for instance (see also [7, 9]). The natural set-up of the HJ problem is the so-called viscosity solution with state-constraints and is well understood. However, the analysis led in this section no longer applies: the measure m develops singularities (not only on the boundary) and one cannot expect uniqueness of the flow m given the vector field $-H_p(x, Du, m)$. To overcome this issue one can device a Lagrangian approach (see [47]). The PDE analysis of this problem is only partially understood (see [49, 50]).

The initial-terminal conditions may also be changed, in what is called the *planning problem*. In that model one wishes to prescribe both the initial and the final distribution, while no terminal condition is assumed on u. This variant of MFG problem fits into the models of optimal transportation theory, since the goal is to transport the density from $m(0) = m_0$ to $m(T) = m_1$ in a way which is optimal for the agents' control. Early results for this problem were given by P.-L. Lions in [149] (Lessons 4-11/12 2009); the second order case was later studied in [163–165], and the deterministic case in [126, 158]. Very recently, the planning problem has been also addressed for the master equation with finite states, see [35].

1.3.8.2   Numerical Methods

The topic will be developed in detail in the contribution of Achdou and Lauriere (see the references therein). Let us just remark here that the computation of the solution of the MFG system is difficult because it involves a forward equation and a backward equation. Let us just quote on this point the pioneering work [2], where a finite difference numerical scheme was proposed in a way that the discretized equations preserve the structure of the MFG system. In some cases one can also take advantage of the fact that the MFG system has a variational structure [29].

1.3.8.3   MFG Systems with Several Populations

MFG models may very naturally involve several populations (say, to fix the ideas, I populations). In this case the system takes the form

$$\begin{cases} (i) \ -\partial_t u_i - \Delta u_i + H_i(x, Du_i, m(t)) = 0 & \text{in } (0, T) \times \mathbb{T}^d \\ (ii) \ \partial_t m_i - \Delta m_i - \text{div} \left(m_i \ D_p H_i(x, Du_i(t, x), m(t)) \right) = 0 & \text{in } (0, T) \times \mathbb{T}^d \\ (iii) \ m_i(0) = m_{i,0}, \ u_i(T, x) = G_i(x, m(T)) & \text{in } \mathbb{T}^d \end{cases}$$

where $i = 1, \dots, I$, u_i denotes the value function of each player in population i and $m = (m_1, \dots, m_I)$ denotes the collection of densities m_i of the population i. The coupling functions H_i and G_i depend on all the densities. Existence of solutions can be proved by fixed point arguments as in Theorem 1.5. Uniqueness, however, is a difficult issue.

The MFG models with several populations were introduced in the early paper by Caines, Huang and Malhamé [132] and revisited by Cirant [86] (for Neumann boundary conditions, see also [24, 155]) and by Kolokoltsov, Li and Yang [137] (for very general diffusions, possibly with jumps). Analysis of segregation phenomena is pursued in [8, 93].

1.3.8.4   MFG of Control

In most application in economics, the coupling between the agents is not only through the distribution of their positions but also of their instantaneous controls. This kind of problem is more subtle than the classical MFG system since it requires, in its general version, a new constraint explaining how to derive the distribution of controls. For instance this new system can take the form (for problems with a constant diffusion):

$$
\begin{cases}
(i) \quad -\partial_t u(t, x) - \Delta u(t, x)) + H(t, x, Du(t, x), \mu(t)) = 0 \text{ in } (0, T) \times \mathbb{R}^d \\
(ii) \quad \partial_t m(t, x) - \Delta m(t, x) - \operatorname{div}\left(m(t, x) H_p(t, x, Du(t, x), \mu(t))\right) = 0 \text{ in } (0, T) \times \mathbb{R}^d \\
(iii) \quad \mu(t) = (id, -H_p(t, x, Du, \mu)) \sharp m(t) \text{ in } (0, T) \\
(iv) \quad m(0, x) = m_0(x), \; u(T, x) = g(x, m(T)) \qquad \text{in } \mathbb{R}^d
\end{cases}
$$

Here $\mu(t)$ is the distribution of states and controls of the players. Its first marginal $m(t)$ is the usual distribution of players. The new relation (iii) explains how to compute $\mu(t)$ from the distribution of the states $m(t)$ and the optimal feedback $-H_p(t, x, Du, \mu)$. Note that (iii) is itself a fixed point problem. In many applications, the players only interact through some moments of the distribution of controls, which simplifies the system. Existence of a solution for MFG of controls can be achieved under rather general conditions (some structure condition, ensuring (iii) to have a solution, is however required). Uniqueness is largely open.

Analysis of such problems can be found, among many other references, in [3, 34, 56, 69, 117, 119].

1.3.8.5   MFG with Common Noise and with a Major Player

Throughout this section we have discussed models in which the agents are subject to individual noises ("idiosyncratic noise") which are independent. However, it is also important to be able to deal with problems in which some random perturbation affects all the players. This perturbation can be quite rough (a white noise) or simply the (random) position of a single player (who, since he/she cannot be considered as an infinitesimal player in this game, is often called a major player). In these setting, the MFG system becomes random. For instance, in the case of MFG with a Brownian common noise, it takes the form

$$
\begin{cases}
du(t, x) = \big[- 2\Delta u(t, x)) + H(x, Du(t, x), m(t)) \\
\qquad\qquad - \sqrt{2}\mathrm{div}(v(t, x))\big]dt + v(t, x) \cdot dW_t & \text{in } (0, T) \times \mathbb{R}^d, \\
dm(t, x) = \big[2\Delta m(t, x)) + \mathrm{div}\big(m(, x)D_p H(x, Du(t, x), m(t))\big)\big]dt \\
\qquad\qquad - \mathrm{div}(m(t, x)\sqrt{2}dW_t), & \text{in } (0, T) \times \mathbb{R}^d, \\
u(T, x) = G(x, m(T)), \quad m(0) = m_0, & \text{in } \mathbb{R}^d
\end{cases}
$$

Here W is the common noise (a Brownian motion). The new variable v is an unknown function which ensures the solution u to the backward HJ to be adapted to the filtration generated by W. Another formulation of this problem involves the master equation and will be discussed below, at the end of Sect. 1.4. Let us just mention now that the analysis of MFG with common noise goes back to [149] (Lessons 12-26/11 2010), where in particular the structure of the master equation with common noise is described. The probabilistic approach of the MFG system is studied in [73] (see also [11, 141]) while the first results on the PDE system above and on the associated master equation are in [65].

MFG problems with a major player have been introduced by Huang in [130]. In a series of papers, Carmona and al. introduced a different notion of solution for the problem [70–72], mainly in a finite state space framework, and they showed that this notion actually corresponds to a Nash equilibrium for infinitely many players. This result is confirmed in [64] where the Nash equilibria for the N-player problem is shown to converge to the corresponding master equation. The master equation for the major problem is also studied in [146] (mostly in finite time horizon and in a finite state space) and in [66] (short-time existence in continuous space). Variants on the major player problem are discussed in [31] (MFG with a dominating player in a Stackelberg equilibrium), [101] (for a principal-agent problem) and in [36].

1.3.8.6 Miscellaneous

Other MFG Systems Let us first mention other variations on the MFG system. Besides the standard continuous-time, continuous-spaces models, the most relevant class of MFG models is probably the MFG on finite state space: see, among main other works, [26, 34, 78, 118]. In these problems the state of a typical player jumps from one state to another. The coupling between the HJ and the FP equations takes a much simpler form of a "forward-backward" system of ordinary differential equations. Another class of MFG problems are MFG on networks [10, 45], in which the state space is a (one dimensional) network. Motivated by knowledge growth models [152], some authors considered MFGs in which the interaction between players leads to a Boltzmann type equation or a Fisher-KPP equation for the distribution function [42, 43, 160, 169]. MFGs involving jump processes, where the diffusion term becomes a fractional Laplacian, have been studied in [44, 83, 89, 103, 137], while MFGs involving dynamics with deterministic jumps have been investigated in [33].

MFGs vs Agent Based Models In MFG theory, agents are assumed to be rational. On the contrary, in Agent Based Models, the agents are supposed to adopt an automatic behavior. The link between the two approach has been discussed in [23, 34, 99] where it is shown that MFG models degenerate into Agent Based Models as the agents become more and more myopic or more and more impatient.

Learning A natural question in Mean Field Games, in view of the complexity of the notion of MFG equilibria, is how players can achieve in practice to play a MFG Nash equilibrium. This kind of problem, also related to the concept of adaptative control [136], has been discussed in particular in the following references: [55, 56, 76, 77, 102, 129].

Efficiency of MFGs In game theory, a classical question is the (in)efficiency of Nash equilibria (the so-called "price of anarchy"): to what extent are Nash equilibria doing socially worse than a global planner? This question has also been addressed for Mean Field Games in [21, 59, 75].

1.4 The Master Equation and the Convergence Problem

In Sect. 1.3 we have explained in detail how the mean field game problem can often be reduced to a system of PDEs, the MFG system. If this MFG system is suitable for the analysis of problems in which players have only independent individual noises (the so-called idiosyncratic noises), it is no longer satisfactory to investigate more complex models (for instance models in which the players are subject to a common randomness, the so-called "MFG models with a common noise"). Nor does it allow to understand the link between N-player differential games and mean field games. To proceed, we need to introduce another equation: the master equation. The master equation is an infinite dimensional hyperbolic equation stated in the space of probability measures. As explained below, it is helpful for the following reasons:

- for standard MFG models, it allows to write the optimal control of a player in feedback form in function of the current time, the current position and *the current distribution of the other players*. This is meaningful since one can expect in practice that players adapt their behavior in function of these data;
- it provides a key tool to investigate the convergence of the N-player game to the MFG system;
- it allows to formalize and investigate more complex MFG models, as MFG with a common noise or MFG with a major player.

In order to discuss the master equation, we first need to have a closer look at the space of probability measures (Sect. 1.4.1) and then understand the notion of derivative in this space (Sect. 1.4.2). Then we present the master equation and state,

almost without proof, the existence and the uniqueness of the solution (Sect. 1.4.3).
We then discuss the convergence of N-player differential games by using the master
equation (Sect. 1.4.4).

1.4.1 The Space of Probability Measures (Revisited)

We have already seen the key role of the space of probability measures in the mean
field game theory. It is now time to investigate the basic properties of this space
more thoroughly. The results are given mostly without proofs, which can be found,
for instance, in the monographs [18, 174, 178, 179].

1.4.1.1 The Monge-Kantorovich Distances

Let X be a Polish space (i.e., a complete separable metric space) and $\mathcal{P}(X)$ be the
set of Borel probability measures on X. There are several ways to metricize the
topology of narrow convergence, at least on some subsets of $\mathcal{P}(X)$. Let us denote
by d the distance on X and, for $p \in [1, +\infty)$, by $\mathcal{P}_p(X)$ the set of probability
measures m such that

$$\int_X d^p(x_0, x)dm(x) < +\infty \qquad \text{for some (and hence for all) point } x_0 \in X.$$

The Monge-Kantorowich distance on $\mathcal{P}_p(X)$ is given by

$$\mathbf{d}_p(m, m') = \inf_{\gamma \in \Pi(m,m')} \left[\int_{X^2} d(x, y)^p d\gamma(x, y) \right]^{1/p} \qquad (1.145)$$

where $\Pi(m, m')$ is the set of Borel probability measures on $X \times X$ such that $\gamma(A \times X) = m(A)$ and $\gamma(X \times A) = m'(A)$ for any Borel set $A \subset X$. In other words, a
Borel probability measure γ on $X \times X$ belongs to $\Pi(m, m')$ if and only if

$$\int_{X^2} \varphi(x)d\gamma(x, y) = \int_X \varphi(x)dm(x) \qquad \text{and}$$

$$\int_{X^2} \varphi(y)d\gamma(x, y) = \int_X \varphi(y)dm'(y) ,$$

for any Borel and bounded measurable map $\varphi : X \to \mathbb{R}$. Note that $\Pi(m, m')$ is
non-empty, because for instance $m \otimes m'$ always belongs to $\Pi(m, m')$. Moreover, by
Hölder inequality, $\mathcal{P}_p(X) \subset \mathcal{P}_{p'}(X)$ for any $1 \leq p' \leq p$ and

$$\mathbf{d}_{p'}(m, m') \leq \mathbf{d}_p(m, m') \qquad \forall m, m' \in \mathcal{P}_p(X) .$$

We now explain that there exists at least an optimal measure in (1.145). This optimal measure is often referred to as *an optimal transport plan* from m to m'.

Lemma 1.18 (Existence of an Optimal Transport Plan) *For any* $m, m' \in \mathcal{P}_p(X)$, *there is at least one measure* $\bar{\gamma} \in \Pi(m, m')$ *with*

$$\mathbf{d}_p(m, m') = \left[\int_{X^2} d(x, y)^p d\bar{\gamma}(x, y) \right]^{1/p} .$$

Proof We first show that $\Pi(m, m')$ is tight and therefore relatively compact for the weak-* convergence. For any $\varepsilon > 0$ there exists a compact set $K_\varepsilon \subset X$ such that $m(K_\varepsilon) \geq 1 - \varepsilon/2$ and $m'(K_\varepsilon) \geq 1 - \varepsilon/2$. Then, for any $\gamma \in \Pi(m, m')$, we have

$$
\begin{aligned}
\gamma(K_\varepsilon \times K_\varepsilon) &\geq \gamma(K_\varepsilon \times X) - \gamma(K_\varepsilon \times (X \backslash K_\varepsilon)) \\
&\geq m(K_\varepsilon) - \gamma(X \times (X \backslash K_\varepsilon)) \\
&\geq 1 - \varepsilon/2 - m'(X \backslash K_\varepsilon) \geq 1 - \varepsilon .
\end{aligned}
$$

This means that $\Pi(m, m')$ is tight. It is also closed for the weak-* convergence. Since the map $\gamma \to \int_{X^2} |x - y|^p d\gamma(x, y)$ is lower semi-continuous for the weak-* convergence, it has a minimum on $\Pi(m, m')$. $\qquad\square$

Let us now check that $\mathbf{d}_p$ is a distance.

Lemma 1.19 *For any* $p \geq 1$, $\mathbf{d}_p$ *is a distance on* $\mathcal{P}_p$.

The proof uses the notion of disintegration of a measure, see Theorem 1.10.

Proof Only the triangle inequality presents some difficulty. Let $m, m', m'' \in \mathcal{P}_p$ and γ, γ' be optimal transport plans from m to m' and from m' to m'' respectively. We disintegrate the measures γ and γ' with respect to m': $d\gamma(x, y) = d\gamma_y(x) dm'(y)$ and $d\gamma'(y, z) = d\gamma'_y(z) dm'(y)$ and we define the measure π on $X \times X$ by

$$\int_{X \times X} \varphi(x, z) d\pi(x, z) = \int_{X \times X \times X} \varphi(x, z) d\gamma_y(x) d\gamma'_y(z) dm'(y) \qquad \forall \phi \in C^0_b(X \times X) .$$

Then one easily checks that $\pi \in \Pi(m, m'')$ and we have, by Hölder inequality,

$$
\begin{aligned}
\left[\int_{X \times X} d^p(x, z) d\pi(x, z) \right]^{1/p} &\leq \\
\left[\int_{X \times X \times X} (d(x, y) + d(y, z))^p d\gamma_y(x) d\gamma'_y(z) dm'(y) \right]^{1/p} & \\
\leq \left[\int_{X \times X} d^p(x, y) d\gamma_y(x) dm'(y) \right]^{1/p} &+ \left[\int_{X \times X} d^p(y, z) d\gamma_y(z) dm'(y) \right]^{1/p} \\
= \mathbf{d}_p(m, m') + \mathbf{d}_p(m', m'')
\end{aligned}
$$

So $\mathbf{d}_p(m, m'') \leq \mathbf{d}_p(m, m') + \mathbf{d}_p(m', m'')$. $\qquad\square$

In these notes, we are mainly interested in two Monge-Kantorovich distances, $\mathbf{d}_1$ and $\mathbf{d}_2$. The distance $\mathbf{d}_2$, which is often called the Wasserstein distance, is particularly useful when X is a Euclidean or a Hilbert space. Its analysis will be the object of the next subsection.

As for the distance $\mathbf{d}_1$, which often takes the name of the Kantorovich-Rubinstein distance, we have already encountered it several times. Let us point out the following equivalent representation, which explains the link with the notion introduced in Sect. 1.2.2:

Theorem 1.18 (Kantorovich-Rubinstein Theorem) *For any $m, m' \in \mathcal{P}_1(X)$,*

$$\mathbf{d}_1(m, m') = \sup \left\{ \int_X f(x)dm(x) - \int_X f(x)dm'(x) \right\}$$

where the supremum is taken over the set of all 1-Lipschitz continuous maps $f : X \to \mathbb{R}$.

Remark 1.17 In fact the above "Kantorovich duality result" holds for much more general costs (i.e., it is not necessary to minimize the power of a distance). The typical assertion in this framework is, for any lower semicontinuous map $c : X \times X \to \mathbb{R}_+ \cup \{+\infty\}$, the following equality holds:

$$\inf_{\gamma \in \Pi(m, m')} \int_{X \times X} c(x, y)d\gamma(x, y) = \sup_{f,g} \int_X f(x)dm(x) + \int_X g(y)dm'(y) ,$$

where the supremum is taken over the maps $f \in L^1_m(X), g \in L^1_{m'}(X)$ such that

$$f(x) + g(y) \leq c(x, y) \qquad \text{for } m - \text{a.e. } x \text{ and } m' - \text{a.e } y.$$

The proof of this result exceeds the scope of these notes and can be found in several textbooks (see [178] for instance).

Let us finally underline the link between convergence for the $\mathbf{d}_p$ distance and narrow convergence:

Proposition 1.4 *A sequence of measures (m_n) of $\mathcal{P}_p(X)$ converges to m for $\mathbf{d}_p$ if and only if (m_n) narrowly converges to m and*

$$\lim_{n \to +\infty} \int_X d^p(x, x_0)dm_n(x) = \int_X d^p(x, x_0)dm(x) \qquad \text{for some (and thus any) } x_0 \in X .$$

The proof for $p = 1$ is a simple consequence of Proposition 1.1 and Theorem 1.18. For the general case, see [178].

1.4.1.2   The Wasserstein Space of Probability Measures on $\mathbb{R}^d$

From now on we work in $X = \mathbb{R}^d$. Let $\mathcal{P}_2 = \mathcal{P}_2(\mathbb{R}^d)$ be the set of Borel probability measures on $\mathbb{R}^d$ with a finite second order moment: m belongs to $\mathcal{P}_2$ if m is a Borel probability measure on $\mathbb{R}^d$ with $\int_{\mathbb{R}^d} |x|^2 m(dx) < +\infty$. The Wasserstein distance is just the Monge-Kankorovich distance when $p = 2$:

$$\mathbf{d}_2(\mu, \nu) = \inf_{\gamma \in \Pi(\mu,\nu)} \left[\int_{\mathbb{R}^{2d}} |x - y|^2 d\gamma(x, y) \right]^{1/2} \qquad (1.146)$$

where $\Pi(\mu, \nu)$ is the set of Borel probability measures on $\mathbb{R}^{2d}$ such that $\gamma(A \times \mathbb{R}^d) = \mu(A)$ and $\gamma(\mathbb{R}^d \times A) = \nu(A)$ for any Borel set $A \subset \mathbb{R}^d$.

An important point, that we shall use sometimes, is the fact that the optimal transport plan can be realized as *an optimal transport map* whenever μ is absolutely continuous.

Theorem 1.19 (Existence of an Optimal Transport Map) *If $\mu \in \mathcal{P}_2$ is absolutely continuous, then, for any $\nu \in \mathcal{P}_2$, there exists a convex map $\Phi : \mathbb{R}^N \to \mathbb{R}$ such that the measure $(id_{\mathbb{R}^d}, D\Phi)\sharp\mu$ is optimal for $\mathbf{d}_2(\mu, \nu)$. In particular $\nu = D\Phi\sharp\mu$.*

Conversely, if the convex map $\Phi : \mathbb{R}^N \to \mathbb{R}$ satisfies $\nu = D\Phi\sharp\mu$, then the measure $(id_{\mathbb{R}^d}, D\Phi)\sharp\mu$ is optimal for $\mathbf{d}_2(\mu, \nu)$.

The proof of this result, due to Y. Brenier [39], exceeds the scope of these notes. It can be found in various places, such as [178].

1.4.2   Derivatives in the Space of Measures

In this section, we discuss different notions of derivatives in the space of probability measures and explain how they are related. This part is, to a large extent, borrowed from [65, 68]. For simplicity, we work in the whole space $\mathbb{R}^d$ and set $\mathcal{P}_2 = \mathcal{P}_2(\mathbb{R}^d)$.

1.4.2.1   The Flat Derivative

Definition 1.9 Let $U : \mathcal{P}_2 \to \mathbb{R}$. We say that U is of class C^1 if there exists a jointly continuous and bounded map $\frac{\delta U}{\delta m} : \mathcal{P}_2 \times \mathbb{R}^d \to \mathbb{R}$ such that

$$U(m') - U(m) = \int_0^1 \int_{\mathbb{R}^d} \frac{\delta U}{\delta m}((1-h)m + hm', y)(m' - m)(dy)dh \qquad \forall m, m' \in \mathcal{P}_2.$$

Moreover we adopt the normalization convention

$$\int_{\mathbb{R}^d} \frac{\delta U}{\delta m}(m, y) m(dy) = 0 \qquad \forall m \in \mathcal{P}_2. \tag{1.147}$$

Remark 1.18 If $U : \mathcal{P}_2(\mathbb{T}^d) \to \mathbb{R}$, then the derivative is defined in the same way, with $\frac{\delta U}{\delta m} : \mathcal{P}_2(\mathbb{T}^d) \times \mathbb{T}^d \to \mathbb{R}$ such that

$$U(m') - U(m) = \int_0^1 \int_{\mathbb{T}^d} \frac{\delta U}{\delta m}((1-h)m + hm', y)(m'-m)(dy)dh \qquad \forall m, m' \in \mathcal{P}_2(\mathbb{T}^d).$$

If U is of class C^1, then the following equality holds for any $m \in \mathcal{P}_2$ and $y \in \mathbb{R}^d$

$$\frac{\delta U}{\delta m}(m, y) = \lim_{h \to 0^+} \frac{1}{h} \left(U((1-h)m + h\delta_y) - U(m) \right).$$

Here is a kind of converse.

Proposition 1.5 *Let $U : \mathcal{P}_2 \to \mathbb{R}$ and assume that the limit*

$$V(m, y) := \lim_{h \to 0^+} \frac{1}{h} \left(U((1-h)m + h\delta_y) - U(m) \right)$$

exists and is jointly continuous and bounded on $\mathcal{P}_2 \times \mathbb{R}^d$. Then U is C^1 and $\frac{\delta U}{\delta m}(m, y) = V(m, y)$.

Proof Although the result can be expected, the proof is a little involved and can be found in [66]. □

Let us recall that, if $\phi : \mathbb{R}^d \to \mathbb{R}^d$ is a Borel measurable map and m is a Borel probability measure on $\mathbb{R}^d$, the image of m by ϕ is the Borel probability measure $\phi \sharp m$ defined by

$$\int_{\mathbb{R}^d} f(x) \phi \sharp m(dx) = \int_{\mathbb{R}^d} f(\phi(y)) m(dy) \qquad \forall f \in C_b^0(\mathbb{R}^d).$$

Proposition 1.6 *Let U be C^1 and be such that $D_y \frac{\delta U}{\delta m}$ exists and is jointly continuous and bounded on $\mathcal{P}_2 \times \mathbb{R}^d$. Then, for any Borel measurable map $\phi : \mathbb{R}^d \to \mathbb{R}^d$ with at most a linear growth, the map $s \to U((id_{\mathbb{R}^d} + s\phi)\sharp m)$ is differentiable at 0 and*

$$\frac{d}{ds} U((id_{\mathbb{R}^d} + s\phi)\sharp m)_{|s=0} = \int_{\mathbb{R}^d} D_y \frac{\delta U}{\delta m}(m, y) \cdot \phi(y) m(dy).$$

Proof Indeed

$$U((id_{\mathbb{R}^d} + s\phi)\sharp m) - U(m)$$

$$= \int_0^1 \int_{\mathbb{R}^d} \frac{\delta U}{\delta m}(m_{h,s}, y)((id_{\mathbb{R}^d} + s\phi)\sharp m) - m)(dy)dh$$

$$= \int_0^1 \int_{\mathbb{R}^d} (\frac{\delta U}{\delta m}(m_{h,s}, y + s\phi(y)) - \frac{\delta U}{\delta m}(m_{h,s}, y))m(dy)dh$$

$$= s \int_0^1 \int_0^1 \int_{\mathbb{R}^d} D_y \frac{\delta U}{\delta m}(m_{h,s}, y + s\tau\phi(y)) \cdot \phi(y)m(dy)dhd\tau,$$

where

$$m_{h,s} = (1 - h)m + h(id_{\mathbb{R}^d} + s\phi)\sharp m.$$

Dividing by s and letting $s \to 0^+$ gives the desired result.   □

Let us recall that, if $m, m' \in \mathcal{P}_2$, the set $\Pi^{opt}(m, m')$ denotes the set of optimal transport plans between m and m' (see Lemma 1.18).

Proposition 1.7 *Under the assumptions of the previous Proposition, let* $m, m' \in \mathcal{P}_2$ *and* $\pi \in \Pi^{opt}(m, m')$. *Then*

$$\left| U(m') - U(m) - \int_{\mathbb{R}^{2d}} D_y \frac{\delta U}{\delta m}(m, x) \cdot (y - x)\pi(dx, dy) \right| \leq o(\mathbf{d}_2(m, m')).$$

Remark 1.19 The same proof shows that, if π is a transport plan between m and m' (not necessarily optimal), then

$$\left| U(m') - U(m) - \int_{\mathbb{R}^{2d}} D_y \frac{\delta U}{\delta m}(m, x) \cdot (y - x)\pi(dx, dy) \right|$$

$$\leq o\left(\left(\int_{\mathbb{R}^{2d}} |x - y|^2 \pi(dx, dy) \right)^{1/2} \right).$$

Proof Let $\phi_t(x, y) = (1 - t)x + ty$ and $m_t = \phi_t \sharp \pi$. Then $m_0 = m$ and $m_1 = m'$ and, for any $t \in (0, 1)$ and any s small we have

$$U(\phi_{t+s}\sharp\pi) - U(\phi_t\sharp\pi)$$

$$= \int_0^1 \int_{\mathbb{R}^d} \frac{\delta U}{\delta m}(m_{s,h}, y)(\phi_{t+s}\sharp\pi - \phi_t\sharp\pi)(dy)dh$$

$$= \int_0^1 \int_{\mathbb{R}^{2d}} \frac{\delta U}{\delta m}(m_{s,h}, (1 - t - s)x + (t + s)y)$$

$$-\frac{\delta U}{\delta m}(m_{s,h}, (1-t)x + ty)\,\pi(dx, dy)dh$$

$$= s\int_0^1\int_0^1\int_{\mathbb{R}^{2d}} D_y\frac{\delta U}{\delta m}$$

$$\times (m_{s,h}, (1-t-\tau s)x + (t+\tau s)y)\cdot(y-x)\,\pi(dx, dy)dhd\tau,$$

where $m_{s,h} = (1-h)\phi_{t+s}\sharp\pi + h\phi_t\sharp\pi$. So, dividing by s and letting $s \to 0$ we find:

$$\frac{d}{dt}U(\phi_t\sharp\pi) = \int_{\mathbb{R}^{2d}} D_y\frac{\delta U}{\delta m}(\phi_t\sharp\pi, (1-t)x + ty)\cdot(y-x)\,\pi(dx, dy).$$

As $D_y\frac{\delta U}{\delta m}$ is continuous and bounded by C, for any $\varepsilon, R > 0$, there exists $r > 0$ such that, if $\mathbf{d}_2(m, m') \le r$ and $|x|, |y| \le R$, then

$$|D_y\frac{\delta U}{\delta m}(\phi_t\sharp\pi, (1-t)x + ty) - D_y\frac{\delta U}{\delta m}(m, x)| \le \varepsilon + 2C\mathbf{1}_{|y-x|\ge r}.$$

So

$$\left|\int_{\mathbb{R}^{2d}} D_y\frac{\delta U}{\delta m}(\phi_t\sharp\pi, (1-t)x + ty)\cdot(y-x)\,\pi(dx, dy)\right.$$

$$\left.-\int_{\mathbb{R}^{2d}} D_y\frac{\delta U}{\delta m}(m, x)\cdot(y-x)\pi(dx, dy)\right|$$

$$\le \delta_R + \int_{(B_R)^2}(\varepsilon + 2C\mathbf{1}_{|x-y|\ge r})|y-x|\pi(dx, dy)$$

$$\le \delta_R + \varepsilon\mathbf{d}_2(m, m') + \frac{2C}{r}\mathbf{d}_2^2(m, m').$$

where

$$\delta_R := \int_{\mathbb{R}^{2d}\setminus(B_R)^2}|D_y\frac{\delta U}{\delta m}(\phi_t\sharp\pi, (1-t)x + ty)\cdot(y-x)|$$

$$+ |D_y\frac{\delta U}{\delta m}(m, x)\cdot(y-x)|\pi(dx, dy)$$

$$\le C\int_{\mathbb{R}^{2d}\setminus(B_R)^2}|y-x|\pi(dx, dy) \le C\mathbf{d}_2(m, m')\pi^{1/2}(\mathbb{R}^{2d}\setminus(B_R)^2)$$

$$= \mathbf{d}_2(m, m')o_R(1).$$

This proves the result. $\qquad\square$

1.4.2.2   W-Differentiability

Next we turn to a more geometric definition of derivative in the space of measures. For this, let us introduce the notion of tangent space to $\mathcal{P}_2$.

Definition 1.10 (Tangent Space) The tangent space $\mathrm{Tan}_m(\mathcal{P}_2)$ of $\mathcal{P}_2$ at $m \in \mathcal{P}_2$ is the closure in $L^2_m(\mathbb{R}^d)$ of $\{D\phi, \ \phi \in C^\infty_c(\mathbb{R}^d)\}$.

Following [17] we define the super and the subdifferential of a map defined on $\mathcal{P}_2$:

Definition 1.11 Let $U : \mathcal{P}_2 \to \mathbb{R}$, $m \in \mathcal{P}_2$ and $\xi \in L^2_m(\mathbb{R}^d, \mathbb{R}^d)$. We say that ξ belongs to the superdifferential $\partial^+ U(m)$ of U at m if, for any $m' \in \mathcal{P}_2$ and any transport plan π from m to m',

$$U(m') \leq U(m) + \int_{\mathbb{R}^d \times \mathbb{R}^d} \xi(x) \cdot (y - x) \pi(dx, dy)$$

$$+ o\left(\left(\int_{\mathbb{R}^{2d}} |x - y|^2 \pi(dx, dy) \right)^{1/2} \right).$$

We say that ξ belongs to the subdifferential $\partial^- U(m)$ of U at m if $-\xi$ belongs to $D^+(-U)(m)$. Finally, we say that the map U is W-differentiable at m if $\partial^+ U(m) \cap \partial^- U(m)$ is not empty.

One easily checks the following:

Proposition 1.8 *If U is W-differentiable at m, then $\partial^+ U(m)$ and $\partial^- U(m)$ are equal and reduce to a singleton, denoted $\{D_m U(m, \cdot)\}$.*

Remark 1.20 On can actually check that $D_m U(m, \cdot)$ belongs to $\mathrm{Tan}_m(\mathcal{P}_2)$.

Proof Let $\xi_1 \in D^+ U(m)$ and $\xi_2 \in D^- U(m)$. We have, for any $m' \in \mathcal{P}_2$ and any transport plan π from m to m',

$$\int_{\mathbb{R}^d \times \mathbb{R}^d} \xi_2(x) \cdot (y - x) \pi(dx, dy) + o\left(\left(\int_{\mathbb{R}^{2d}} |x - y|^2 \pi(dx, dy) \right)^{1/2} \right)$$

$$\leq U(m') - U(m) \leq \int_{\mathbb{R}^d \times \mathbb{R}^d} \xi_1(x) \cdot (y - x) \pi(dx, dy)$$

$$+ o\left(\left(\int_{\mathbb{R}^{2d}} |x - y|^2 \pi(dx, dy) \right)^{1/2} \right).$$

In particular, if we choose $m' = (1 + h\phi)\sharp m$ and $\pi = (Id, Id + h\phi)\sharp m$ for some $\phi \in L^2_m(\mathbb{R}^d, \mathbb{R}^d)$ and $h > 0$ small, we obtain

$$h \int_{\mathbb{R}^d} \xi_2(x) \cdot \phi(x) m(dx) + o(h) \leq U(m') - U(m) \leq h \int_{\mathbb{R}^d} \xi_1(x) \cdot \phi(x) m(dx) + o(h),$$

from which we easily infer that $\xi_1 = \xi_2$ in $L^2_m(\mathbb{R}^d)$.   □

Remark 1.19 implies that, if U is C^1 with $D_y \delta U/\delta m$ continuous and bounded, then U is W-differentiable. In this case it is obvious that $D_y \delta U/\delta m$ belongs to $\mathrm{Tan}_m(\mathcal{P}_2)$ by definition and that $D_m U(m, \cdot) = D_y \delta U/\delta m$. From now on we systematically use the notation $D_m U(m, \cdot) = D_y \delta U/\delta m$ in this case.

1.4.2.3   Link with the L-Derivative

Another possibility for the notion of derivative is to look at probability measures as the law of random variables with values in $\mathbb{R}^d$ and to use the fact that the set of random variables, under suitable moment conditions, is a Hilbert space.

Let $(\Omega, \mathcal{F}, \mathbb{P})$ an atomless probability space (meaning that, for any $E \in \mathcal{F}$ with $\mathbb{P}[E] > 0$, there exists $E' \in \mathcal{F}$ with $E' \subset E$ and $0 < \mathbb{P}[E'] < \mathbb{P}[E]$). Given a map $U : \mathcal{P}_2 \to \mathbb{R}$, we consider its extension $\tilde{U}$ to the set of random variables $L^2(\Omega, \mathbb{R}^d)$:

$$\tilde{U}(X) = U(\mathcal{L}(X)) \qquad \forall X \in L^2(\Omega, \mathbb{R}^d).$$

(recall that $\mathcal{L}(X)$ is the law of X, i.e., $\mathcal{L}(X) := X\sharp\mathbb{P}$. Note that $\mathcal{L}(X)$ belongs to $\mathcal{P}_2$ because $X \in L^2(\Omega)$). The main point is that $L^2(\Omega, \mathcal{F}, \mathbb{P})$ is a Hilbert space, in which the notion of Frechet differentiability makes sense.

For instance, if U is a map of the form

$$U(m) = \int_{\mathbb{R}^d} \phi(x) m(dx) \qquad \forall m \in \mathcal{P}_2, \tag{1.148}$$

where $\phi \in C^0_c(\mathbb{R}^d)$ is given, then

$$\tilde{U}(X) = \mathbb{E}[\phi(X)] \qquad \forall X \in L^2(\Omega, \mathbb{R}^d).$$

Definition 1.12 The map $U : \mathcal{P}_2 \to \mathbb{R}$ is L-differentiable at $m \in \mathcal{P}_2$ if there exists $X \in L^2(\Omega, \mathbb{R}^d)$ such that $\mathcal{L}(X) = m$ and the extension $\tilde{U}$ of U is Frechet differentiable at X.

The following result says that the notion of L-differentiability coincides with that of W-differentiability and is independent of the probability space and of the representative X. The first statement in that direction goes back to Lions [149] (Lesson 31/10 2008), the version given here can be found in [109] (see also [19], from which the sketch of proof of Lemma 1.21 is largely inspired).

Theorem 1.20 *The map U is W-differentiable at $m \in \mathcal{P}_2$ if and only if $\tilde{U}$ is Frechet differentiable at some (or thus any) $X \in L^2(\Omega, \mathbb{R}^d)$ with $\mathcal{L}(X) = m$. In this case*

$$\nabla \tilde{U}(X) = D_m U(m, X).$$

The result can be considered as a structure theorem for the L-derivative.

For instance, if U is as in (1.148) for some map $\phi \in C_c^1(\mathbb{R}^d)$, then it is almost obvious that

$$\nabla \tilde{U}(X) = D\phi(X)$$

and thus

$$D_m U(m, x) = D\phi(x).$$

The proof of Theorem 1.20 is difficult and we only sketch it briefly. Complete proofs can be found in [109] or [19]. The first step is the fact that, if X and X' have the same law, then so do $\nabla \tilde{U}(X)$ and $\nabla \tilde{U}(X')$:

Lemma 1.20 *Let $U : \mathcal{P}_2 \to \mathbb{R}$ and $\tilde{U}$ be its extension. Let X, X' be two random variables in $L^2(\Omega, \mathbb{R}^d)$ with $\mathcal{L}(X) = \mathcal{L}(X')$. If $\tilde{U}$ is Frechet differentiable at X, then $\tilde{U}$ is differentiable at X' and $(X, \nabla \tilde{U}(X))$ has the same law as $(X', \nabla \tilde{U}(X'))$.*

(Sketch of) Proof The idea behind this fact is that, if X and X' have the same law, then one can "almost" find a bi-measurable and measure-preserving transformation $\tau : \Omega \to \Omega$ such that $X = X' \circ \tau$. Admitting this statement for a while, we have, for any $H' \in L^2$ small,

$$\tilde{U}(X' + H') = \tilde{U}((X' + H') \circ \tau) = \tilde{U}(X + H' \circ \tau)$$

$$= \tilde{U}(X) + \mathbb{E}\left[\nabla \tilde{U}(X) \cdot H' \circ \tau\right] + o(\|H' \circ \tau\|_2)$$

$$= \tilde{U}(X') + \mathbb{E}\left[\nabla \tilde{U}(X) \circ \tau^{-1} \cdot H'\right] + o(\|H'\|_2).$$

This shows that $\tilde{U}$ is differentiable at X' with differential given by $\nabla \tilde{U}(X) \circ \tau^{-1}$. Thus $(X', \nabla \tilde{U}(X')) = (X, \nabla \tilde{U}(X)) \circ \tau^{-1}$, which shows that $(X, \nabla \tilde{U}(X))$ and $(X', \nabla \tilde{U}(X'))$ have the same law.

In fact the existence of τ does not hold in general. However, one can show that, for any $\varepsilon > 0$, there exists $\tau : \Omega \to \Omega$ bi-measurable and measure preserving and such that $\|X' - X \circ \tau\|_\infty \le \varepsilon$. A (slightly technical) adaptation of the proof above then gives the result (see [51] or [68] for the details). □

Next we show that $\nabla \tilde{U}(X)$ is a function of X:

Lemma 1.21 *Assume that $\tilde{U}$ is differentiable at $X \in L^2(\Omega, \mathbb{R}^d)$. Then there exists a Borel measurable map $g : \mathbb{R}^d \to \mathbb{R}^d$ such that $\nabla \tilde{U}(X) = g(X)$ a.s..*

(Sketch of) Proof To prove the claim, we just need to check that $\nabla \tilde{U}(X)$ is $\sigma(X)$-measurable (see Theorem 20.1 in [37]), which can be recasted into the fact that $\nabla \tilde{U}(X) = \mathbb{E}\left[\nabla \tilde{U}(X)|X\right]$. Let $\mu = \mathcal{L}(X, \nabla \tilde{U}(X))$ and let $\mu(dx, dy) = (\delta_x \otimes \nu_x(dy))\mathbb{P}_X(dx)$ be its disintegration with respect to its first marginal $\mathbb{P}_X$. Let λ be the restriction of the Lebesgue measure to $Q_1 := [0, 1]^d$. Then, as λ has an L^1 density, the optimal transport from λ to ν_x is unique and given by the gradient of a convex map $\psi_x(\cdot)$ (Brenier's Theorem, see [179]). So we can find[2] a measurable map $\psi : \mathbb{R}^d \times \mathbb{R}^d \to \mathbb{R}^d$ such that, for $\mathbb{P}_X$-a.e. $x \in \mathbb{R}^d$, $\psi_x(\cdot)\sharp\lambda = \nu_x$. Let Z be a random variable with law λ and independent of $(X, \nabla \tilde{U}(X))$.

Note that $\mu = \mathcal{L}(X, \nabla \tilde{U}(X)) = \mathcal{L}(X, \psi_X(Z))$ because, for any $f \in C_b^0(\mathbb{R}^d \times \mathbb{R}^d)$,

$$
\begin{aligned}
\mathbb{E}[f(X, \psi_X(Z))] &= \int_{\mathbb{R}^d} \int_{Q_1} f(x, \psi_x(z))\lambda(dz)\mathbb{P}_X(dx) \\
&= \int_{\mathbb{R}^d} \int_{\mathbb{R}^d} f(x, y)(\psi_x \sharp \lambda)(dy)\mathbb{P}_X(dx) \\
&= \int_{\mathbb{R}^d} \int_{\mathbb{R}^d} f(x, y))\nu_x(dy)\mathbb{P}_X(dx) = \int_{\mathbb{R}^{2d}} f(x, y)\mu(dx, dy).
\end{aligned}
$$

So, for any ε,

$$
\tilde{U}(X + \varepsilon\nabla\tilde{U}(X)) = \tilde{U}(X + \varepsilon\psi_X(Z)),
$$

from which we infer, taking the derivative with respect to ε at $\varepsilon = 0$:

$$
\mathbb{E}\left[\left|\nabla\tilde{U}(X)\right|^2\right] = \mathbb{E}\left[\nabla\tilde{U}(X) \cdot \psi_X(Z)\right].
$$

Note that, as Z is independent of $(X, \nabla\tilde{U}(X))$, we have

$$
\mathbb{E}\left[\nabla\tilde{U}(X) \cdot \psi_X(Z)\right] = \mathbb{E}\left[\nabla\tilde{U}(X) \cdot \mathbb{E}[\psi_x(Z)]_{x=X}\right],
$$

where, for $\mathbb{P}_X$-a.e. x,

$$
\begin{aligned}
\mathbb{E}[\psi_x(Z)] &= \int_{Q_1} \psi_x(z)\lambda(dz) = \int_{Q_1} y\,(\psi_x\sharp\lambda)(dy) \\
&= \int_{\mathbb{R}^d} y\,\nu_x(dy) = \mathbb{E}\left[\nabla\tilde{U}(X)|X = x\right].
\end{aligned}
$$

---

[2] Warning: here the proof is sloppy and the possibility of a measurable selection should be justified.

So, by the tower property of the conditional expectation, we have

$$\mathbb{E}\left[\left|\nabla\tilde{U}(X)\right|^2\right] = \mathbb{E}\left[\nabla\tilde{U}(X)\cdot\mathbb{E}\left[\nabla\tilde{U}(X)|X\right]\right] = \mathbb{E}\left[\left|\mathbb{E}\left[\nabla\tilde{U}(X)|X\right]\right|^2\right].$$

Using again standard properties of the conditional expectation we infer the equality $\nabla\tilde{U}(X) = \mathbb{E}\left[\nabla\tilde{U}(X)|X\right]$, which shows the result.   □

Proof of Theorem 1.20 Let us first assume that U is W-differentiable at some $m \in \mathcal{P}_2$. Then there exists $\xi := D_m U(m, \cdot) \in L^2_m(\mathbb{R}^d)$ such that, for any $m' \in \mathcal{P}_2$ and any transport plan π between m and m' we have

$$\left|U(m') - U(m) - \int_{\mathbb{R}^d\times\mathbb{R}^d} \xi(x)\cdot(y-x)\pi(dx, dy)\right|$$
$$\leq o\left(\left(\int_{\mathbb{R}^{2d}} |x-y|^2\pi(dx, dy)\right)^{1/2}\right).$$

Therefore, for any $X \in L^2$ such that $\mathcal{L}(X) = m$, for any $H \in L^2$, if we denote by m' the law of $X + H$ and by π the law of $(X, X+H)$, we have

$$\left|\tilde{U}(X+H) - \tilde{U}(X) - \mathbb{E}[\xi(X)\cdot H]\right| = \left|U(m') - \tilde{U}(m) - \int_{\mathbb{R}^{2d}} \xi(x)\cdot(y-x)\pi(x, y)\right|$$
$$\leq o\left(\left(\int_{\mathbb{R}^{2d}} |x-y|^2\pi(dx, dy)\right)^{1/2}\right)$$
$$= o\left(\mathbb{E}\left[|X-Y|^2\right]^{1/2}\right).$$

This shows that U is L-differentiable.

Conversely, let us assume that U is L-differentiable at m. We know from Lemma 1.21 that, for any $X \in L^2$ such that $\mathcal{L}(X) = m$, $\tilde{U}$ is differentiable at X and $\nabla\tilde{U}(X) = \xi(X)$ for some Borel measurable map $\xi : \mathbb{R}^d \to \mathbb{R}^d$. In view of Lemma 1.20, the map ξ does not depend on the choice of X. So, for any $\varepsilon > 0$, there exists $r > 0$ such that, for any X with $\mathcal{L}(X) = m$ and any $H \in L^2$ with $\|H\| \leq r$, one has

$$\left|\tilde{U}(X+H) - \tilde{U}(X) - \mathbb{E}[\xi(X)\cdot H]\right| \leq \varepsilon.$$

Let now $m' \in \mathcal{P}_2$ and π be a transport plan between m and m' such that $\int_{\mathbb{R}^{2d}} |x - y|^2 \pi(dx, dy) \leq r^2$. Let (X, Y) with law π. We set $H = Y - X$ and note that $\|H\|_2 \leq r$. So we have

$$\left| U(m') - \tilde{U}(m) - \int_{\mathbb{R}^{2d}} \xi(x) \cdot (y - x)\pi(x, y) \right|$$

$$= \left| \tilde{U}(X + H) - \tilde{U}(X) - \mathbb{E}\left[\xi(X) \cdot H\right] \right| \leq \varepsilon.$$

This proves the W-differentiability of U. $\square$

1.4.2.4   Higher Order Derivatives

We say that U is partially C^2 if U is C^1 and if $D_y \delta U / \delta m$ and $D_{yy}^2 \delta U / \delta m$ exist and are continuous and bounded on $\mathcal{P}_2 \times \mathbb{R}^d$.

We say that U is C^2 if $\frac{\delta U}{\delta m}$ is C^1 in m with a continuous and bounded derivative: namely $\frac{\delta^2 U}{\delta m^2} = \frac{\delta}{\delta m}(\frac{\delta U}{\delta m}) : \mathcal{P}_2 \times \mathbb{R}^d \times \mathbb{R}^d \to \mathbb{R}$ is continuous in all variables and bounded. We say that U is twice L-differentiable if the map $D_m U$ is L-differentiable with respect to m with a second order derivative $D_{mm}^2 U = D_{mm}^2 U(m, y, y')$ which is continuous and bounded on $\mathcal{P}_2 \times \mathbb{R}^d \times \mathbb{R}^d$ with values in $\mathbb{R}^{d \times d}$. One can check that this second order derivative enjoys standard properties of derivatives, such as the symmetry:

$$D_{mm}^2 U(m, y, y') = D_{mm}^2 U(m, y', y).$$

See [65, 68].

1.4.2.5   Comments

For a general description of the notion of derivatives and the historical background, we refer to [68], Chap. V. The notion of flat derivative is very natural and has been introduced in several contexts and under various assumptions. We follow here [65]. Let us note however that these notions of derivatives can be traced back to [14], while the construction of Proposition 1.5 has already a counterpart in [159].

The initial definition of sub and super differential in the space $\mathcal{P}_2$, introduced in [18], is the following: ξ belongs to $\partial^+ U(m)$ if $\xi \in \mathrm{Tan}_m(\mathcal{P}_2)$ and

$$U(m') \leq U(m) + \inf_{\pi \in \Pi^{opt}(m,m')} \int_{\mathbb{R}^d \times \mathbb{R}^d} \xi(x) \cdot (y - x)\pi(dx, dy) + o(\mathbf{d}_2(m, m')).$$

It is proved in [109] that this definition coincides with the one introduced in Definition 1.11.

The notion of L-derivative and the structure of this derivative has been first discussed by Lions in [149] (see also [51] for a proof of Theorem 1.20 in which the function is supposed to be continuously differentiable). The proof of Theorem 1.20, without the extra continuity condition, is due to Gangbo and Tudorascu [109] (see also [19], revisited here in a loose way).

1.4.3 The Master Equation

In this section we investigate the partial differential equation:

$$
\begin{cases}
-\partial_t U - \Delta_x U + H(x, D_x U) - \displaystyle\int_{\mathbb{T}^d} \mathrm{div}_y \, D_m U(t, x, m, y) \, dm(y) \\
\qquad + \displaystyle\int_{\mathbb{T}^d} D_m U(t, x, m, y) \cdot H_p(y, D_x U) \, dm(y) = F(x, m) \qquad (1.149) \\
\qquad\qquad \text{in } (0, T) \times \mathbb{T}^d \times \mathcal{P}_2 \\
U(T, x, m) = G(x, m) \qquad \text{in } \mathbb{T}^d \times \mathcal{P}_2
\end{cases}
$$

In this equation, $U = U(t, x, m)$ is the unknown. As explained below, $U(t, x, m)$ can be interpreted as the minimal cost, in the mean field problem, for a small player at time t in position x, if the distribution of the other players is m. Equation (1.149) is often called the *first order master equation* since it only involves first order derivatives with respect to the measure. This is in contrast with what happens for MFG problems with a common noise, for which the corresponding master equation also involves second order derivatives (see Sect. 1.4.3.3). After explaining the existence and the uniqueness of a solution for (1.149) (Sect. 1.4.3.1), we discuss other frameworks for the master equation: the case of finite state space (Sect. 1.4.3.2) and the MFG problem with a common noise (Sect. 1.4.3.3).

Throughout this part, we work in the torus $\mathbb{T}^d$ and in the space $\mathcal{P}_2 = \mathcal{P}_2(\mathbb{T}^d)$ of Borel probability measures on $\mathbb{T}^d$ endowed with the Wasserstein distance $\mathbf{d}_2$. The notion of derivative is the one discussed in the previous part (with the minor difference explained in Remark 1.18).

1.4.3.1 Existence and Uniqueness of a Solution for the Master Equation

Definition 1.13 We say that a map $V : [0, T] \times \mathbb{T}^d \times \mathcal{P}_2 \to \mathbb{R}$ is a classical solution to the Master equation (1.149) if

- V is continuous in all its arguments (for the $\mathbf{d}_1$ distance on $\mathcal{P}_2$), is of class C^2 in x and C^1 in time,
- V is of class C^1 with respect to m with a derivative $\frac{\delta V}{\delta m} = \frac{\delta V}{\delta m}(t, x, m, y)$ having globally continuous first and second order derivatives with respect to the space variables.

- The following relation holds for any $(t, x, m) \in (0, T) \times \mathbb{T}^d \times \mathcal{P}_2$:

$$\begin{cases} -\partial_t V(t, x, m) - \Delta_x V(t, x, m) + H(x, D_x V(t, x, m)) - \int_{\mathbb{T}^d} \operatorname{div}_y D_m V(t, x, m, y) \, dm(y) \\ + \int_{\mathbb{T}^d} D_m V(t, x, m, y) \cdot H_p(y, D_x V(t, y, m)) \, dm(y) = F(x, m) \end{cases}$$

and $V(T, x, m) = G(x, m)$ in $\mathbb{T}^d \times \mathcal{P}_2$.

Throughout the section, $H : \mathbb{T}^d \times \mathbb{R}^d \to \mathbb{R}$ is smooth, globally Lipschitz continuous and satisfies the coercivity condition:

$$C^{-1} \frac{I_d}{1 + |p|} \leq H_{pp}(x, p) \leq C I_d \qquad \text{for } (x, p) \in \mathbb{T}^d \times \mathbb{R}^d. \tag{1.150}$$

We also always assume that the maps $F, G : \mathbb{T}^d \times \mathcal{P}_1 \to \mathbb{R}$ are globally Lipschitz continuous and monotone:

$$F \text{ and } G \text{ are monotone.} \tag{1.151}$$

Note that assumption (1.151) implies that $\frac{\delta F}{\delta m}$ and $\frac{\delta G}{\delta m}$ satisfy the following monotonicity property (explained for F):

$$\int_{\mathbb{T}^d} \int_{\mathbb{T}^d} \frac{\delta F}{\delta m}(x, m, y) \mu(x) \mu(y) \, dx \, dy \geq 0$$

for any smooth map $\mu : \mathbb{T}^d \to \mathbb{R}$ with $\int_{\mathbb{T}^d} \mu = 0$.

Let us fix $n \in \mathbb{N}^*$ and $\alpha \in (0, 1/2)$. We set

$$\operatorname{Lip}_n\left(\frac{\delta F}{\delta m}\right) := \sup_{m_1 \neq m_2} (\mathbf{d}_1(m_1, m_2))^{-1} \left\| \frac{\delta F}{\delta m}(\cdot, m_1, \cdot) - \frac{\delta F}{\delta m}(\cdot, m_2, \cdot) \right\|_{C^{n+2\alpha} \times C^{n-1+2\alpha}}$$

and use the symmetric notation for G. We call $(\mathbf{HF(n)})$ the following regularity conditions on F:

$(\mathbf{HF(n)})$ $\displaystyle \sup_{m \in \mathcal{P}_1} \left(\|F(\cdot, m)\|_{C^{n+2\alpha}} + \left\| \frac{\delta F(\cdot, m, \cdot)}{\delta m} \right\|_{C^{n+2\alpha} \times C^{n+2\alpha}} \right) + \operatorname{Lip}_n\left(\frac{\delta F}{\delta m}\right) < \infty.$

and $(\mathbf{HG(n)})$ the symmetric condition on G:

$(\mathbf{HG(n)})$ $\displaystyle \sup_{m \in \mathcal{P}_1} \left(\|G(\cdot, m)\|_{C^{n+2\alpha}} + \left\| \frac{\delta G(\cdot, m, \cdot)}{\delta m} \right\|_{C^{n+2\alpha} \times C^{n+2\alpha}} \right) + \operatorname{Lip}_n\left(\frac{\delta G}{\delta m}\right) < \infty.$

In order to explain the existence of a solution to the master equation, we need to introduce the solution of the MFG system: for any $(t_0, m_0) \in [0, T) \times \mathcal{P}_2$, let (u, m) be the solution to:

$$\begin{cases} -\partial_t u - \Delta u + H(x, Du) = F(x, m(t)) \\ \partial_t m - \Delta m - \operatorname{div}(m H_p(x, Du)) = 0 \\ u(T, x) = G(x, m(T)), \ m(t_0, \cdot) = m_0 \end{cases} \qquad (1.152)$$

Thanks to the monotonicity condition (1.151), we know that the system admits a unique solution, see Theorem 1.4. Then we set

$$U(t_0, x, m_0) := u(t_0, x) \qquad (1.153)$$

Theorem 1.21 *Assume that* **(HF(n))** *and* **(HG(n))** *hold for some* $n \geq 4$. *Then the map* U *defined by* (1.153) *is the unique classical solution to the master equation* (1.149).
Moreover, U *is globally Lipschitz continuous in the sense that*

$$\|U(t_0, \cdot, m_0) - U(t_0, \cdot, m_1)\|_{C^{n+\alpha}} \leq C_n \mathbf{d}_1(m_0, m_1) \qquad (1.154)$$

with Lipschitz continuous derivatives:

$$\|D_m U(t_0, \cdot, m_0, \cdot) - D_m U(t_0, \cdot, m_1, \cdot)\|_{C^{n+\alpha} \times C^{n+\alpha}} \leq C_n \mathbf{d}_1(m_0, m_1) \qquad (1.155)$$

for any $t_0 \in [0, T]$, $m_0, m_1 \in \mathcal{P}_1$.

Relation (1.153) says that the solutions of the MFG system (1.152) can be considered as characteristics of the master equation (1.149). As it will be transparent in the analysis of the MFG problem on a finite state space (Sect. 1.4.3.2 below), this means that the master equation is a kind of transport in the space of measures. The difficulty is that it is nonlinear, nonlocal (because of the integral terms) and without a comparison principle.

The proof of Theorem 1.21, although not very difficult in its principle, is quite technical and will be mostly omitted here. The main issue is to check that the map U defined by (1.153) satisfies (1.154), (1.155). This exceeds the scope of these notes and we refer the reader to [65] for a proof. Once we know that U is quite smooth, the conclusion follows easily:

Sketch of Proof of Theorem 1.21 (Existence) Let $m_0 \in \mathcal{P}(\mathbb{T}^d)$ with a C^1, positive density. Let $t_0 > 0$, (u, m) be the solution of the MFG system (1.152) starting from m_0 at time t_0. Then

$$\frac{U(t_0 + h, x, m_0) - U(t_0, x, m_0)}{h} = \frac{U(t_0 + h, x, m_0) - U(t_0 + h, x, m(t_0 + h))}{h}$$
$$+ \frac{U(t_0 + h, x, m(t_0 + h)) - U(t_0, x, m_0)}{h}.$$

As

$$\partial_t m - \text{div}[m(D(\ln(m)) + H_p(x, Du))] = 0,$$

Lemma 1.22 below says that

$$\mathbf{d}_1(m(t_0 + h), (id - h\Phi)\sharp m_0) = o(h)$$

where

$$\Phi(x) := D(\ln(m_0(x))) + H_p(x, Du(t_0, x))$$

and $o(h)/h \to 0$ as $h \to 0$. So, by Lipschitz continuity of U and then differentiability of U,

$$\begin{aligned}
U(t_0 + h, x, m(t_0 + h)) &= U(t_0 + h, x, (id - h\Phi)\sharp m_0) + o(h) \\
&= U(t_0 + h, x, m_0) - h \int_{\mathbb{T}^d} D_m U(t_0 + h, x, m_0, y) \\
&\quad \cdot \Phi(y) m_0(y) dy + o(h),
\end{aligned}$$

and therefore, by continuity of U and $D_m U$,

$$\begin{aligned}
\lim_{h \to 0} & \frac{U(t_0 + h, x, m(t_0 + h)) - U(t_0 + h, x, m_0)}{h} \\
&= - \int_{\mathbb{T}^d} \left(D_m U(t_0, x, m_0, y) \cdot [D(\ln(m_0)) + H_p(y, Du(t_0))]\right) m_0(y) dy.
\end{aligned}$$

On the other hand, for $h > 0$,

$$U(t_0+h, x, m(t_0+h)) - U(t_0, x, m_0) = u(t_0+h, x) - u(t_0, x) = h\partial_t u(t_0, x) + o(h),$$

so that

$$\lim_{h \to 0^+} \frac{U(t_0 + h, x, m(t_0 + h)) - U(t_0, x, m_0)}{h} = \partial_t u(t_0, x).$$

Therefore $\partial_t U(t_0, x, m_0)$ exists and is equal to

$$
\begin{aligned}
\partial_t U(t_0, x, m_0) &= \int_{\mathbb{T}^d} (D_m U(t_0, x, m_0, y) \cdot [D(\ln(m_0)) + H_p(y, Du(t_0))]) \\
&\quad \times m_0(y) dy + \partial_t u(t_0, x) \\
&= - \int_{\mathbb{T}^d} \operatorname{div}_y D_m U(t_0, x, m_0, y) m_0(y) dy \\
&\quad + \int_{\mathbb{T}^d} D_m U(t_0, x, m_0, y) \cdot H_p(y, Du(t_0)) \, m_0(y) dy \\
&\quad - \Delta u(t_0, x) + H(x, Du(t_0, x)) - F(x, m_0) \\
&= - \int_{\mathbb{T}^d} \operatorname{div}_y D_m U(t_0, x, m_0, y) m_0(y) dy \\
&\quad + \int_{\mathbb{T}^d} D_m U(t_0, x, m_0, y) \cdot H_p(y, D_x U(t_0, y, m_0)) \, m_0(y) dy \\
&\quad - \Delta_{xx} U(t_0, x, m_0) + H(x, D_x U(t_0, x, m_0)) - F(x, m_0)
\end{aligned}
$$

This means that U satisfies (1.149) at (t_0, x, m_0). By continuity, U satisfies the equation everywhere.   □

Lemma 1.22 *Let $V = V(t, x)$ be a C^1 vector field, $m_0 \in \mathcal{P}_2$ and m be the weak solution to*

$$
\begin{cases}
\partial_t m + \operatorname{div}(mV) = 0 \\
m(0) = m_0 .
\end{cases}
$$

Then

$$
\lim_{h \to 0^+} \mathbf{d}_1(m(h), (id + hV(0, \cdot))\sharp m_0) / h = 0.
$$

Proof Recall that $m(h) = X^{\cdot}(h)\sharp m_0$, where $X^x(h)$ is the solution to the ODE

$$
\begin{cases}
\frac{d}{dt} X^x(t) = V(t, X^x(t)) \\
X^x(0) = x .
\end{cases}
$$

Let ϕ be a Lipschitz test function. Then

$$
\begin{aligned}
\int_{\mathbb{T}^d} \phi(x)(m(h) &- (id + hV(0, \cdot))\sharp m_0)(dx) \\
&= \int_{\mathbb{T}^d} (\phi(X^x(h)) - \phi(x + hV(0, x))) m_0(dx) \\
&\leq \|D\phi\|_\infty \int_{\mathbb{T}^d} |X^x(h) - x - hV(0, x)| m_0(dx) = \|D\phi\|_\infty o(h),
\end{aligned}
$$

which proves that $\mathbf{d}_1(m(h), (id + hV(0, \cdot))\sharp m_0) = o(h)$.   □

Proof of Theorem 1.21 (Uniqueness) We use a technique introduced in [149] (Lesson 5/12/2008), consisting at looking at the MFG system (1.152) as a system of characteristics for the master equation (1.149). We reproduce here this argument for the sake of completeness. Let V be another solution to the master equation. The main point is that, by definition of solution $D^2_{xy} \frac{\delta V}{\delta m}$ is bounded, and therefore $D_x V$ is Lipschitz continuous with respect to the measure variable.

Let us fix (t_0, m_0). In view of the Lipschitz continuity of $D_x V$, one can easily uniquely solve the PDE (by standard fixed point argument):

$$\begin{cases} \partial_t \tilde{m} - \Delta \tilde{m} - \operatorname{div}(\tilde{m} H_p(x, D_x V(t, x, \tilde{m}))) = 0 \\ \tilde{m}(t_0) = m_0 \end{cases}$$

Then let us set $\tilde{u}(t, x) = V(t, x, \tilde{m}(t))$. By the regularity properties of V, $\tilde{u}$ is at least of class $C^{2,1}$ with

$$\begin{aligned} \partial_t \tilde{u}(t, x) &= \partial_t V(t, x, \tilde{m}(t)) + \langle \frac{\delta V}{\delta m}(t, x, \tilde{m}(t), \cdot), \partial_t \tilde{m}(t) \rangle_{C^2, (C^2)'} \\ &= \partial_t V(t, x, \tilde{m}(t)) \\ &\quad + \langle \frac{\delta V}{\delta m}(t, x, \tilde{m}(t), \cdot), \Delta \tilde{m} + \operatorname{div}(\tilde{m} H_p(\cdot, D_x V(t, \cdot, \tilde{m}))) \rangle_{C^2, (C^2)'} \\ &= \partial_t V(t, x, \tilde{m}(t)) + \int_{\mathbb{T}^d} \operatorname{div}_y D_m V(t, x, \tilde{m}(t), y) \, d\tilde{m}(t)(y) \\ &\quad - \int_{\mathbb{T}^d} D_m V(t, x, \tilde{m}(t), y) \cdot H_p(y, D_x V(t, y, \tilde{m})) \, d\tilde{m}(t)(y) \end{aligned}$$

Recalling that V satisfies the master equation we get

$$\begin{aligned} \partial_t \tilde{u}(t, x) &= -\Delta_x V(t, x, \tilde{m}(t)) + H(x, D_x V(t, x, \tilde{m}(t))) - F(x, \tilde{m}(t)) \\ &= -\Delta \tilde{u}(t, x) + H(x, D\tilde{u}(t, x)) - F(x, \tilde{m}(t)) \end{aligned}$$

with terminal condition $\tilde{u}(T, x) = V(T, x, \tilde{m}(T)) = G(x, \tilde{m}(T))$. Therefore the pair $(\tilde{u}, \tilde{m})$ is a solution of the MFG system (1.152). As the solution of this system is unique, we get that $V(t_0, x, m_0) = U(t_0, x, m_0)$ is uniquely defined. $\qquad\square$

1.4.3.2 The Master Equation for MFG Problems on a Finite State Space

We consider here a MFG problem on a finite state space: let $I \in \mathbb{N}$, $I \geq 2$ be the number of states. Players control their jump rate from one state to another; their cost depends on the jump rate they choose and on the distribution of the other players on the states. In this finite state setting, this distribution is simply an element of the simplex $\mathcal{S}_{I-1}$ with

$$\mathcal{S}_{I-1} := \left\{ m \in \mathbb{R}^I, \; m = (m_i)_{i=1,\dots,I}, \; m_i \geq 0, \; \forall i, \; \sum_i m_i = 1 \right\}.$$

Given $m = (m_i) \in S_{I-1}$, m_i is the proportion of players in state i.

The MFG System In this setting the MFG system takes the form of a coupled system of ODEs: for $i = 1, \ldots, I$,

$$
\begin{cases}
-\dfrac{d}{dt}u^i(t) + H^i((u^j(t) - u^i(t))_{j \neq i}, m(t)) = 0 & \text{in } (0, T) \\[2mm]
\dfrac{d}{dt}m_i(t) - \displaystyle\sum_{j \neq i} m_j(t) \dfrac{\partial H^j}{\partial p_i}((u^k(t) - u^j(t))_{k \neq j}, m(t)) \\[4mm]
\qquad\qquad + m_i(t) \displaystyle\sum_{j \neq i} \dfrac{\partial H^i}{\partial p_j}((u^k(t) - u^i(t))_{k \neq i}, m(t)) = 0 & \text{in } (0, T) \\[4mm]
m_i(t_0) = m_{i,0}, \ u^i(T) = g^i(m(T)).
\end{cases}
$$

$$(1.156)$$

In the above system, the unknown is $(u, m) = (u^i(t), m^i(t))$, where $u^i(t)$ is the value function of a player at time t and in position i while $m(t)$ is the distribution of players at time t, with $m(t) \in S_{I-1}$ for any t. The map $H^i : \mathbb{R}^{I-1} \times S_{I-1} \to \mathbb{R}$ is the Hamiltonian of the problem in state i while $m_0 = (m_{i,0}) \in S_{I-1}$ is the initial distribution at time $t_0 \in [0, T)$ and $g^i : S_{I-1} \to \mathbb{R}$ is the terminal cost in state i. As usual, this is a forward-backward system.

The Structure for Uniqueness As for standard MFG systems, the existence of a solution is relatively easy; the uniqueness relies on a specific structure of the coupling and on a monotonicity condition which become here:

$$H^i(z, m) = h^i(z) - f^i(m) \tag{1.157}$$

where h^i is strictly convex in z and

$$\sum_{i=1}^{I}(f^i(m) - f^i(m'))(m^i - (m')^i) \geq 0, \ \sum_{i=1}^{I}(g^i(m) - g^i(m'))(m^i - (m')^i) \geq 0,$$

$$\forall m, m' \in S_{I-1}. \tag{1.158}$$

The Master Equation To find a solution of this MFG problem in feedback form (i.e., such that the control of a players depends on the state of this player and on the distribution of the other players), one can proceed as in the continuous space case and set $U^i(t, m_0) = u^i(t_0)$, where m_0 is the initial distribution of the players at time t_0 and (u, m) is the solution to (1.156). Then U solves the following hyperbolic system, for $i = 1, \ldots, I$,

$$
\begin{cases}
\begin{aligned}
-\partial_t U^i(t, m) + H^i((U^j(t, m) - U^i(t, m))_{j \neq i}, m) & \\
- \sum_{j=1}^{I} \frac{\partial U^i}{\partial p_j}(t, m) \Big(\sum_{k \neq j} m_k \frac{\partial H^k}{\partial p_j}((U^l(t, m) - U^k(t, m))_{l \neq k}, m) & \\
- m_j \sum_{k \neq j} \frac{\partial H^j}{\partial p_k}((U^l(t, m) - U^j(t, m))_{l \neq j}, m) \Big) & = 0 \text{ in } (0, T) \times \mathcal{S}_{I-1} \\
U^i(T, m) = g^i(m) \qquad \text{in } \mathcal{S}_{I-1} &
\end{aligned}
\end{cases}
$$

This is the master equation in the framework of the finite state space problem. It can be rewritten in a more compact way in the form

$$
\partial_t U + (F(m, U) \cdot D) U = G(m, U) \tag{1.159}
$$

where $F, G : \mathcal{S}_{I-1} \times \mathbb{R}^I \to \mathbb{R}^I$ are defined by

$$
F(m, U) = \Big(\sum_{k \neq j} m_k \frac{\partial H^k}{\partial p_j}((U^l - U^k)_{l \neq k}, m) - m_j \sum_{k \neq j} \frac{\partial H^j}{\partial p_k}((U^l - U^j)_{l \neq j}, m) \Big)_j
$$

and $G(m, U) = -(H^j(U, m))_j$. Equation (1.159) has to be understood as follows: for any $i \in \{1, \ldots, I\}$,

$$
\partial_t U^i + (F(m, U) \cdot D) U^i = -H^i(U, m).
$$

Link Between Two Notions of Monotonicity The monotonicity condition stated in (1.158) is equivalent with the fact that the pair (G, F) is monotone (in the classical sense) from $\mathbb{R}^{2d}$ into itself. Indeed, recalling the structure condition (1.157), we have

$$
\langle (G, F)(m, U) - (G, F)(m', U'), (m, U) - (m', U') \rangle
$$

$$
= \sum_{j} (h^j((U^k - U^j)_{k \neq j}) - h^j((U'^k - U'^j)_{k \neq j}))(m_j - m'_j)
$$

$$
- \sum_{j} (f^j(m) - f^j(m'))(m_j - m'_j)
$$

$$
+ \sum_{j \neq k} \Big(m_k \frac{\partial h^k}{\partial p_j}((U^l - U^k)_{l \neq k}) - m'_k \frac{\partial h^k}{\partial p_j}((U'^l - U'^k)_{l \neq k}) \Big)(U^j - U'^j)
$$

$$
- \sum_{j=1}^{I} \Big(m_j \sum_{k \neq j} \frac{\partial h^j}{\partial p_k}((U^l - U^j)_{l \neq j}) - m'_j \sum_{k \neq j} \frac{\partial h^j}{\partial p_k}((U'^l - U'^j)_{l \neq j}) \Big)
$$

$$\times (U^j - U'^j)$$

$$= -\sum_{j}(f^j(m) - f^j(m'))(m_j - m'_j)$$

$$-\sum_{j=1}^{I} m_j \left(h^j((U'^k - U'^j)_{k\neq j}) - h^j((U^k - U^j)_{k\neq j}) \right)$$

$$-\sum_{k\neq j} \frac{\partial h^j}{\partial p_k}((U^l - U^j)_{l\neq j})(U'^k - U^k - U'^j - U^j) \bigg)$$

$$-\sum_{j=1}^{I} m'_j \left(h^j((U^k - U^j)_{k\neq j}) - h^j((U'^k - U'^j)_{k\neq j}) \right)$$

$$-\sum_{k\neq j} \frac{\partial h^j}{\partial p_k}((U'^l - U'^j)_{l\neq j})(U^k - U'^k - U^j - U'^j) \bigg),$$

which is nonnegative since (1.158) holds and h is convex.

The finite state space is very convenient in the analysis of MFGs: it makes complete sense in terms of modeling and, in addition, it simplifies a lot the analysis of the master equation. First of all, this is a finite dimensional problem. Secondly, under the monotonicity condition, the solution of the master equation is also monotone and it is known that monotone maps are BV in open sets in which they are finite: so some regularity is easily available.

1.4.3.3 The MFG Problem with a Common Noise

The aim of this part is to say a few words about the MFGs in which all agents are subject to a common source of randomness. This kind of models are often met in macro-economy, after the pioneering work of Krusell-Smith [138]. We start with a toy example, in which the agents are subject to a single shock. Then we describe the more delicate model where the shock is a Brownian motion.

An Illustrative Example

We consider here a problem in which the agents face a common noise which, in this elementary example, is a random variable Z on which the coupling costs F and G depend: $F = F(x, m, Z)$ and $G = G(x, m, Z)$. The game is played in finite horizon T and the exact value of Z is revealed to the agents at time $T/2$ (to fix the ideas).

To fix the ideas, we assume that the agents directly control their drift:

$$dX_t = \alpha_t dt + \sqrt{2}dB_t$$

(where (α_t) is the control with values in $\mathbb{R}^d$ and B a Brownian motion). In contrast with the previous discussions, the control α_t is now adapted to the filtration generated by B *and to the noise* Z when $t \geq T/2$. The cost is now of the form

$$J(\alpha) = \mathbb{E}\left[\int_0^T \frac{1}{2}|\alpha_t|^2 + F(X_t, m(t), Z)\, dt + G(X_T, m(T), Z)\right],$$

where F and G depend on the position of the player, on the distribution of the agents and on the common noise Z. As all the agents will choose their optimal control in function of the realization of Z (of course after time $T/2$), one expect the distribution of players to be random after $T/2$ and to depend on the noise Z.

On the time interval $[T/2, T]$, the agents have to solve a classical control problem (which depends on Z and on $(m(t))$):

$$u(t, x) := \inf_\alpha \mathbb{E}\left[\int_t^T \frac{1}{2}|\alpha_t|^2 + F(X_t, m(t), Z)\, dt + G(X_T, m(T)) \mid Z\right]$$

which depends on the realization of Z and solves the HJ equation (with random coefficients):

$$\begin{cases} -\partial_t u - \Delta u + \frac{1}{2}|Du|^2 = F(x, m(t), Z) \text{ in } (T/2, T) \times \mathbb{R}^d \\ u(T, x, Z) = G(x, m(T), Z) \text{ in } \mathbb{R}^d. \end{cases} \tag{1.160}$$

On the other hand, on the time interval $[0, T/2)$, the agent has no information on Z and, by dynamic programming, one expects to have

$$u(t, x) := \inf_\alpha \mathbb{E}\left[\int_t^{T/2} \frac{1}{2}|\alpha_t|^2 + \bar{F}(X_t, m(t))\, dt + u(T/2^+, X_{T/2})\right],$$

where $\bar{F}(x, m) = \mathbb{E}[F(x, m, Z)]$ (recall that $m(t)$ is deterministic on $[0, T/2]$). Thus, on the time interval $[0, T/2]$, u solves

$$\begin{cases} -\partial_t u - \Delta u + \frac{1}{2}|Du|^2 = \bar{F}(x, m(t)) \text{ in } (0, T/2) \times \mathbb{R}^d \\ u(T/2^-, x) = \mathbb{E}\left[u(T/2^+, x)\right] \text{ in } \mathbb{R}^d \end{cases} \tag{1.161}$$

As for the associated Kolmogorov equation, on the time interval $[0, T/2]$ (where the optimal feedback $-Du$ is purely deterministic) we have as usual:

$$\partial_t m - \Delta m - \operatorname{div}(m\, Du(t, x)) = 0 \text{ in } (0, T/2) \times \mathbb{R}^d, \qquad m(0) = m_0. \tag{1.162}$$

while on the time interval $[T/2, T]$, m becomes random (as the control $-Du$) and solves

$$\partial_t m - \Delta m - \text{div}\,(m\,Du(t, x, Z)) = 0 \text{ in } (T/2) \times \mathbb{R}^d, \qquad m(T/2^-) = m(T/2^+).$$
$$(1.163)$$

Note the relation: $m(T/2^-) = m(T/2^+)$, which means that the dynamics of the crowd is continuous in time.

Let us point out some remarkable features of the problem. Firstly, the pairs (u, m) are no longer deterministic, and are adapted to the filtration generated by the common noise (here this filtration is trivial up to time $T/2$ and is the σ-algebra generated by Z after $T/2$). Secondly, the map u is discontinuous: this is due to the shock of information at time $T/2$.

The existence of a solution to the MFG system (1.160)–(1.163) can be obtained in two steps. First one solves the MFG system on $[T/2, T]$: given any measure $m_0 \in \mathcal{P}_1(\mathbb{R}^d)$, let (u, m) be the solution to

$$\begin{cases} -\partial_t u(t, x, Z) - \Delta u(t, x, Z) + \dfrac{1}{2}|Du(t, x, Z)|^2 = F(x, m(t), Z) \text{ in } (T/2, T) \times \mathbb{R}^d \\ \partial_t m(t, x, Z) - \Delta m(t, x, Z) - \text{div}(m(t, x, Z)Du(t, x, Z)) = 0 \text{ in } (T/2, T) \times \mathbb{R}^d \\ m(T/2, dx, Z) = m_0(dx), \qquad u(T, x, Z) = G(x, m(T, x, Z), Z) \text{ in } \mathbb{R}^d \end{cases}$$

Note that u and m depend of course on m_0. If we require F and G to be monotone, then this solution is unique and we can set $U(x, m_0, Z) = u(T/2, x, Z)$ (with the notation of Sect. 1.4.3.1, it should be $U(T/2^+, x, m_0, Z)$, but we omit the $T/2$ for simplicity). It is not difficult to check that, if the couplings F, G are smoothing, then U is continuous in m (uniformly in (x, Z)), measurable in Z and C^2 in x uniformly in (m, Z). In addition, it is a simple exercise to prove that U is monotone as well. Therefore, if we set $\bar{U}(x, m) = \mathbb{E}[U(x, m, Z)]$, then $\bar{U}$ is also continuous in m and C^2 in x and monotone. So the system

$$\begin{cases} -\partial_t u(t, x) - \Delta u(t, x) + \dfrac{1}{2}|Du(t, x)|^2 = \bar{F}(x, m(t)) \text{ in } (0, T/2) \times \mathbb{R}^d \\ \partial_t m(t, x) - \Delta m(t, x) - \text{div}(m(t, x)Du(t, x)) = 0 \text{ in } (0, T/2) \times \mathbb{R}^d \\ m(0, dx) = m_0(dx), \qquad u(T, x) = \bar{U}(x, m(T/2)) \text{ in } \mathbb{R}^d \end{cases}$$

has a unique solution. Note that u is a discontinuous function of time, but the discontinuity

$$u(T/2^+, x) - u(T/2^-, x) = u(T/2^+, x) - \mathbb{E}\left[u(T/2^+, x)\right]$$

has zero mean, that is, it is a "one-step martingale".

Common Noise of Brownian Type
In general, MFG with a common noise involve much more complex randomness than a single shock that occurs at a given time. We discuss here very briefly a case

in which the common noise is a Brownian motion. As before, we just consider an elementary model in order to fix the ideas.

The game is played in finite horizon T. The agents control directly their drift: their state solves therefore the SDE

$$dX_t = \alpha_t dt + \sqrt{2} dB_t + \sqrt{2\beta} dW_t,$$

where (α_t) is the control with values in $\mathbb{R}^d$, B the idiosyncratic noise (a Brownian motion, independent for each player) and W is the common noise (a Brownian motion, the same for each player), $\beta \geq 0$ denoting the intensity of this noise. The control α_t is now adapted to the filtration generated by B and W. The cost is of the (standard) form

$$J(\alpha) = \mathbb{E}\left[\int_0^T \frac{1}{2}|\alpha_t|^2 + F(X_t, m(t))\, dt + G(X_T, m(T))\right],$$

where F and G depend on the position of the player and on the distribution of the agents.

The main difference with the classical case is that now the flow of measures m is random and adapted to the filtration generated by W. To understand why it should be so, let us come back to the setting with finitely many agents (in which one sees better the difference between B and W). If there are N agents, controlling their state with a feedback control $\alpha = \alpha_t(x)$ (possibly random), then the state of player i, for $i \in \{1, \ldots, N\}$, solves

$$dX_t^i = \alpha_t(X_t^i) dt + \sqrt{2} dB_t^i + \sqrt{2\beta} dW_t.$$

Note that the B^i are independent (idiosyncratic noise) and independent of the common noise W. Let m_t^N be the empirical measure associated to the X^i:

$$m_t^N = \frac{1}{N}\sum_{i=1}^N \delta_{X_t^i}.$$

Let us assume that m^N converges to some m (formally) and let us try to guess the equation for m. We have, for any smooth test function $\phi = \phi(t, x)$ with a compact support,

$$\int_{\mathbb{R}^d} \phi(t, x) m_t(dx) = \lim_N \int_{\mathbb{R}^d} \phi(t, x) m_t^N(dx),$$

where, by Itô's formula,

$$\int_{\mathbb{R}^d} \phi(t,x) m_t^N(dx) = \frac{1}{N} \sum_{i=1}^{N} \phi(t, X_t^i)$$

$$= \frac{1}{N} \sum_{i=1}^{N} \phi(t, X_0^i)$$

$$+ \frac{1}{N} \sum_{i=1}^{N} \int_0^t (\partial_t \phi(s, X_s^i) + D\phi(s, X_s^i) \cdot \alpha_t(X_s^i) + (1+\beta) \Delta \phi(s, X_s^i)) ds$$

$$+ \frac{1}{N} \sum_{i=1}^{N} \int_0^t D\phi(s, X_s^i) \cdot (dB_s^i + dW_s)$$

$$= \int_{\mathbb{R}^d} \phi(t,x) m^N(0, dx)$$

$$+ \int_0^t \int_{\mathbb{R}^d} (\partial_t \phi(s, x) + D\phi(s, x) \cdot \alpha_t(x) + (1+\beta) \Delta \phi(s, x)) m_s^N(dx) ds$$

$$+ \beta \int_0^t (\int_{\mathbb{R}^d} D\phi(s, x) m_s^N(dx)) \cdot dW_s + \frac{1}{N} \sum_{i=1}^{N} \int_0^t D\phi(s, X_s^i) \cdot dB_s^i.$$

As $N \to +\infty$, the last term vanishes because, by Itô's isometry,

$$\lim_{N \to +\infty} \mathbb{E} \left[\left| \frac{1}{N} \sum_{i=1}^{N} \int_0^t D\phi(s, X_s^i) \cdot dB_s^i \right|^2 \right]$$

$$= \lim_{N \to +\infty} \frac{1}{N^2} \sum_{i=1}^{N} \mathbb{E} \left[\int_0^t |D\phi(s, X_s^i)|^2 ds \right] = 0.$$

So we find

$$\int_{\mathbb{R}^d} \phi(t,x) m_t(dx) = \int_{\mathbb{R}^d} \phi(t,x) m(0, dx)$$

$$+ \int_0^t \int_{\mathbb{R}^d} (\partial_t \phi(s, x) + D\phi(s, x) \cdot \alpha_t(x)$$

$$+ (1+\beta) \Delta \phi(s, x)) m_s(dx) ds$$

$$+ \beta \int_0^t (\int_{\mathbb{R}^d} D\phi(s, x) m_s(dx)) \cdot dW_s.$$

This means that m solves in the sense of distributions the stochastic Kolmogorov equation:

$$dm_t = [(1 + \beta)\Delta m_t - \operatorname{div}(m_t \alpha)] \, dt - \sqrt{2\beta}\operatorname{div}(m_t \, dW_t).$$

As the flow m is stochastic and adapted to the filtration generated by W, the value function u is stochastic as well and is adapted to the filtration generated by W. It turns out that u solves a backward Hamilton-Jacobi equation. The precise form of this equation is delicate because, as it is random and backward, it has to involve an extra unknown vector field $v = v_t(x)$ which ensures the solution u to be adapted to the filtration generated by W (see, on that subject, the pioneering work by Peng [162] and the discussion in [65] (Chapter 4) or in [68] (Part II, Section 1.4.2)). The stochastic MFG system associated with the problem becomes (if the initial distribution of the players is $\bar{m}_0$):

$$\begin{cases} du_t = \left[-(1 + \beta)\Delta u_t + \frac{1}{2}|Du_t|^2 - F(x, m_t) - \sqrt{2\beta}\operatorname{div}(v_t) \right] dt - \sqrt{2\beta}v_t \cdot dW_t \\ dm_t = [(1 + \beta)\Delta m_t + \operatorname{div}(m_t \, Du_t)] \, dt - \sqrt{2\beta}\operatorname{div}(m_t \, dW_t) \\ m_0 = \bar{m}_0, \ u_T = G(\cdot, m_T) \end{cases}$$

Finally, one can associate with the problem a master equation, which plays the same role as without common noise. It takes the form of a second order (in measure) equation on the space of measures:

$$\begin{cases} -\partial_t U - (1 + \beta)\Delta_x U + \frac{1}{2}|D_x U|^2 - (1 + \beta)\int_{\mathbb{R}^d} \operatorname{div}_y D_m U(t, x, m, y) \, m(dy) \\ \qquad + \int_{\mathbb{R}^d} D_m U(t, x, m, y) \cdot D_x U(t, y, m) \, m(dy) - 2\beta \int_{\mathbb{R}^d} \operatorname{div}_x D_m U(t, x, m, y) \, m(dy) \\ \qquad - \beta \int_{\mathbb{R}^d \times \mathbb{R}^d} \operatorname{Tr}(D^2_{mm} U(t, x, m, y, y'))m(dy)m(dy') = F(x, m) \\ \qquad\qquad\qquad \text{in } (0, T) \times \mathbb{R}^d \times \mathcal{P}_2 \\ U(T, x, m) = G(x, m) \qquad \text{in } \mathbb{R}^d \times \mathcal{P}_2 \end{cases}$$

where the unknown is $U = U(t, x, m)$.

1.4.3.4   Comments

Most formal properties of the Master equation have been introduced and discussed by Lions in [149] (Lesson 5/12/2008 and the Course 2010-'11), including of course the representation formula (1.153). The actual proof of the existence of a solution of the master equation is a tedious verification that (1.153) actually gives a solution. This has required several steps: the first paper in this direction is [41], where a master equation is studied for linear Hamiltonian and without coupling terms ($F = G = 0$); [108] analyzes the master equation in short time and without the diffusion term; [85] obtains the existence and uniqueness for the master equation (1.149); [65]

establishes the existence and uniqueness of solutions for the master equation with common noise under the Lasry-Lions monotonicity condition (see also [68]). There has been few works since then on the subject outside the above references and the analysis on finite state space in [27, 34]: see [11, 12, 66]. Another approach, not discussed in these notes, is the so-called "Hilbertian approach" developed by Lions in [149] (see e.g. Lesson 31/10 2008, and later the seminar 08/11/2013): the idea is to write the master equation (or, more precisely, its space derivative) in the Hilbert space of square integrable random variables and use this Hilbert structure to obtain existence and uniqueness results.

The reader may notice that we have worked here under the monotonicity assumption. We could have also considered the problem in short time, or with a "small coupling". All these settings correspond to situation in which the MFG system has a unique solution for any initial measure. When this does not hold, the solution of the master equation is expected to be discontinuous. One knows almost nothing on the definition of the master equation outside of the smooth set-up: this remains one of the major issues of the topic. To overcome this difficulty, an idea would be to add a common noise to smoothen the solution. Although this approach is not understood in the whole space, there are now a few results in this direction in the finite state space: we discuss this point now.

The MFG problem on finite state space has been first described by Lions [149] (Lesson 14/1 2011 and the Course 2011-'12). The probabilistic interpretation is carefully explained in [78], while the well-posedness of the master equation (and its use for the convergence of the Nash system) is discussed in this setting in [26] and [79]. The addition of a common noise to the master equation in finite state space is described in [34] and [27]. In particular, [27] provides the existence of smooth solutions even without the monotonicity assumption (see also [146], on problems with a major player). Finally, for the master equation on finite state space we definitively refer to the contribution by F. Delarue in the present volume.

1.4.4   Convergence of the Nash System

In this section, we study the convergence of Nash equilibria in differential games with a finite number of players, as the number of players tends to infinity. We would like to know if the limit is a MFG model. Let us recall that, in Sect. 1.3.3 we explained how to use the MFG system to find an ε-Nash equilibrium in a N-player game. So here we consider the converse problem. As we will see, this question is much more subtle and, in fact, not completely understood.

On one hand, this problem depends on the structure of information one allows to the players in the finite player game. If in this game players observe only their own position (but they are aware of the controls played by the other players and hence their average distribution), then the limit problem is (almost always) a MFG game (see the notes below). On the other hand, if players observe each other closely and remember all the past actions, the convergence cannot be expected because a deviating player can always be punished in the game with finitely many players (this is the so-called Folk Theorem), while it is not the case in Mean Field Games. This kind of strategy, however, is not always convincing because a player is often led to punish him/herself in order to punish a deviation. So the most interesting case is when players play in closed loop strategies (in function of the current position of the other players): indeed, this kind of strategy is time consistent (and is associated with a PDE, the Nash system). However, the answer to the convergence problem is then much more complicated and we only have a partial picture.

We consider here a very smooth case, in which the Nash equilibrium in the N-player game satisfies a time-consistency condition. More precisely, we assume that the Nash equilibrium is given through the solution $(v^{N,i})$ of the so-called Nash system:

$$
\begin{cases}
-\partial_t v^{N,i} - \sum_j \Delta_{x_j} v^{N,i} + H(x_i, D_{x_i} v^{N,i}) \\
\qquad + \sum_{j \neq i} H_p(x_j, D_{x_j} v^{N,j}) \cdot D_{x_j} v^{N,i} = F(x_i, m_X^{N,i}) \qquad \text{in } (0, T) \times \mathbb{T}^{Nd} \\
v^{N,i}(T, x) = G(x_i, m_X^{N,i}) \qquad \text{in } \mathbb{T}^{Nd}
\end{cases}
$$

(1.164)

where we set, for $X = (x_1, \ldots, x_N) \in (\mathbb{T}^d)^N$, $m_X^{N,i} = \dfrac{1}{N-1} \sum_{j \neq i} \delta_{x_j}$. We explain below how this system is associated with a Nash equilibrium.

Assuming that the coupling functions F and G are monotone, our aim is to show that the solution $(v^{N,i})$ converges, in a suitable sense, to the solution of the master equation without a common noise.

Throughout this part we denote by $U = U(t, x, m)$ the solution of the master equation built in Theorem 1.21 which satisfies (1.154) and (1.155). It solves

$$
\begin{cases}
-\partial_t U - \Delta_x U + H(x, D_x U) - \displaystyle\int_{\mathbb{T}^d} \operatorname{div}_y D_m U \, dm(y) \\
\qquad + \displaystyle\int_{\mathbb{T}^d} D_m U(t, x, m, y) \cdot H_p(y, D_x U(t, y, m)) \, dm(y) = F(x, m) \qquad (1.165) \\
\qquad\qquad \text{in } (0, T) \times \mathbb{T}^d \times \mathcal{P}_2 \\
U(T, x, m) = G(x, m) \qquad \text{in } \mathbb{T}^d \times \mathcal{P}_2
\end{cases}
$$

Throughout the section, we suppose that the assumptions of the previous section are in force.

1.4.4.1 The Nash System

Let us first explain the classical interpretation of the Nash system (1.164):

The game consists, for each player $i = 1, \ldots, N$ and for any initial position $x_0 = (x_0^1, \ldots, x_0^N)$, in minimizing

$$J_i(t_0, x_0, (\alpha^j)) = \mathbb{E}\left[\int_{t_0}^T L(X_t^i, \alpha_t^i) + F(X_t^i, m_{X_t}^{N,i})\, dt + G(X_t^i, m_{X_t}^{N,i})\right]$$

where, for each $i = 1, \ldots, N$,

$$dX_t^i = \alpha_t^i dt + \sqrt{2} dB_t^i, \qquad X_{t_0}^i = x_0^i$$

We have set $X_t = (X_t^1, \ldots, X_t^N)$. The Brownian motions (B_t^i) are independent, but the controls (α^i) are supposed to depend on the filtration $\mathcal{F}$ generated by all the Brownian motions.

Proposition 1.9 (Verification Theorem) *Let $(v^{N,i})$ be a classical solution to the above system. Then the N-uple of maps $(\alpha^{i,*})_{i=1,\ldots,d} := (-H_p(x_i, D_{x_i} v^{N,i}))_{i=1,\ldots,d}$ is a Nash equilibrium in feedback form of the game: for any $i = 1, \ldots, d$, for any initial condition $(t_0, x_0) \in [0, T] \times \mathbb{T}^{Nd}$, for any control α^i adapted to the whole filtration $\mathcal{F}$, one has*

$$J_i(t_0, x_0, (\alpha^{j,*})) \leq J_i(t_0, x_0, \alpha^i, (\alpha^{j,*})_{j\neq i})$$

Proof The proof relies on a standard verification argument and is left to the reader.
□

1.4.4.2 Finite Dimensional Projections of U

Let U be the solution to the master equation (1.165). For $N \geq 2$ and $i \in \{1, \ldots, N\}$ we set

$$u^{N,i}(t, X) = U(t, x_i, m_X^{N,i}) \quad \text{where}$$

$$X = (x_1, \ldots, x_N) \in (\mathbb{T}^d)^N, \quad m_X^{N,i} = \frac{1}{N-1}\sum_{j\neq i}\delta_{x_j}. \tag{1.166}$$

Note that the $u^{N,i}$ are at least C^2 with respect to the x_i variable because so is U. Moreover, $\partial_t u^{N,i}$ exists and is continuous because of the equation satisfied by U.

The next statement says that $u^{N,i}$ is actually globally $C^{1,1}$ in the space variables:

Proposition 1.10 *For any $N \geq 2$, $i \in \{1, \ldots, N\}$, $u^{N,i}$ is of class $C^{1,1}$ in the space variables, with*

$$D_{x_j} u^{N,i}(t, X) = \frac{1}{N-1} D_m U(t, x_i, m_X^N, x_j) \qquad (j \neq i)$$

and

$$\left\| D_{x_k, x_j} u^{N,i}(t, \cdot) \right\|_\infty \leq \frac{C}{N} \qquad (k \neq i, \ j \neq i).$$

Proof Let $X = (x_j) \in (\mathbb{T}^d)^N$ be such that $x_j \neq x_k$ for any $j \neq k$. Let $\varepsilon :=$ $\min_{j \neq k} |x_j - x_k|$. For $V = (v_j) \in (\mathbb{R}^d)^N$ with $v_i = 0$, we consider a smooth vector field $\phi : \mathbb{T}^d \to \mathbb{R}^d$ such that

$$\phi(x) = v_j \qquad \text{if } x \in B(x_j, \varepsilon/4).$$

Then, as U satisfies (1.154), (1.155), we can apply Proposition 1.6 which says that, (omitting the dependence with respect to t for simplicity)

$$
\begin{aligned}
u^{N,i}(X + V) - u^{N,i}(X) &= U((\text{id} + \phi) \sharp m_X^{N,i}) - U(m_X^{N,i}) \\
&= \int_{\mathbb{T}^d} D_m U(m_X^{N,i}, y) \cdot \phi(y) \, dm_X^{N,i}(y) + O(\|\phi\|_{L^2(m_X^{N,i})}^2) \\
&= \frac{1}{N-1} \sum_{j \neq i} D_m U(m_X^{N,i}, x_j) \cdot v_j + O(\sum_{j \neq i} |v_j|^2)
\end{aligned}
$$

This shows that $u^{N,i}$ has a first order expansion at X with respect to the variables $(x_j)_{j \neq i}$ and that

$$D_{x_j} u^{N,i}(t, X) = \frac{1}{N-1} D_m U(t, x_i, m_X^N, x_j) \qquad (j \neq i).$$

As $D_m U$ is continuous with respect to all its variables, $u^{N,i}$ is C^1 with respect to the space variables in $[0, T] \times \mathbb{T}^{Nd}$.

The second order regularity of the $u^{N,i}$ can be established in the same way. □

We now show that $(u^{N,i})$ is "almost" a solution to the Nash system (1.164). More precisely, next Proposition states that the $(u^{N,i})$ solve the Nash system (1.164) up to an error of size $1/N$.

Proposition 1.11 *One has, for any* $i \in \{1, \ldots, N\}$,

$$
\begin{cases}
-\partial_t u^{N,i} - \sum_j \Delta_{x_j} u^{N,i} + H(x_i, D_{x_i} u^{N,i}) \\
\qquad + \sum_{j \neq i} D_{x_j} u^{N,i}(t, X) \cdot H_p(x_j, D_{x_j} u^{N,j}(t, X)) = F(x_i, m_X^{N,i}) + r^{N,i}(t, X) \\
\qquad\qquad\qquad\qquad\qquad\qquad \text{in } (0, T) \times \mathbb{T}^{Nd} \\
u^{N,i}(T, X) = G(x_i, m_X^{N,i}) \qquad \text{in } \mathbb{T}^{Nd}
\end{cases}
$$

$$(1.167)$$

where $r^{N,i} \in L^\infty((0, T) \times \mathbb{T}^{dN})$ *with*

$$\|r^{N,i}\|_\infty \leq \frac{C}{N}.$$

Proof As U solves (1.165), one has at a point $(t, x_i, m_X^{N,i})$:

$$
-\partial_t U - \Delta_x U + H(x_i, D_x U) - \int_{\mathbb{T}^d} \operatorname{div}_y D_m U(t, x_i, m_X^{N,i}, y) \, dm_X^{N,i}(y)
$$
$$
+ \int_{\mathbb{T}^d} D_m U(t, x_i, m_X^{N,i}, y) \cdot H_p(y, D_x U(t, y, m_X^{N,i})) \, dm_X^{N,i}(y) = F(x_i, m_X^{N,i})
$$

So $u^{N,i}$ satisfies:

$$
-\partial_t u^{N,i} - \Delta_{x_i} u^{N,i} + H(x_i, D_{x_i} u^{N,i}) - \frac{1}{N-1} \sum_{j \neq i} \operatorname{div}_y D_m U(t, x_i, m_X^{N,i}, y_j)
$$
$$
+ \frac{1}{N-1} \sum_{j \neq i} D_{x_j} u^{N,i}(t, X) \cdot H_p(x_j, D_x U(t, x_j, m_X^{N,i})) = F(x_i, m_X^{N,i})
$$

By the Lipschitz continuity of $D_x U$ with respect to m, we have

$$
\left| D_x U(t, x_j, m_X^{N,i}) - D_x U(t, x_j, m_X^{N,j}) \right| \leq C \mathbf{d}_1(m_X^{N,i}, m_X^{N,j}) \leq \frac{C}{N-1},
$$

so that, by Proposition 1.10,

$$
\left| \frac{1}{N-1} D_x U(t, x_j, m_X^{N,i}) - D_{x_j} u^{N,j}(t, X) \right| \leq \frac{C}{N^2}
$$

and

$$
\frac{1}{N-1} \sum_{j \neq i} D_{x_j} u^{N,i}(t, X) \cdot H_p(x_j, D_x U(t, x_j, m_X^{N,i}))
$$
$$
= \sum_{j \neq i} D_{x_j} u^{N,i}(t, X) \cdot H_p(x_j, D_{x_j} u^{N,j}(t, X)) + O(1/N).
$$

On the other hand,

$$\sum_j \Delta_{x_j} u^{N,i} = \Delta_{x_i} u^{N,i} + \sum_{j \neq i} \Delta_{x_j} u^{N,i}$$

where, using Proposition 1.10 and the Lipschitz continuity of $D_m U$ with respect to m,

$$\sum_{j \neq i} \Delta_{x_j} u^{N,i} = \int_{\mathbb{T}^d} \mathrm{div}_y \, D_m U(t, x_i, m_X^{N,i}, y) dm_X^{N,i}(y) + O(1/N) \qquad \text{a.e.}$$

Therefore

$$-\partial_t u^{N,i} - \sum_j \Delta_{x_j} u^{N,i} + H(x_i, D_{x_i} u^{N,i})$$
$$+ \sum_{j \neq i} D_{x_j} u^{N,i}(t, X) \cdot H_p(x_j, D_{x_j} u^{N,j}(t, X)) + O(1/N) = F(x_i, m_X^{N,i}).$$

$\square$

1.4.4.3 Convergence

We are now ready to state the main convergence results of [65]: the convergence of the value function and the convergence of the optimal trajectories. Let us strongly underline that we have to work here under the restrictive assumption that there exists a classical solution to the master equation. This solution is known to exist only on short time intervals or under the Lasry-Lions monotonicity assumption. Outside this framework, a recent (and beautiful) result of Lacker [140] states that the limit problem is a weak solution of a MFG model (i.e., involving some extra randomness), provided the idiosyncratic noise is non degenerate.

Let us start with the convergence of the value function:

Theorem 1.22 *Let* $(v^{N,i})$ *be the solution to* (1.164) *and* U *be the classical solution to the master equation* (1.165)*. Fix* $N \geq 1$ *and* $(t_0, m_0) \in [0, T] \times \mathcal{P}_1$.

(i) *For any* $\mathbf{x} \in (\mathbb{T}^d)^N$, *let* $m_{\mathbf{x}}^N := \frac{1}{N} \sum_{i=1}^N \delta_{x_i}$. *Then*

$$\sup_{i=1,\cdots,N} \left| v^{N,i}(t_0, \mathbf{x}) - U(t_0, x_i, m_{\mathbf{x}}^N) \right| \leq C N^{-1}.$$

(ii) *For any* $i \in \{1, \dots, N\}$ *and* $x_i \in \mathbb{T}^d$, *let us set*

$$w^{N,i}(t_0, x_i, m_0) := \int_{\mathbb{T}^d} \cdots \int_{\mathbb{T}^d} v^{N,i}(t_0, \mathbf{x}) \prod_{j \neq i} m_0(dx_j),$$

where $\mathbf{x} = (x_1, \dots, x_N)$. *Then,*

$$\left\| w^{N,i}(t_0, \cdot, m_0) - U(t_0, \cdot, m_0) \right\|_{L^1(m_0)} \leq \begin{cases} CN^{-1/d} & \text{if } d \geq 3 \\ CN^{-1/2} \log(N) & \text{if } d = 2 \\ CN^{-1/2} & \text{if } d = 1 \end{cases}.$$

In (i) and (ii), the constant C does not depend on t_0, m_0, i nor N.

Theorem 1.22 says, in two different ways, that the $(v^{N,i})_{i \in \{1, \dots, N\}}$ are close to U. In the first statement, one compares $v^{N,i}(t, \mathbf{x})$ with the solution of the master equation evaluated at the empirical measure $m_{\mathbf{x}}^N$ while, in the second statement, the averaged quantity $w^{N,i}$ can directly be compared with the solution of the MFG system (1.152) thanks to the representation formula (1.153) for the solution U of the master equation.

The proof of Theorem 1.22 consists in comparing the "optimal trajectories" for $v^{N,i}$ and for $u^{N,i}$, for any $i \in \{1, \dots, N\}$. For this, let us fix $t_0 \in [0, T)$, $m_0 \in \mathcal{P}_2$ and let $(Z_i)_{i \in \{1, \dots, N\}}$ be an i.i.d family of N random variables of law m_0. We set $\mathbf{Z} = (Z_i)_{i \in \{1, \dots, N\}}$. Let also $((B_t^i)_{t \in [0,T]})_{i \in \{1, \dots, N\}}$ be a family of N independent d-dimensional Brownian motions which is also independent of $(Z_i)_{i \in \{1, \dots, N\}}$. We consider the systems of SDEs with variables $(\mathbf{X}_t = (X_{i,t})_{i \in \{1, \dots, N\}})_{t \in [0,T]}$ and $(\mathbf{Y}_t = (Y_{i,t})_{i \in \{1, \dots, N\}})_{t \in [0,T]}$ (the SDEs being set on $\mathbb{R}^d$ with periodic coefficients):

$$\begin{cases} dX_{i,t} = -H_p\big(X_{i,t}, D_{x_i} u^{N,i}(t, \mathbf{X}_t)\big) dt \\ \qquad\qquad\qquad + \sqrt{2} d B_t^i, \quad t \in [t_0, T], \\ X_{i,t_0} = Z_i, \end{cases} \tag{1.168}$$

and

$$\begin{cases} dY_{i,t} = -H_p\big(Y_{i,t}, D_{x_i} v^{N,i}(t, \mathbf{Y}_t)\big) dt \\ \qquad\qquad\qquad + \sqrt{2} d B_t^i, \quad t \in [t_0, T], \\ Y_{i,t_0} = Z_i. \end{cases} \tag{1.169}$$

Note that the (Y_i) are the optimal solutions for the Nash system, while, by the mean field theory, the (X_i) are close to the optimal solutions in the mean field limit.

Since the $(u^{N,i})_{i \in \{1, \dots, N\}}$ are symmetrical, the processes $((X_{i,t})_{t \in [t_0,T]})_{i \in \{1, \dots, N\}}$ are exchangeable. The same holds for the $((Y_{i,t})_{t \in [t_0,T]})_{i \in \{1, \dots, N\}}$ and, actually, the N $\mathbb{R}^{2d}$-valued processes $((X_{i,t}, Y_{i,t})_{t \in [t_0,T]})_{i \in \{1, \dots, N\}}$ are also exchangeable.

Theorem 1.23 *We have, for any $i \in \{1, \dots, N\}$,*

$$\mathbb{E}\Big[\sup_{t \in [t_0,T]} |Y_{i,t} - X_{i,t}| \Big] \leq \frac{C}{N}, \qquad \forall t \in [t_0, T], \tag{1.170}$$

$$\mathbb{E}\Big[\int_{t_0}^T |D_{x_i} v^{N,i}(t, \mathbf{Y}_t) - D_{x_i} u^{N,i}(t, \mathbf{Y}_t)|^2 dt \Big] \leq CN^{-2}, \tag{1.171}$$

and, $\mathbb{P}$-almost surely, for all $i = 1, \ldots, N$,

$$|u^{N,i}(t_0, \mathbf{Z}) - v^{N,i}(t_0, \mathbf{Z})| \le CN^{-1}, \tag{1.172}$$

where C is a (deterministic) constant that does not depend on t_0, m_0 and N.

The main step of the proof of Theorem 1.22 and Theorem 1.23 consists in comparing the maps $v^{N,i}$ and $u^{N,i}$ along the optimal trajectory Y_i. Using the presence of the idiosyncratic noises B^i and Proposition 1.11 gives (1.171), from which one derives that the X_i and the Y_i solve almost the same SDE, whence (1.170). We refer to [65] for details.

1.4.4.4 Comments

The question of the convergence of N-player games to the MFG system has been and is still one of the most puzzling questions of the MFG theory (together with the notion of discontinuous solution for the master equation). In their pioneering works [143–145] Lasry and Lions first discussed the convergence for open-loop problems in a Markovian setting, because in this case the Nash equilibrium system reduces to a coupled system of N equations in $\mathbb{R}^d$ (instead of N equations in $\mathbb{R}^{Nd}$), and in short time, where the estimates on the derivatives of the $v^{N,i}$ propagate from the initial condition.

The convergence of open-loop Nash equilibria (in a general setting) is now completely understood thanks to the works of Fischer [105] and Lacker [139], who identified completely the possible limits: these limits are always MFG equilibria. If these results are technically subtle, they are not completely surprising because at the limit players actually play open-loop controls: so there is not a qualitative difference between the game with finitely many players and the mean field game.

The question of convergence of closed-loop equilibria is more subtle. As shows a counter-example in [68, I.7.2.5], this convergence does not hold in full generality: however, the conditions under which it holds are still not clear. We have presented above what happens in MFG problems for monotone coupling and nondegenerate idiosyncratic noise. The result also holds for MFG problems with a common noise: see [65]. The convergence is quite strong, and there is a convergence rate. In that same setting, [97] and [96] study the central limit theorem and the large deviation. Lacker's result [140], on the other hand, allows to prove the convergence towards (weak) solutions of MFG equilibria without using the master equation, under the assumption of nondegeneracy of the idiosyncratic noise only. The result relies on the fact that, in some average sense, the deviation of a player barely affects the distribution of the players when N is large. Heuristically, this is due to the presence of the noise, which prevents the players to guess if another player has deviated or not. One of the drawbacks of Lacker's paper is that there might be a lot of (weak) MFG equilibria, outside of the monotone case where it is unique. It is possible that

actually only one of these equilibria is selected at the limit: this is what happens in
the examples discussed in [80, 95].

Appendix: P.-L. Lions' Courses on Mean Field Games at the Collège de France

Mean Field Game theory has been largely developed from Lions's ideas on the topic
as presented in his courses at the Collège de France during the period 2007–2012.
These courses have been recorded and can be found at the address:
   http://www.college-de-france.fr/site/pierre-louis-lions/_course.htm
   To help the reader to navigate between the different years, we collect here some
informal notes on the organization of the courses. We will use brackets to link some
of the topics below to the content of the previous sections.

Organization 2007–2008

(Symmetric functions of many variables; differentiability on the Wasserstein space)

- 09/11/2007
  Behavior as $N \to \infty$ of symmetric functions of N variables. Distances on
  spaces of measures. Eikonal equation in the space of measures (by Lax-Oleinik
  formula). Monomial on the space of measures. Hewitt-Savage theorem.
- 16/11/2007
  A proof of Hewitt-Savage theorem by the use of monomials on the space of
  measures.
- 23/11/2007

  1st hour: A remark on quantum mechanics (antisymmetric functions of N
      variables).
  2nd hour: Extensions on the result about the behavior as $N \to \infty$ of
      symmetric functions of N variables.

  - other moduli of continuity ($|u^N(X) - u^N(Y)| \le C \inf_\sigma \max_i |x_i - y_{\sigma(i)}|$).
  - relaxation of the symmetry assumption: symmetry by blocs.
  - distances with weights (replacing $1/N$ by weights (λ_i)).

  Discussion on the differential calculus on $\mathcal{P}_2$: functions C^1 over $\mathcal{P}_2$ defined
  through conditions on their restriction to measures with finite support.
- 07/12/2007

  1st hour: Back to the differential calculus on $\mathcal{P}_2$; application to linear
      transport equation, to 1st order HJ equations (discussion on scaling

$(1/N) \sum_i H(ND_{x_i} u^N)$ - discussion on the restriction to subquadratic hamiltonians).

2nd hour: Second order equations. Heat equations (independent noise, common noise); case of diffusions depending on the measure.

- 14/12/2007
Discussion about differentiability, C^1, $C^{1,1}$ on the Wasserstein space [cfr. Sect. 1.4.2]. Wasserstein distance computed by random variables.

Organization 2008–2009

(Hamilton-Jacobi equation in the Wasserstein space - Derivation and analysis of the MFG system)

- 24/10/08
Nash equilibria in one shot symmetric games as the number of players tends to infinity (example of the towel on the beach).
Characterization of the limit of Nash equilibria.
Existence - Discussion on the uniqueness through an example.
Nash equilibria (in the game with infinitely many players) as optima of a functional (efficiency principle).
- 31/10/2008
Differentiability on $\mathcal{P}_2$ through the representation as a function of random variables. Definition of C^1, link with the differentiability of functions of many variables. Structure of the derivative: law independent of the choice of the representative, derivative as a function of the random variable [cfr. Sect. 1.4.2].
First order Hamilton-Jacobi equations in the space of measures. Definition with test functions in $L^2(\Omega)$. Lax-Oleinik formula. Uniqueness of the solution.
- 07/11/2008
First order Hamilton-Jacobi equations in the space of measures: comparison. Limit of HJ with many variables: Eikonal equation, extension to general Hamiltonians, weak coupling.
Discussion about the choice of the test function: is it possible to take test functions on $L^2(\Omega)$ which depend on the law only?
- 14/11/2008
1st hour: 2nd order equations in probability spaces. Back to the limit of equations (A) $\partial_t u^N - \Delta u^N = 0$ and (B) $\partial_t u^N - \sum_{i,j} \frac{\partial^2 u^N}{\partial x_i \partial x_j} = 0$: different expressions for the limit.
2nd hour: strategies for the proof of uniqueness for the limit equation (A): (1) by verification—restricted to linear equation, (2) in $L^2(\mathbb{R}^d)$—requires coercivity conditions which are missing here, (3) Feng-Katsoulakis technique—works mostly for the heat equation and relies on the contracting properties of the heat equation in the Wasserstein space.

- 21/11/2008

  (Digression: Back to the family of polynomials: restriction to $U(m) = \Pi_k \int_{\mathbb{R}^d} \phi_k(x)m(x)$.)

  Analysis of the "limit heat equation" in the Wasserstein space (case (A)): explanation of the fact that it is a first order equation - interpretation as a geometric equation.

  Back to uniqueness: use of HJ in Hilbert spaces (cf. Lions, Swiech). Key point: diffusion almost in finite dimension. Proof of uniqueness by using formulation in $L^2(\Omega)$.

  Nonlinear equations of the form

  $$(*) \qquad \partial_t u^N - \frac{1}{N} \sum_i F(N \, D^2 u_i^N) = 0.$$

  Heuristics for the limit by polynomials.

  Limit equation of ($*$): $\partial_t U - \mathbb{E}_1[F(\mathbb{E}_2[U''(G, G)])] = 0$. Uniqueness: as before.

  Beginning of the analysis of the case of complete correlation.

- 28/11/2008

  Analysis of "limit heat equation" in the Wasserstein space (case (B)). Discussion on the well-posedness.

  Remark on the dual equation.

- 05/12/2008

  Derivation of the MFG system from the N-player game [cfr. Sect. 1.4.4].

  Back to the system of N equations and link with Nash equilibria. Ref. Bensoussan-Frehse. Uniqueness of smooth solutions; existence: more difficult, requires conditions in x of the Hamiltonian (growth of $\frac{\partial H}{\partial x}$).

  Problem: understand what happens as $N \to +\infty$.

  Key point: one needs to have $|\frac{\partial u_j^N}{\partial x_j}| \le C$ and $|\frac{\partial u_j^N}{\partial x_i}| \le C/N$. Known for T small or special structure of H. Open in general.

  One then expects that $u_i^N \to U(x_i, m, t)$. Derivation of the Master equation for U (without common noise, [cfr. Sect. 1.4.3]).

  Discussion on the Master equation; uniqueness. No maximum principle.

  Derivation of the MFG system from the Master equation.

  Direct derivation of the MFG system from the Nash system: evolution of the density of the players in the $\mathbb{R}^{Nd}$ system for the Nash equilibrium with N players when starting from an initial density m_0; cost of a player with respect to the averaged position of the other players. Propagation of chaos under the assumption $|\frac{\partial^2 u_j^N}{\partial x_j \partial x_k}| \le C/N^2$.

- 19/12/2008

  Analysis of the MFG system for time dependent problems: second order [cfr. Thm 1.4 and Thm 1.11].

  Existence: H Lipschitz or regularizing coupling.

  Discussion on the coupling: local or nonlocal, regularizing.

Case H Lipschitz + coupling of the form $g = g(m, \nabla m)$ with a polynomial growth in ∇m. A priori estimates for (m, u) and its derivatives.

Case of a regularizing coupling $F = F(m)$ without condition on H (here $H = H(\nabla u)$): a priori estimates by Bernstein method.

- 09/01/2009

  Existence of solutions for the MFG system: by strategy of fixed point and approximation.

  Starting point: H Lipschitz and regularizing coupling.

  Other cases by approximation.

  Description of "la ola".

  Discussion on the uniqueness for the system MFG. Two regimes: monotone coupling versus small time horizon.

- 16/01/2009

  1st hour: Interpretation of the MFG system (with a local coupling and planning problem setting) as an optimal control problem of the Fokker-Planck equation [cfr. Thm 1.17].

  Comment on the existence of a minimum, on the uniqueness (counter-example to uniqueness when the monotonicity is lost).

  Loss of uniqueness by analysis of the linearized system (when existence of a trivial solution): the linearized problem is well-posed only if the horizon is small.

  2nd hour: Use of the Hopf-Cole transform for quadratic Hamiltonians [cfr. Remark 1.13].

  Back on the existence of the solution to the MFG system [cfr. Remark 1.12]:

  $$\begin{cases} -\partial_t u - \Delta u + H(p) = f(m) \\ \partial_t m - \Delta m - \operatorname{div}(m H_p(Du)) = 0 \end{cases}$$

  - if f is bounded and H is subquadratic, existence of smooth solutions (e.g., $H(p) = p^\alpha, \alpha \le 2$). (works also for $f(m) = c_0 m^p$ for p small).
  - if H is superquadratic and f is nonincreasing: open problem.
  - if $f(m) = cm^\beta$ with $c > 0$, $H(p) = c_0|p|^\gamma$ with $\gamma > 1$. First a priori estimate on $\int\int m^{1+\beta} + m|Du|^\gamma \le C$. Second a priori estimate obtained by multiplying by Δm the equation for u, and by Δu the equation of m and adding the resulting quantities (computation for $\gamma = 2$): one gets $\frac{d}{dt}\int Du\, Dm = \int |D^2 u|^2 m + f'(m)|Dm|^2$.

Organization 2009–2010

(Analysis of the MFG system: the local coupling - Variational approach to MFGs)

- 06/11/2009

  Presentation of the MFG system.

1st hour: Maximum principle in the deterministic case for smooth solutions: if $u_0 \leq v_0$, then $u \leq v$.

Proof by reduction to a time-space elliptic equation with boundary conditions Dirichlet and nonlinear Neumann (+ discussion on the link with Euler equation). Proof that this is an elliptic equation.

2nd hour: generalization to the case where the initial condition on u is a function of m. Discussion of the maximum principle when the running cost f grows: not true in general.

Discussion of the maximum principle when the continuity equation has a right-hand side.

- 13/11/2009

  Comparison principle in the second order setting with a quadratic hamiltonian.

  Quadratic Hamiltonian: change of variable (Hopf-Cole transform, [cfr. Remark 1.13]) and algorithm to build solutions.

  Conjecture: no comparison principle for more general Hamiltonians.

- 20/11/2009

  Comparison principle: second order setting with a quadratic Hamiltonian and stationary MFG systems.

  Comments on the convergence of the MFG system as $T \to +\infty$ [cfr. Sect. 1.3.6]: convergence of $m^T(t)$, $u^T(t)- < u^T(t) >$, and $< u^T(t) > /T$. Claim that $u^T(t) - \bar{\lambda}(T - t)$ converges.

  Ergodic problem: comparison in the deterministic setting: if $f_1 \leq f_2$, then $\bar{\lambda}_1 \leq \bar{\lambda}_2$. When $H(x, \xi) \geq H(x, 0)$ for all ξ, then $m = [f^{-1}(x, \lambda)]_+$ where λ is such that $\int m = 1$. Then $u = constant$ in $\{m > 0\}$; solve $H(x, Du) = \lambda$ in $\{m = 0\}$ with boundary conditions. Justification by $\nu \to 0^+$ for instance.

  Comparison in the second order setting: quadratic H.

  Planification problems. Approach by penalization. Link with Wasserstein.

- 27/11/2009

  Link between MFG with optimal control of (backward) Fokker-Plank equation:

  $$\partial_t m + \Delta m + div(m\alpha) = 0, \qquad m(T, x) = m_1(x)$$

  where $\alpha = \alpha(x, t)$ and the cost is of the form

  $$\int_0^T \int_Q mL(x, \alpha)dxdt + \Psi(m) + \int_Q \Phi(x, m(0, x))dx$$

  Planing pb: $\Phi = \frac{1}{2\varepsilon}\|m - m_0\|_2^2$.

  Derivation of the optimality conditions. Generalization to the case $L(x, \alpha, m)$ which is a functional of m. Approach by optimal control to the planning problem. Leads to controllability issues. Discussion of the polynomial case.

  2nd hour: First order planning problem: existence of a smooth solution.

  Step 1: link with quasilinear elliptic equations with nonlinear boundary conditions [cfr. Remark 1.16].

Step 2: L^∞ estimates on $w := \partial_t u + H(x, Du)$ (i.e., estimate on m): extension of Bernstein method by looking at the equation satisfied by w.

Step 3: L^∞ estimate on u. Indeed u is smooth and solves $\partial_t u + H(Du) = f(m)$ where $f(m)$ is bounded. So it is a forward and backward solution which gives the result.

- 04/12/2009

Planning problem (without diffusion): link with quasilinear elliptic equation (in time-space) with nonlinear boundary conditions. Lipschitz estimates on u: Bernstein method again. Difficulties: constants are subsolutions and boundary conditions.

- 11/12/2009

First hour: Back to the first order planning problem.

Dual problem, i.e., optimal control of HJ equation [cfr. Sect. 1.3.7.2]. Namely

$$\inf_u \int \int G(\frac{\partial u}{\partial t} + H(Du)) - \int (m_1 u(T) - m_0 u(0)).$$

Computation of the first variation, and link with the MFG system. Comment on the fact that $f = f(m)$ has to be strictly increasing. Generalization to second order problems.

Counter-examples:

(i) (reminder when H at most linear (first or second order): existence of solutions). In this case there is no existence of solution for the dual problem (at least for small time).

(ii) Regularity? Normalization: $H(0) = 0$, $H'(0) = 0$, $f(1) = 0$, $A = H''(0) > 0$, $f'(1) = a > 0$. Then $m = 1$, $u = 0$ is the unique solution for $m_0 = m_T = 0$. One linearizes to get $\partial_t v - \nu \Delta v = an$, $\partial_t n + \nu \Delta n + div(A Dv) = 0$ with $n(0) = n_0$ and $n(T) = n_T$ where $\int n_0 = \int n_1$. Stability requires that $A > 0$. Proposition: the linearized (periodic) problem is well-posed iif $A > 0$, $a \geq 0$, $\nu \geq 0$. Proof for first order, straightforward; for second order, Fourier.

Second hour: end of the proof.

Second Order Planning Problem Approach by optimization (optimal control of Fokker-Planck equation) yields the existence and uniqueness of very weak solutions. Main issue: regularity. Understood when $H = \frac{1}{2}|p|^2$. Theorem: when $H = \frac{1}{2}|p|^2$, and f non decreasing with polynomial growth, then there is a unique smooth solution. Generalization to the case $|H''(p) - I| \leq \frac{C}{\sqrt{1+|p|^2}}$ (conj. could be generalized to the case $cI \leq H'' \leq CI$). Proof by the Hopf-Cole transformation.

- 18/12/2009

MFG problems with congestion terms [cfr. Example 1.1]: minimize $\mathbb{E}\left[\int_t^T q^{-1}|\alpha_s|^q (m(s, X_s))^a ds + u_0(X_T)\right]$ with $dX_s = \sigma dW_s - \alpha_s ds$ where

$q > 1$ and $a > 0$. Leads to the MFG system of the form

$$\begin{cases} \partial_t u - \nu \Delta u + \frac{1}{p} \frac{|Du|^p}{m^b} = 0 \\ -\partial_t m - \nu \Delta m + \mathrm{div}(\frac{|Du|^{p-2}Du}{m^b}m) = 0 \\ m(T) = m_T, \ u(0) = u_0 \end{cases} \qquad (1.173)$$

Discussion of the (lack of) link with the optimal control of the Fokker-Planck equation. Uniqueness condition for the MFG system (for $p = 2$ and $0 < b \le 2$).

- 08/01/2010
  Back to the congestion problem. Uniqueness of the solution of (1.173) in the case (1) where the Hamiltonian is of the form $|Du|^2/(2f(m))$ (and the term in the divergence by $m\,Du/f(m)$) and (2) $p > 1$ and $0 < b \le 4/p'$.
  Discussion on the existence of a solution for $\nu = 0$ by using the fact that the equation of u is an elliptic equation in time-space: bounds on u, m and on Du. Regularity issue if m vanishes.
  Analysis of the case $\nu > 0$, $p = 2$, $0 < b(\le 2)$: a priori estimates and notion of solution.
- 15/01/2010
  Back to to the congestion problem (1.173) when $p = 2$, $b = 1$. A priori estimates continued (bounds on u, on $\int \int |Du|^2(1 + m^{-1})$, on $\int \int |D^2u|^2$ and on $\int \int |Du|^2|Dm|^2/m^2$). Existence of a solution by approximation (replacing $|Du|^2/m$ by $|Du|^2/(\delta + m)$ for $\delta > 0$).

Organization 2010–2011

(the master equation in infinite and finite dimension)

- 05/11/2010
  Uniqueness for the MFG system when $H = H(Du, m)$ [cfr. Thm 1.13]. Different approaches: monotonicity, continuation, reduction to an elliptic equation.
- 12/11/2010
  Uniqueness for the MFG system when $H = H(Du, m)$ (continued): linearization, problems with actualization rate.
  On the Master equation (MFGf)[3]:

  1. Heuristics: Master equation as a limit system of Nash equilibria with N players as $N \to +\infty$
  2. The Master Equation contains the MFG equation (when $\beta = 0$)
  3. Back to the uniqueness proof: U is monotone
  4. Back to $N \to +\infty$: MFGf contains the Nash system without individual noise.

---

[3] Warning: missing term in the MFGf.

- 19/11/2010
  1st hour: Back to the Master equation.[4] Check that when $\nu \neq 0$ the equation does not match with Nash eq for N players. Link with optimal control problems in the case of separate variables (discussion of the case of non separate variables).
  2nd hour: Hamilton-Jacobi equation associated with an optimal control of Fokker-Planck equation. Derivation of the master equation by taking the derivative of the Hamilton-Jacobi equation.
- 26/11/2010
  Erratum on the master equation. Interpretation of the Master Equation as a limit as $N \to +\infty$: explanation of the second order terms [cfr. Sect. 1.4.3.3].

  (1) Interpretation in terms of optimal control problem ($\beta = 0$)
  (2) Uniqueness related to the convexity of F and Φ
  (3) General principle for the link between optimal control and the Master Equation in infinite dimension.

- 03/12/2010
  System derived from Hamilton-Jacobi: propagation of monotonicity.
- 10/12/2010
  System derived from Hamilton-Jacobi:

  - Propagation of monotonicity for second order systems.
  - Propagation of smoothness, method of characteristics.

- 17/12/2010
  Propagation of monotonicity for $\frac{\partial U}{\partial t} + (H'(DU)D)U = f(x) + \sum a_{\alpha,\beta} \frac{\partial^2 U}{\partial x_\alpha x_\beta}$.
- 07/01/2011
  Existence and uniqueness of a monotone solution for $\frac{\partial U}{\partial t} + (H'(DU)D)\dot{U} = f(x)$.
  Remarks on semi-concavity for HJ equations.
- 14/01/2011
  1st hour: Structure of the master equation in the discrete setting (without diffusion):

$$\partial_t U_i + \left(\sum_j x_j H_j'(x, \nabla U)\nabla\right)U_i + H_i(x, \nabla U) = 0.$$

Propagation of Monotonicity
2nd hour: Propagation of monotonicity for independent noises (in the infinite dimensional setting). Finite dimensional setting, in which the noise yields a term of the form $\sum_{k,l} a_{kl}x_l \partial_k U_i + \sum_k a_{ki}U_k$.
Monotonicity for the common noise (in the infinite dimensional setting; the finite dimensional setting being open).

---

[4]Warning: missing term in the MFGf.

Organization 2011–2012

(Analysis of the master equation for MFG in the finite state space, [cfr. Sect. 1.4.3.2])

- 28/10/2011
  Analysis of equation: $\frac{\partial U}{\partial t} + (U.\nabla)U = 0$ (where $U : \mathbb{R}^n \times (0, +\infty) \to \mathbb{R}^n$).

  - case $U_0 = \nabla\phi_0$: then $U = \nabla\phi$ with ϕ sol of HJ equation.
  - case U_0 monotone, bounded and Lipschitz continuous: existence and uniqueness of a monotone, bounded and Lipschitz continuous sol, which is smooth if U_0 is smooth.

  Generalization to $\frac{\partial V}{\partial t} + (F(V).\nabla)V = 0$, provided F and V_0 monotone (since $U = F(V)$ the initial equation)
  Explicit formula: linear case, method of characteristics: solution is given by $U = (U_0^{-1} + t I_d)^{-1}$ as long as there is no shock. Quid in general?
  Propagation of the condition $\frac{\partial U_i}{\partial x_j} \leq 0$, $j \neq i$.
- 04/11/2011
  Back to the system $\frac{\partial U}{\partial t} + (U.\nabla)U = 0$.
  Propagation of the condition $\frac{\partial U_i}{\partial x_j} \leq 0$, $j \neq i$. Consequence: $\frac{\partial U_i}{\partial x_i}$ is a bounded measure.
  A striking identity: if U is a classical solution of $\frac{\partial U}{\partial t} + (F(U).\nabla)U = 0$, then
  $\frac{\partial}{\partial t} \det(\nabla U) + \operatorname{div}(F(U) \det(\nabla U)) = 0$.
- 25/11/2011
  Application to non-convex HJ equations: examples of smooth solutions.
- 09/12/2011
  Propagation of monotonicity with second order terms.
- 16/12/2011
  Analysis of $\frac{\partial U}{\partial t} + (F(U).\nabla)U = 0$.
  Following Krylov idea: introduce $W(x, \eta, t) = U(x, t) \cdot \eta$.
- 06/01/2012
  Analysis of $\frac{\partial U}{\partial t} + (F(U).\nabla)U = f(x)$: existence of a smooth global solution under monotonicity assumptions.
  A priori estimates when U_0 satisfies $U_0'(z)\xi \cdot \xi \geq \alpha|U_0'(z)\xi|^2$ for some $\alpha > 0$ and any z, ξ.
- 13/01/2012
  Analysis of $\frac{\partial U}{\partial t} + (F(U).\nabla)U = 0$ with U_0 and F monotone (continued). A priori estimates on ∇U under the assumption that there exists $\alpha > 0$ such that $F'(z)\xi \cdot \xi \geq \alpha|F'(z)\xi|^2$ for any z, ξ.
  Generalization to the case with a right-hand side of the form $a_{kl}\partial_{kl}U^i + b_{kl}^i\partial_l U^k$ where $a_{\alpha\beta}$ symmetric ≥ 0.

Additional Notes

- 08/11/2013
  Seminar: on the differentiability in Wasserstein space, point of view of the random variables. MFGs in the finite state case: the master equation as a first order hyperbolic system. Back to the infinite dimensional case, the Hilbertian approach: if $U(t, x, X)$ is the solution of the classical master equation, one sets $V(t, X) = U(t, X, \mathcal{L}(X))$. Discussion of the monotonicity in the Hilbertian framework.

References

1. Y. Achdou, Finite difference methods for mean field games, in *Hamilton-Jacobi Equations: Approximations, Numerical Analysis and Applications*, ed. by P. Loreti, N.A. Tchou. Lecture Notes in Mathematics, vol. 2074 (Springer, Heidelberg, 2013), pp. 1–47
2. Y. Achdou, I. Capuzzo Dolcetta, Mean field games: numerical methods. SIAM J. Numer. Anal. **48**, 1136–1162 (2010)
3. Y. Achdou, Z. Kobeissi, *Mean Field Games of Controls: Finite Difference Approximations* (2020). Preprint arXiv:2003.03968
4. Y. Achdou, A. Porretta, Convergence of a finite difference scheme to weak solutions of the system of partial differential equation arising in mean field games. SIAM J. Numer. Anal. **54**, 161–186 (2016)
5. Y. Achdou, A. Porretta, Mean field games with congestion. Ann. I. H. Poincaré **35**, 443–480 (2018)
6. Y. Achdou, F. Camilli, I. Capuzzo Dolcetta, Mean field games: numerical methods for the planning problem. SIAM J. Control Opt. **50**, 77–109 (2012)
7. Y. Achdou, F.J. Buera, J.-M. Lasry, P.-L. Lions, B. Moll, Partial differential equation models in macroeconomics. Philos. Trans. R. Soc. A **372**, 20130397 (2014)
8. Y. Achdou, M. Bardi, M. Cirant, Mean field games models of segregation. Math. Models Methods Appl. Sci. **27**, 75–113 (2017)
9. Y. Achdou, J. Han, J.-M. Lasry, P.-L. Lions, B. Moll, Income and wealth distribution in macroeconomics: a continuous-time approach. Technical report, National Bureau of Economic Research, 2017
10. Y. Achdou, M.K. Dao, O. Ley, N. Tchou, A class of infinite horizon mean field games on networks. Netw. Heterog. Media **14**, 537–566 (2019)
11. S. Ahuja, Well-posedness of mean field games with common noise under a weak monotonicity condition. SIAM J. Control Optim. **54**, 30–48 (2016)
12. S. Ahuja, W. Ren, T.-W. Yang, Asymptotic analysis of mean field games with small common noise. Asymptot. Anal. **106**, 205–232 (2018)
13. S.R. Aiyagari, Uninsured idiosyncratic risk and aggregate saving. Quart. J. Econ. **109**, 659–684 (1994)
14. S. Albeverio, Y.G. Kondratiev, M. Röckner, Analysis and geometry on configuration spaces. J. Funct. Anal. **154**, 444–500 (1998)
15. L. Ambrosio, Transport equation and cauchy problem for BV vector fields. Invent. Math. **158**, 227–260 (2004)
16. L. Ambrosio, Transport equation and cauchy problem for non-smooth vector fields, in *Calculus of Variations and Nonlinear Partial Differential Equations*. Lecture Notes in Mathematics, vol. 1927 (Springer, Berlin, 2008), pp. 1–41

17. L. Ambrosio, W. Gangbo, Hamiltonian odes in the Wasserstein space of probability measures. Commun. Pure Appl. Math.: J. Issued Courant Inst. Math. Sci. **61**, 18–53 (2008)
18. L. Ambrosio, N. Gigli, G. Savaré, *Gradient Flows: In Metric Spaces and in the Space of Probability Measures* (Springer, New York, 2006)
19. A. Alfonsi, B. Jourdain, Lifted and geometric differentiability of the squared quadratic Wasserstein distance (2018). Preprint. arXiv:1811.07787
20. R.J. Aumann, Markets with a continuum of traders. Econometrica: J. Econ. Soc. **1964**, 39–50 (1964)
21. M. Balandat, C.J. Tomlin, On efficiency in mean field differential games, in *2013 American Control Conference*, June 2013 (IEEE, New York, 2013), pp. 2527–2532
22. J.M. Ball, *A Version of the Fundamental Theorem for Young Measures, PDE's and Continuum Models of Phase Transitions*, ed. by Rascle, M., Serre, D., Slemrod, M. Lecture Notes in Physics, vol. 344 (Springer, Berlin, 1989), pp. 207–215
23. M. Bardi, P. Cardaliaguet, Convergence of some Mean Field Games systems to aggregation and flocking models (2020). Preprint. arXiv:2004.04403
24. M. Bardi, M. Cirant, Uniqueness of solutions in mean field games with several populations and Neumann conditions, in *PDE Models for Multi-agent Phenomena* (Springer, New York, 2018), pp. 1–20
25. M. Bardi, M. Fischer, On non-uniqueness and uniqueness of solutions in finite-horizon mean field games. ESAIM: Control Optim. Calculus Var. **25**, 44 (2019)
26. E. Bayraktar, A. Cohen, Analysis of a finite state many player game using its master equation. SIAM J. Control Optim. **56**, 3538–3568 (2018)
27. E. Bayraktar, A. Cecchin, A. Cohen, F. Delarue, Finite state mean field games with Wright-Fisher common noise (2019). Preprint. arXiv:1912.06701
28. J.-D. Benamou, Y. Brenier, A computational fluid mechanics solution to the Monge-Kantorovich mass transfer problem. Numer. Math. **84**, 375–393 (2000)
29. J.D. Benamou, G. Carlier, F. Santambrogio, Variational mean field games, in *Active Particles*, vol. 1 (Birkhäuser, Cham, 2017), pp. 141–171
30. A. Bensoussan, J. Frehse, P. Yam, *Mean Field Games and Mean Field Type Control Theory*, vol. 101 (Springer, New York, 2013)
31. A. Bensoussan, M. Chau, S. Yam, Mean field games with a dominating player. Appl. Math. Optim. **74**, 91–128 (2016)
32. C. Bertucci, Optimal stopping in mean field games, an obstacle problem approach. J. Math. Pures Appl. **120**, 165–194 (2018)
33. C. Bertucci, Fokker-Planck equations of jumping particles and mean field games of impulse control. In Annales de l'Institut Henri Poincaré C, Analyse non linéaire (2020)
34. C. Bertucci, J.-M. Lasry, P.-L. Lions, Some remarks on mean field games. Comm. Part. Differ. Equ. **44**, 205–227 (2019)
35. C. Bertucci, J.-M. Lasry, P.-L. Lions, Master equation for the finite state space planning problem (2020). Preprint arXiv:2002.09330
36. C. Bertucci, J.-M. Lasry, P.-L. Lions, Strategic advantages in mean field games with a major player (2020). Preprint arXiv:2002.07034
37. P. Billingsley, *Probability and Measure* (Wiley, New York, 2008)
38. G. Bouveret, R. Dumitrescu, P. Tankov, Mean-field games of optimal stopping: a relaxed solution approach. Siam J. Control Optim. **58**, 1795–1821 (2020)
39. Y. Brenier, Polar factorization and monotone rearrangement of vector-valued functions. Commun. Pure Appl. Math. **44**, 375–417 (1991)
40. A. Briani, P. Cardaliaguet, Stable solutions in potential mean field game systems. Nonlinear Differ. Equ. Appl. **25**, 1 (2018)
41. R. Buckdahn, J. Li, S. Peng, C. Rainer, Mean-field stochastic differential equations and associated PDEs. Ann. Probab. **45**, 824–878 (2017)
42. M. Burger, A. Lorz, M.T. Wolfram, On a Boltzmann mean field model for knowledge growth. SIAM J. Appl. Math. **76**(5), 1799–1818 (2016)

43. M. Burger, A. Lorz, M.T. Wolfram, Balanced growth path solutions of a Boltzmann mean field game model for knowledge growth. Kinet. Relat. Models **10**(1),117–140 (2017)
44. F. Camilli, R. De Maio A time-fractional mean field game. Adv. Differ. Equ. **24**(9/10), 531–554 (2019)
45. F. Camilli, C. Marchi, Stationary mean field games systems defined on networks. SIAM J. Control Optim. **54**(2), 1085–1103 (2016)
46. L. Campi, M. Fischer, *n*-player games and mean-field games with absorption. Ann. Appl. Probab. **28**, 2188–2242 (2018)
47. P. Cannarsa, R. Capuani, Existence and uniqueness for mean field games with state constraints, in *PDE Models for Multi-agent Phenomena* (Springer, Cham, 2018), pp. 49–71
48. P. Cannarsa, C. Sinestrari, Semiconcave functions, in *Hamilton-Jacobi Equations and Optimal Control* (Birkhauser, Boston, 2004)
49. P. Cannarsa, R. Capuani, P. Cardaliaguet, $C^{1,1}$-smoothness of constrained solutions in the calculus of variations with application to mean field games. Math. Eng. **1**, 174–203 (2019)
50. P. Cannarsa, R. Capuani, P. Cardaliaguet, Mean field games with state constraints: from mild to pointwise solutions of the pde system (2018). Preprint. arXiv:1812.11374
51. P. Cardaliaguet, Notes on mean field games, Technical report, 2010
52. P. Cardaliaguet, Long time average of first order mean field games and weak KAM theory. Dynam. Games Appl. **3**, 473–488 (2013)
53. P. Cardaliaguet, Weak solutions for first order mean field games with local coupling, in *Analysis and Geometry in Control Theory and Its Applications* (Springer, Cham, 2015), pp. 111–158
54. P. Cardaliaguet, P.J. Graber, Mean field games systems of first order. ESAIM: Control Optim. Calc. Var. **21**, 690–722 (2015)
55. P. Cardaliaguet, S. Hadikhanloo, Learning in mean field games: the fictitious play. ESAIM: Control Optim. Calc. Var. **23**(2), 569–591 (2017)
56. P. Cardaliaguet, C.-A. Lehalle, Mean field game of controls and an application to trade crowding. Math. Finan. Econ. **12**, 335–363 (2018)
57. P. Cardaliaguet, M. Masoero, Weak KAM theory for potential MFG. J. Differ. Equ. **268**, 3255–3298 (2020)
58. P. Cardaliaguet, A. Porretta, Long time behavior of the master equation in mean field game theory. Anal. PDE **12**, 1397–1453 (2019)
59. P. Cardaliaguet, C. Rainer, On the (in) efficiency of MFG equilibria. SIAM J. Control Optim. **57**(4), 2292–2314 (2019)
60. P. Cardaliaguet, J.-M. Lasry, P.-L. Lions, A. Porretta, Long time average of mean field games. Netw. Heterog. Media **7**, 279–301 (2012)
61. P. Cardaliaguet, J.-M. Lasry, P.-L. Lions, A. Porretta, Long time average of mean field games with a nonlocal coupling. SIAM J. Control Optim. **51**, 3558–3591 (2013)
62. P. Cardaliaguet, P.J. Graber, A. Porretta, D. Tonon, Second order mean field games with degenerate diffusion and local coupling. Nonlinear Differ. Equ. Appl. **22**, 1287–1317 (2015)
63. P. Cardaliaguet, A.R. Mészáros, F. Santambrogio, First order mean field games with density constraints: pressure equals price. SIAM J. Control Optim. **54**, 2672–2709 (2016)
64. P. Cardaliaguet, M. Cirant, A. Porretta, Remarks on Nash equilibria in mean field game models with a major player. Proceedings of the American Math. Society **148**, 4241–4255 (2020)
65. P. Cardaliaguet, F. Delarue, J.-M. Lasry, P.-L. Lions, *The Master Equation and the Convergence Problem in Mean Field Games:(AMS-201)*, vol. 201 (Princeton University Press, Princeton, 2019)
66. P. Cardaliaguet, M. Cirant, A. Porretta, Splitting methods and short time existence for the master equations in mean field games (2020). Preprint. arXiv:2001.10406
67. R. Carmona, F. Delarue, Probabilistic analysis of mean-field games. SIAM J. Control Optim. **51**, 2705–2734 (2013)
68. R. Carmona, F. Delarue, *Probabilistic Theory of Mean Field Games with Applications* (Springer Verlag, New York, 2017)

69. R. Carmona, D. Lacker, A probabilistic weak formulation of mean field games and applications. Ann. Appl. Probab. **25**, 1189–1231 (2015)
70. R. Carmona, P. Wang, Finite state mean field games with major and minor players (2016). Preprint. arXiv:1610.05408
71. R. Carmona, P. Wang, An alternative approach to mean field game with major and minor players, and applications to herders impacts. Appl. Math. Optim. **76**, 5–27 (2017)
72. R. Carmona, X. Zhu, A probabilistic approach to mean field games with major and minor players. Ann. Appl. Probab. **26**, 1535–1580 (2016)
73. R. Carmona, F. Delarue, D. Lacker, Mean field games with common noise. Ann. Probab. **44**, 3740–3803 (2016)
74. R. Carmona, F. Delarue, D. Lacker, Mean field games of timing and models for bank runs. Appl. Math. Optim. **76**, 217–260 (2017)
75. R. Carmona, C.V. Graves, Z. Tan, Price of anarchy for mean field games. ESAIM: Proc. Surv. **65**, 349–383 (2019)
76. R. Carmona, M. Laurière, Z. Tan, Linear-quadratic mean-field reinforcement learning: convergence of policy gradient methods (2019). Preprint. arXiv:1910.04295
77. R. Carmona, M. Laurière, Z. Tan, Model-free mean-field reinforcement learning: mean-field MDP and mean-field Q-learning (2019). Preprint. arXiv:1910.12802
78. A. Cecchin, M. Fischer, Probabilistic approach to finite state mean field games. Appl. Math. Optim. **2018**, 1–48 (2018)
79. A. Cecchin, G. Pelino, Convergence, fluctuations and large deviations for finite state mean field games via the master equation. Stoch. Process. Appl. **129**, 4510–4555 (2019)
80. A. Cecchin, P.D. Pra, M. Fischer, G. Pelino, On the convergence problem in mean field games: a two state model without uniqueness. SIAM J. Control Optim. **57**, 2443–2466 (2019)
81. A. Cesaroni, M. Cirant, Concentration of ground states in stationary MFG systems. Analysis PDE (2019)
82. A. Cesaroni, N. Dirr, C. Marchi, Homogenization of a mean field game system in the small noise limit. SIAM J. Math. Anal. **48**, 2701–2729 (2016)
83. A. Cesaroni, M. Cirant, S. Dipierro, M. Novaga, E. Valdinoci, On stationary fractional mean field games. J. Math. Pures Appl. **122**, 1–22 (2019)
84. P. Chan, R. Sircar, Fracking, renewables, and mean field games. SIAM Rev. **59**(3), 588–615 (2017)
85. J.-F. Chassagneux, D. Crisan, F. Delarue, Classical solutions to the master equation for large population equilibria. Preprint arXiv:1411.3009
86. M. Cirant, Multi-population mean field games systems with Neumann boundary conditions. J. Math. Pures Appl. **103**, 1294–1315 (2015)
87. M. Cirant, Stationary focusing mean-field games. Commun. Part. Differ. Equ. **41**, 1324–1346 (2016)
88. M. Cirant, On the existence of oscillating solutions in non-monotone mean-field games. J. Differ. Equ. **266**, 8067–8093 (2019)
89. M. Cirant, A. Goffi, On the existence and uniqueness of solutions to time-dependent fractional MFG. SIAM J. Math. Anal. **51**(2), 913–954 (2019)
90. M. Cirant, L. Nurbekyan, The variational structure and time-periodic solutions for mean-field games systems. Minimax Theory Appl. **3**, 227–260 (2018)
91. M. Cirant, A. Porretta, Long time behavior and turnpike solutions in non monotone mean field games. Preprint
92. M. Cirant, D. Tonon, Time-dependent focusing mean-field games: the sub-critical case. J. Dyn. Differ. Equ. **31**, 49–79 (2019)
93. M. Cirant, G. Verzini, Bifurcation and segregation in quadratic two-populations mean field games systems. ESAIM: Control Optim. Calc. Var. **23**, 1145–1177 (2017)
94. F.H. Clarke, *Optimization and Nonsmooth Analysis*, 2nd edn. Classics in Applied Mathematics, vol. 5 (Society for Industrial and Applied Mathematics (SIAM), Philadelphia, PA, 1990)
95. F. Delarue, R.F. Tchuendom, Selection of equilibria in a linear quadratic mean-field game. Stoch. Process. Appl. **130**, 1000–1040 (2020)

96. F. Delarue, D. Lacker, K. Ramanan, From the master equation to mean field game limit theory: a central limit theorem. Electron. J. Probab. **24**, 54 pp. (2019)
97. F. Delarue, D. Lacker, K. Ramanan, From the master equation to mean field game limit theory: large deviations and concentration of measure. Ann. Probab. **48**, 211–263 (2020)
98. R.J. DiPerna, P.-L. Lions, Ordinary differential equations, transport theory and sobolev spaces. Invent. Math. **98**, 511–547 (1989)
99. P. Degond, J.G. Liu, C. Ringhofer, Large-scale dynamics of mean-field games driven by local Nash equilibria. J. Nonlinear Sci. **24**(1), 93–115 (2014)
100. I. Ekeland, R. Témam, R. *Convex Analysis and Variational Problems*, English ed., vol. 28. Classics in Applied Mathematics (Society for Industrial and Applied Mathematics (SIAM), Philadelphia, PA, 1999). Translated from the French
101. R. Elie, T. Mastrolia, D. Possamaï, A tale of a principal and many, many agents. Math. Oper. Res. **44**, 40–467 (2019)
102. R. Elie, J. Pérolat, M. Laurière, M. Geist, O. Pietquin, Approximate fictitious play for mean field games (2019). Preprint. arXiv:1907.02633
103. O. Ersland, E.R. Jakobsen, On classical solutions of time-dependent fractional mean field game systems (2020). Preprint. arXiv:2003.12302
104. R. Ferreira, D. Gomes, Existence of weak solutions to stationary mean-field games through variational inequalities. SIAM J. Math. Anal. **50**, 5969–6006 (2018)
105. M. Fischer, On the connection between symmetric n-player games and mean field games. Ann. Appl. Probab. **27**, 757–810 (2017)
106. W.H. Fleming, R.W. Rishel, *Deterministic and Stochastic Optimal Control*, vol. 1 (Springer Science & Business Media, Berlin, 2012)
107. W.H. Fleming, H.M. Soner, *Controlled Markov Processes and Viscosity Solutions*, vol. 25 (Springer Science & Business Media, Berlin, 2006)
108. W. Gangbo, A. Swiech, Existence of a solution to an equation arising from the theory of mean field games. J. Differ. Equ. **259**, 6573–6643 (2015)
109. W. Gangbo, A. Tudorascu, On differentiability in the Wasserstein space and well-posedness for Hamilton–Jacobi equations. J. Math. Pures Appl. **125**, 119–174 (2019)
110. D. Gilbarg, N.S. Trudinger, *Elliptic Partial Differential Equations of Second Order* (Springer, Berlin, 2015)
111. F. Golse, On the dynamics of large particle systems in the mean field limit, in *Macroscopic and Large Scale Phenomena: Coarse Graining, Mean Field Limits and Ergodicity* (Springer, Cham, 2016), pp. 1–144
112. D. Gomes, S. Patrizi, Obstacle mean-field game problem. Interfaces Free Bound. **17**, 55–68 (2015)
113. D.A. Gomes, H. Mitake, Existence for stationary mean-field games with congestion and quadratic Hamiltonians. Nonlinear Differ. Equ. Appl. **22**, 1897–1910 (2015)
114. D.A. Gomes, E.A. Pimentel, Time dependent mean-field games with logarithmic nonlinearities. SIAM J. Math. Anal. **47**, 3798–3812 (2015)
115. D. Gomes, J. Saùde, Mean field games models - a brief survey. Dyn. Games Appl. **4**, 110–154 (2014)
116. D.A. Gomes, V.K. Voskanyan, Short-time existence of solutions for mean-field games with congestion. J. Lond. Math. Soc. **92**, 778–799 (2015)
117. D.A. Gomes, V.K. Voskanyan, Extended deterministic mean-field games. SIAM J. Control Optim. **54**, 1030–1055 (2016)
118. D.A. Gomes, J. Mohr, R.R. Souza, Discrete time, finite state space mean field games. J. Math. Pures Appl. **93**, 308–328 (2010)
119. D.A. Gomes, S. Patrizi, V. Voskanyan, On the existence of classical solutions for stationary extended mean field games. Nonlinear Anal. Theory Methods Appl. **99**, 49–79 (2014)
120. D.A. Gomes, E.A. Pimentel, H. Sànchez-Morgado, Time-dependent mean-field games in the subquadratic case. Commun. Part. Differ. Eq. **40**, 40–76 (2015)
121. D.A. Gomes, E.A. Pimentel, H. Sànchez-Morgado, Time-dependent mean-field games in the superquadratic case. ESAIM Control. Optim. Calc. Var. **22**, 562–580 (2016)

122. D.A. Gomes, E.A. Pimentel, V. Voskanyan, *Regularity Theory for Mean-Field Game Systems* (Springer, Berlin, 2016)
123. P.J. Graber, Weak solutions for mean field games with congestion (2015). ArXiv e-print 1503.04733
124. P.J. Graber, A.R. Mészáros, Sobolev regularity for first order mean field games. Annales de l'Institut Henri Poincaré C, Analyse non linéaire **35**, 1557–1576 (2018)
125. P.J. Graber, C. Mouzouni, On mean field games models for exhaustible commodities trade. ESAIM: Control Optim. Calc. Var. **26**, 11 (2020)
126. P.J. Graber, A.R. Mészáros, F.J. Silva, D. Tonon, The planning problem in mean field games as regularized mass transport. Calc. Var. Part. Differ. Equ. **58**, 115 (2019)
127. O. Gueant, A reference case for mean field games models. J. Math. Pures Appl. (9) **92**, 76–294 (2009)
128. O. Gueant, J.-M. Lasry, P.-L. Lions, Mean field games and applications, in *Paris-Princeton Lectures on Mathematical Finance 2010* (Springer, Berlin, Heidelberg, 2011), pp. 205–266
129. S. Hadikhanloo, F.J. Silva, Finite mean field games: fictitious play and convergence to a first order continuous mean field game. J. Math. Pures Appl. **132**, 369–397 (2019)
130. M. Huang, Large-population LQG games involving a major player: the Nash certainty equivalence principle. SIAM J. Control Optim. **48**, 3318–3353 (2010)
131. M. Huang, P.E. Caines, R.P. Malhamé, Individual and mass behaviour in large population stochastic wireless power control problems: centralized and Nash equilibrium solutions, in *42nd IEEE Conference on Decision and Control, 2003. Proceedings*, vol. 1 (IEEE, New York, 2003), pp. 98–103
132. M. Huang, R.P. Malhamé, P.E. Caines, Large population stochastic dynamic games: closed-loop Mckean-Vlasov systems and the nash certainty equivalence principle. Commun. Inf. Syst. **6**, 221–252 (2006)
133. M. Huang, R.P. Malhamé, P.E. Caines, An invariance principle in large population stochastic dynamic games. J. Syst. Sci. Complex. **20**, 162–172 (2007)
134. M. Huang, R.P. Malhamé, P.E. Caines, Large-population cost-coupled LQG problems with nonuniform agents: individual-mass behavior and decentralized ε-Nash equilibria. IEEE Trans. Autom. Control **52**, 1560–1571 (2007)
135. M. Huang, R.P. Malhamé, P.E. Caines, The Nash certainty equivalence principle and Mckean-Vlasov systems: an invariance principle and entry adaptation, in *2007 46th IEEE Conference on Decision and Control* (IEEE, New York, 2007), pp. 121–126
136. A.C. Kizilkale, P.E. Caines, Mean field stochastic adaptive control. IEEE Trans. Autom. Control **58**(4), 905–920 (2012)
137. V.N. Kolokoltsov, J. Li, W. Yang, Mean field games and nonlinear Markov processes (2011). Preprint. arXiv:1112.3744
138. P. Krusell, A.A. Smith Jr, Income and wealth heterogeneity in the macroeconomy. J. Polit. Econ. **106**, 867–896 (1998)
139. D. Lacker, A general characterization of the mean field limit for stochastic differential games. Probab. Theory Relat. Fields **165**, 581–648 (2016)
140. D. Lacker, On the convergence of closed-loop Nash equilibria to the mean field game limit. Ann. Appl. Probab. **30**, 1693–1761 (2020)
141. D. Lacker, K. Webster, Translation invariant mean field games with common noise. Electron. Commun. Probab. **20**, 1–13 (2015)
142. O.A. Ladyzhenskaia, V.A. Solonnikov, N. Uraltseva, *Linear and Quasi-Linear Equations of Parabolic Type*, vol. 23 (American Mathematical Society, Providence, RI, 1998)
143. J.-M. Lasry, P.-L. Lions, Jeux à champ moyen. I –le cas stationnaire. Comptes Rendus Mathématique **343**, 619–625 (2006)
144. J.-M. Lasry, P.-L. Lions, Jeux à champ moyen. II –horizon fini et contrôle optimal. Comptes Rendus Mathématique **343**, 679–684 (2006)
145. J.-M. Lasry, P.-L. Lions, Mean field games. Jpn. J. Math. **2**, 229–260 (2007)
146. J.-M. Lasry, P.-L. Lions, Mean-field games with a major player. Comptes Rendus Mathematique **356**, 886–890 (2018)

147. H. Lavenant, F. Santambrogio, Optimal density evolution with congestion: l^∞ bounds via flow interchange techniques and applications to variational mean field games. Commun. Part. Differ. Equ. **43**, 1761–1802 (2018)
148. H. Lavenant, F. Santambrogio, New estimates on the regularity of the pressure in density-constrained mean field games. J. Lond. Math. Soc. **100**, 644–667 (2019)
149. P.-L. Lions, Cours au college de france, 2007–2012
150. P.-L. Lions, J.-M. Lasry, Large investor trading impacts on volatility, in *Paris-Princeton Lectures on Mathematical Finance 2004* (Springer, Berlin, 2007), pp. 173–190
151. P.-L. Lions, P.E. Souganidis, Homogenization of the backward-forward mean-field games systems in periodic environments (2019). Preprint arXiv:1909.01250
152. R.E. Lucas Jr, B. Moll, Knowledge growth and the allocation of time. J. Polit. Econ. **122**(1), 1–51 (2014)
153. M. MASOERO, On the long time convergence of potential MFG. Nonlinear Differ. Equ. Appl. **26**, 15 (2019)
154. G. Mazanti, F. Santambrogio, Minimal-time mean field games. Math. Models Methods Appl. Sci. **29**, 1413–1464 (2019)
155. A.R. Mészáros, F.J. Silva, On the variational formulation of some stationary second-order mean field games systems. SIAM J. Math. Anal. **50**, 1255–1277 (2018)
156. M. Nutz, A mean field game of optimal stopping. SIAM J. Control Optim. **56**, 1206–1221 (2018)
157. M. Nutz, J. San Martin, X. Tan, Convergence to the mean field game limit: a case study. Ann. Appl. Probab. **30**, 259–286 (2020)
158. C. Orrieri, A. Porretta, G. Savaré, A variational approach to the mean field planning problem. J. Funct. Anal. **277**, 1868–1957 (2019)
159. L. Overbeck, M. Rockner, B. Schmuland, An analytic approach to Fleming-Viot processes with interactive selection. Ann. Probab. **23**, 1–36 (1995)
160. G. Papanicolaou, L. Ryzhik, K. Velcheva, Travelling waves in a mean field learning model (2020). Preprint. arXiv:2002.06287
161. P. Pedregal, Optimization, relaxation and Young measures. Bull. Am. Math. Soc. **36**, 27–58 (1999)
162. S. Peng, Stochastic Hamilton–Jacobi–Bellman equations. SIAM J. Control Optim. **30**, 284–304 (1992)
163. A. Porretta, On the planning problem for a class of Mean Field Games. C. R. Acad. Sci. Paris, Ser. I **351**, 457–462 (2013)
164. A. Porretta, On the planning problem for the Mean Field Games system. Dyn. Games Appl. **4**, 231–256 (2014)
165. A. Porretta, Weak solutions to Fokker–Planck equations and mean field games. Arch. Ration. Mech. Anal. **216**, 1–62 (2015)
166. A. Porretta, On the weak theory for mean field games systems. Boll. Unione Mat. Ital. **10**, 411–439 (2017)
167. A. Porretta, On the turnpike property in mean field games. Minimax Theory Appl. **3**, 285–312 (2018)
168. A. Porretta, M. Ricciardi, Mean field games under invariance conditions for the state space. Commun. Part. Differ. Equ. **45**, 146–190 (2020)
169. A. Porretta, L. Rossi, Traveling waves for a nonlocal KPP equation and mean-field game models of knowledge diffusion. preprint arxiv:2010.10828v1
170. A. Porretta, E. Zuazua, Long time versus steady state optimal control. Siam J. Control Optim. **51**, 4242–4273 (2013)
171. S.T. Rachev, L. Rüschendorf, Mass Transportation Problems: Volume I: Theory, vol. 1 (Springer Science & Business Media, Berlin, 1998)
172. D. Revuz, M. Yor, *Continuous Martingales and Brownian Motion*, vol. 293 (Springer Science & Business Media, Berlin, 2013)
173. F. Santambrogio, A modest proposal for MFG with density constraints. Netw. Heterog. Media **7**, 337–347 (2012)

174. F. Santambrogio, *Optimal Transport for Applied Mathematicians* (Birkäuser, New York, NY, 2015), pp. 99–102
175. F. Santambrogio, Regularity via duality in calculus of variations and degenerate elliptic PDEs. J. Math. Anal. Appl. **457**, 1649–1674 (2018)
176. H. Spohn, *Large Scale Dynamics of Interacting Particles* (Springer Science & Business Media, Berlin, 2012)
177. A.-S. Sznitman, Topics in propagation of chaos, in *Ecole d'été de probabilités de Saint-Flour XIX—1989* (Springer, Berlin, 1991), pp. 165–251
178. C. Villani, *Topics in Optimal Transportation*, vol. 58 (American Mathematical Society, Providence, 2003)
179. C. Villani, *Optimal Transport: Old and New*, vol. 338 (Springer Science & Business Media, Berlin, 2008)
180. J. Yong, X.Y. Zhou, Stochastic controls: Hamiltonian systems and HJB equations, vol. 43 (Springer Science & Business Media, Berlin, 1999)
181. L.C. Young, *Lectures on Calculus of Variations and Optimal Control Theory* (W. B. Saunders, Philadelphia, 1969). Reprinted by Chelsea, 1980

Chapter 2
Lecture Notes on Variational Mean Field Games

Filippo Santambrogio

Abstract These lecture notes aim at giving the details presented in the short course (6 h) given in Cetraro, in the CIME School about MFG of June 2019. The topics which are covered concern first-order MFG with local couplings, and the main goal is to prove that minimizers of a suitably expressed global energy are equilibria in the sense that a.e. trajectory solves a control problem with a running cost depending on the density of all the agents. Both the case of a cost penalizing high densities and of an L^∞ constraint on the same densities are considered. The details of a construction to prove that minimizers actually define equilibria are presented under a boundedness assumption of the running cost, which is proven in the relevant cases.

2.1 Introduction and Modeling

The theory of Mean Field Games was introduced around 2006 at the same time by Lasry and Lions, [23–25], and by Huang et al. [21], in order to describe the evolution of a population of rational agents, each one choosing (or controlling) a path in a state space, according to some preferences which are affected by the presence of other agents nearby in a way physicists call mean-field effect. The evolution is described through a Nash equilibrium in a game with a continuum of players. This can be interpreted as a limit as $N \to \infty$ of a game with N indistinguishable players, each one having a negligible effect as $N \to \infty$ on the mean-field. The class of games we consider, called Mean Field Games (MFG for short), are very particular differential games: typically, in a differential game the role of the time variable is crucial since if a player decides to deviate from a given strategy (a notion which is at the basis of the Nash equilibrium definition),

F. Santambrogio (✉)
Institut Camille Jordan, Université Claude Bernard - Lyon 1, Villeurbanne, France

Institut Universitaire de France (IUF), Paris, France
e-mail: santambrogio@math.univ-lyon1.fr

© The Editor(s) (if applicable) and The Author(s), under exclusive license
to Springer Nature Switzerland AG 2020

P. Cardaliaguet, A. Porretta (eds.), *Mean Field Games*,
Lecture Notes in Mathematics 2281, https://doi.org/10.1007/978-3-030-59837-2_2

the other can react to this change, so that the choice of a strategy is usually not defined as the choice of a path, but of a function selecting a path according to the information the player has at each given time. Yet, when each player is considered as negligible, any deviation he/she performs will have no effect on the other players, so that they will not react. In this way we have a static game where the space of strategies is a space of paths. Because of indistinguishability, the main tool to describe such equlibria will be the use of measures on paths, and in this setting we will use the terminology of *Lagrangian equilibria*. In fluid mechanics, indeed, the Lagrangian formulation consists in "following" each particle and providing for each of them the corresponding trajectories. On the other hand, fluid mechanics also uses another language, the so-called Eulerian formulation, where certain quantities, and in particular the density and the velocity of the particles, are given as a function of time and space. MFG equilibria can also be described through a system of PDEs in *Eulerian variables*, where the key ingredients are the density ρ and the value function φ of the control problem solved by each player, the velocity $\mathbf{v}(t, x)$ of the agents at (t, x) being, by optimality, related to the gradient $\nabla\varphi(t, x)$.

The MFG theory is now studied by may scholars in many countries, with a quickly growing set of references. For a general overview of the theory, it is possible to refer to the 6-years course given by P.-L. Lions at Collège de France, for which videorecording is available in French [28] or to the lecture notes by P. Cardaliaguet [9], based on the same course. In the present lecture notes we will only be concerned with a sub-class of MFG, those which are deterministic, have a variational structure, and are in some sense *congestion games*, where the cost for an agent passing through a certain point depends, in an increasing way, on the density $\rho(t, x)$ at such a point. We will also see a variant of this class of problems where the penalization on ρ is replaced by a constraint on it (of the form $\rho(t, x) \leq 1$, for instance), which does not fit exactly this framework but shares most of the ideas and the properties. The topic of this course and these lecture notes were already presented in [5], so that there will be some superposition with such a survey paper, but in these notes we will focus on some more particular cases so as to be able to provide more technical details and proofs. Moreover, not all regularity results were available when [5] was written, and some proofs are simplified here.

2.1.1 A Coupled System of PDEs

Let us describe in a more precise way the simplest MFG models and the sub-class that we consider. First, we look at a population of agents moving inside a domain Ω (which can be a bounded domain in $\mathbb{R}^d$ or, for instance, the flat torus $\mathbb{T}^d :=$ $\mathbb{R}^d/\mathbb{Z}^d\dots$), and we suppose that every agent chooses his own trajectory solving a minimization problem

$$\min \int_0^T \left(\frac{|x'(t)|^2}{2} + h(t, x(t))\right) \, dt + \Psi(x(T)),$$

with given initial point $x(0)$. The mean-field effect will be modeled through the fact that the function $h(t, \cdot)$ depends on the density ρ_t of the agents at time t. The dependence of the cost on the velocity x' could of course be more general than a simple quadratic function, but in all these lecture notes we will focus on the quadratic case (some results that we present could be generalized, while for some parts of the analysis, in particular the regularity obtained via optimal transport methods, the use of the quadratic cost is important).

For the moment, we consider the evolution of the density ρ_t as an input, i.e. we suppose that agents know it. Hence, we can suppose the function h to be given, and we want to study the above optimization problem. The main tool to analyze it, coming from optimal control theory, is the value function. The value function φ is in this case defined via

$$\varphi(t_0, x_0) := \min \left\{ \int_{t_0}^{T} \left(\frac{|x'(t)|^2}{2} + h(t, x) \right) \, dt + \Psi(x(T)), \ x : [t_0, T] \to \Omega, x(t_0) = x_0 \right\}$$

(2.1)

and it has some important properties. First, it solves the *Hamilton–Jacobi equation*

$$\text{(HJ)} \qquad \begin{cases} -\partial_t \varphi + \frac{1}{2}|\nabla \varphi|^2 = h, \\ \varphi(T, x) = \Psi(x) \end{cases}$$

(in the viscosity sense, but we will not pay attention to this technicality, so far); second, the optimal trajectories $x(t)$ can be computed using φ, since they are the solutions of

$$x'(t) = -\nabla \varphi(t, x(t)).$$

Now if we call $\mathbf{v}$ the velocity field which advects the density ρ (which means that ρ is the density of a bunch of particles each following a trajectory $x(t)$ solving $x'(t) = \mathbf{v}_t(x(t))$), fluid mechanics tells us that the pair $(\rho, \mathbf{v})$ solves the *continuity equation*

$$\text{(CE)} \qquad \partial_t \rho + \nabla \cdot (\rho \mathbf{v}) = 0$$

in the weak sense, together with no-flux boundary conditions $\rho \mathbf{v} \cdot n = 0$, modeling the fact that no mass enters or exits Ω.

In MFG we look for an equilibrium in the sense of Nash equilibria: a configuration where no player would spontaneously decide to change his choice if he/she assumes that the choices of the others are fixed. This means that we can consider the densities ρ_t as an input, compute $h[\rho]$, then compute the optimal trajectories through the (HJ) equation, then the solution of (CE) and get new densities as an output: we have an equilibrium if and only if the output densities coincide with the input. This means solving the following coupled (HJ)+(CE) system: the function φ

solves (HJ) with a right-hand side depending on ρ, which on turn evolves according to (CE) with a velocity field depending on $\nabla \varphi(t, x)$.

$$
\begin{cases}
-\partial_t \varphi + \frac{|\nabla \varphi|^2}{2} = h[\rho] \\
\partial_t \rho - \nabla \cdot (\rho \nabla \varphi) = 0, \\
\varphi(T, x) = \Psi(x), \quad \rho(0, x) = \overline{\rho_0}(x).
\end{cases}
\tag{2.2}
$$

To be more general, it is possible to conside a stochastic case, where agents follow controlled stochastic differential equations of the form $dX_t = \alpha_t \, dt + \sqrt{2\nu} \, dW_t$ and minimize

$$
\mathbb{E}\left[\int_0^T \left(\frac{1}{2}\alpha_t^2 + h[\rho_t](X_t) \right) \, dt + \Psi(X(T)) \right].
$$

In this case, a Laplacian appears both in the (HJ) and in the (CE) equations:

$$
\begin{cases}
-\partial_t \varphi - \nu \Delta \varphi + \frac{|\nabla \varphi|^2}{2} - h[\rho] = 0 \\
\partial_t \rho - \nu \Delta \rho - \nabla \cdot (\rho \nabla \varphi) = 0.
\end{cases}
\tag{2.3}
$$

2.1.2 Questions and Difficulties

Let us be precise now about the kind of questions that one would like to attack, at the interface between mathematical analysis and modeling.

From the analysis and PDE point of view, the most natural questions to ask in MFG is the existence (and possibly the uniqueness and the regularity properties) of the solutions of systems like the above ones. This is an Eulerian question, as the objects which are involved, the density and the velocity, which is related to the value function, are defined for time-space points (t, x). This question can be intuitively attacked via fixed points methods: given ρ, compute $h[\rho]$, φ, then the solution of the evolution equation on ρ, thus getting a new density evolution $\tilde{\rho}$, and look for $\tilde{\rho} = \rho$. Yet, this requires strong continuity properties of this sequence of operators (and, by the way, uniqueness of those solutions if we want the operators to be univalued) which corresponds to uniqueness, regularity, and stability properties of the solutions of the corresponding PDEs. These properties are not always easy to get, but can be usually obtained when

- either we have $\nu > 0$ in System (2.3), i.e. the equations are parabolic and the regularization effect of the Laplacian provides the desired estimates (this applies quite easily to the present quadratic case, where a change-of-variable $u = e^{-\varphi/2}$ transforms the Hamilton–Jacobi equation of (2.3) into a linear parabolic equation; the general case is harder and require to use uniqueness and stability properties which are valid for the Fokker–Planck equations under milder regularity assumptions, and which have been recently proven in [36]);

- or the correspondence $\rho \mapsto h[\rho]$ is strongly regularizing (in particular this happens for non-local operators of the form $h[\rho](x) = \int \eta(x - y) \, d\rho(y)$ for a smooth kernel η). Indeed, if h is guaranteed to be smooth, then φ satisfies semiconcavity properties implying BV estimates on the drift $\nabla\varphi$; this, in turn, provides uniqueness and stability for the continuity equation thanks to the DiPerna Lions theory [16] (this is more or less the point of view presented in [9]).

One of the main interesting cases which is left out is the case where $\nu = 0$ and $h[\rho] = g \circ \rho$ (the local case, where h at a point directly depends only on ρ at the same point). Whenever g is an increasing function this is a very natural model to describe aversion to overcrowding and recalls in a striking way the models about Wardrop equilibria (see [14, 15, 44]).

From the point of view of modeling and game theory, the other natural question is to provide the existence of an equilibrium in the sense of finding which trajectories are followed by each players (or, since players are considered to be indistinguishable, just finding a measure on possible trajectories). This is on the contrary a Lagrangian question, as individual trajectories are involved. The unknown is then a probability measure on a suitable space of paths, which induces measures ρ_t at each instant of time. From these measures we deduce the function $h(t, \cdot)$, which is an ingredient for the optimization problem solved by every agent. The goal is then to choose such a probability on paths so that a.e. path is optimal for the cost built upon the function h which is induced by such probability. Again, there is a difficulty in the local case with $\nu = 0$. Indeed, if $\rho_t(\cdot)$ is just the density of a measure, it is defined only a.e. and so will be $h(t, \cdot)$. Hence, there will be no meaning in integrating it on a path, unless we choose a precise representative, which is a priori arbitrary unless ρ_t is continuous. Of course this difficulty does not exist whenever h is defined via convolution, and in many cases it can also be overcome in the local case for $\nu > 0$ since parabolic equations have a regularization effect and one can expect ρ_t to be smooth.

For both the Eulerian and the Lagrangian question, an answer comes from a variational interpretation: it happens that a solution to the equilibrium system (2.2) can be found by an overall minimization problem as first outlined in the seminal work by Lasry and Lions [24]. This allows to prove existence of a solution in a suitable sense, and the optimality conditions go in the direction of a Lagrangian equilibrium, as we will see in Sects. 2.2.2 and 2.2.3.

2.2 Variational Formulation

As we said, solutions to the equilibrium system (2.2) can be found by an overall minimization problem.

The description that we give below will be focused on the case $h[\rho](x) = V(x)+g(\rho(x))$, where we identify the measure ρ with its density w.r.t. the Lebesgue measure on Ω. The function $V : \Omega \to \mathbb{R}_+$ is a potential taking into account different local costs of different points in Ω.

For the variational formulation, we consider all the possible population evolutions, i.e. pairs $(\rho, \mathbf{v})$ satisfying $\partial_t \rho + \nabla \cdot (\rho \mathbf{v}) = 0$ (note that this is the Eulerian way of describing such a movement; in Sect. 2.2.2 we will see how to express it in a Lagrangian language) and we minimize the following energy

$$\mathcal{A}(\rho, \mathbf{v}) := \int_0^T \int_\Omega \left(\frac{1}{2} \rho_t |\mathbf{v}_t|^2 + \rho_t V + G(\rho_t) \right) \, \mathrm{d}x \, \mathrm{d}t + \int_\Omega \Psi \, \mathrm{d}\rho_T,$$

where G is the anti-derivative of g, i.e. $G'(s) = g(s)$ for $s \in \mathbb{R}^+$ with $G(0) = 0$. We fix by convention $G(s) = +\infty$ for $\rho < 0$. Note in particular that G is convex, as its derivative is the increasing function g.

The above minimization problem recalls, in particular when $V = 0$, the Benamou–Brenier dynamic formulation for optimal transport (see [4]). The main difference with the Benamou–Brenier problem is that here we add to the kinetic energy a congestion cost G; also note that usually in optimal transport the target measure ρ_T is fixed, and here it is part of the optimization (but this is not a crucial difference). Finally, note that the minimization of a Benamou–Brenier energy with a congestion cost was already present in [8] where the congestion term was used to model the motion of a crowd with panic.

As is often the case in congestion games, the quantity $\mathcal{A}(\rho, \mathbf{v})$ is not the total cost for all the agents. Indeed, the term $\int \int \frac{1}{2} \rho |\mathbf{v}|^2$ is exactly the total kinetic energy, and the last term $\int \Psi \, \mathrm{d}\rho_T$ is the total final cost, as well as the cost $\int V \, \mathrm{d}\rho_t$ exactly coincides with the total cost enduced by the potential V; yet, the term $\int G(\rho)$ is not the total congestion cost, which should be instead $\int \rho g(\rho)$. This means that the equilibrium minimizes an overall energy (we have what is called a potential game), but not the total cost; this gives rise to the so-called *price of anarchy*.

Another important point is the fact that the above minimization problem is convex, which was by the way the key idea of [4]. Indeed, the problem is not convex in the variables $(\rho, \mathbf{v})$, because of the product term $\rho |\mathbf{v}|^2$ in the functional and of the product $\rho \mathbf{v}$ in the differential constraint. But if one changes variable, defining $\mathbf{w} = \rho \mathbf{v}$ and using the variables $(\rho, \mathbf{w})$, then the constraint becomes linear and the functional convex. We will write $\bar{\mathcal{A}}(\rho, \mathbf{w})$ for the functional $\mathcal{A}(\rho, \mathbf{v})$ written in these variables. The important point for convexity is that the function

$$\mathbb{R} \times \mathbb{R}^d \ni (s, \mathbf{w}) \mapsto \begin{cases} \frac{|\mathbf{w}|^2}{2s} & \text{if } s > 0, \\ 0 & \text{if } (s, \mathbf{w}) = (0, 0), \\ +\infty & \text{otherwise} \end{cases}$$

is convex (and it is actually obtained as $\sup\{as + b \cdot \mathbf{w} \ : \ a + \frac{1}{2}|b|^2 \le 0\}$).

2.2.1   Convex Duality

In order to convince the reader of the connection between the minization of $\mathcal{A}(\rho, \mathbf{v})$ (or of $\bar{\mathcal{A}}(\rho, \mathbf{w})$) and the equilibrium system (2.2), we will use some formal argument from convex duality. A rigorous equivalence between optimizers and equilibria will be, instead, presented in the Lagrangian framework in Sect. 2.2.3.

In order to formally produce a dual problem to $\min \mathcal{A}$, we will use a min-max exchange procedure. First, we write the constraint $\partial_t \rho + \nabla \cdot (\rho \mathbf{v}) = 0$ in weak form, i.e.

$$\int_0^T \int_\Omega (\rho \partial_t \phi + \nabla \phi \cdot \rho \mathbf{v}) + \int_\Omega \phi_0 \overline{\rho_0} - \int_\Omega \phi_T \rho_T = 0 \qquad (2.4)$$

for every function $\phi \in C^1([0, T] \times \Omega)$ (note that we do not impose conditions on the values of ϕ on $\partial \Omega$, hence this is equivalent to completing (CE) with a no-flux boundary condition $\rho \mathbf{v} \cdot n = 0$). Equation (2.4) requires, in order to make sense, that we give a meaning at ρ_t for every instant of time t (and in particular for $t = T$), which is possible because whenever the kinetic term is finite then the curve ρ_t is a(n absolutely) continuous curve in the space of measures (continuous for the weak convergence, and absolutely continuous for the W_2 Wasserstein distance, see Sect. 2.4.1). However, we do not insist on this now, as the presentation stays quite formal.

Using (2.4), we can re-write our problem as

$$\min_{\rho, \mathbf{v}} \ \mathcal{A}(\rho, \mathbf{v}) + \sup_\phi \int_0^T \int_\Omega (\rho \partial_t \phi + \nabla \phi \cdot \rho \mathbf{v}) + \int_\Omega \phi_0 \overline{\rho_0} - \int_\Omega \phi_T \rho_T,$$

since the sup in ϕ takes value 0 if the constraint is satisfied and $+\infty$ if not. We now switch the inf and the sup and get

$$\sup_\phi \int_\Omega \phi_0 \overline{\rho_0} + \inf_{\rho, \mathbf{v}} \int_\Omega (\Psi - \phi_T) \rho_T + \int_0^T \int_\Omega \left(\frac{1}{2} \rho_t |\mathbf{v}_t|^2 + \rho_t V + G(\rho_t) + \rho \partial_t \phi + \nabla \phi \cdot \rho \mathbf{v} \right) dx \, dt.$$

First, we minimize w.r.t. $\mathbf{v}$, thus obtaining $\mathbf{v} = -\nabla \phi$ (on $\{\rho_t > 0\}$) and we replace $\frac{1}{2} \rho |\mathbf{v}|^2 + \nabla \phi \cdot \rho \mathbf{v}$ with $-\frac{1}{2} \rho |\nabla \phi|^2$. Then we get, in the double integral,

$$\inf_\rho \{ G(\rho) - \rho(-V - \partial_t \phi + \frac{1}{2} |\nabla \phi|^2) \} = -\sup_\rho \{ p\rho - G(\rho) \} = -G^*(p),$$

where we set $p := -V - \partial_t \phi + \frac{1}{2} |\nabla \phi|^2$ and G^* is defined as the Legendre transform of G. Then, we observe that the minimization in the final cost simply gives as a result 0 if $\Psi \geq \phi_T$ (since the minimization is only performed among positive ρ_T) and $-\infty$

otherwise. Hence we obtain a dual problem of the form

$$\sup\left\{-\mathcal{B}(\phi, p) := \int_\Omega \phi_0\overline{\rho_0} - \int_0^T \int_\Omega G^*(p) \;:\; \phi_T \leq \Psi, \; -\partial_t\phi + \frac{1}{2}|\nabla\phi|^2 = V + p\right\}.$$

Note that the condition $G(s) = +\infty$ for $s < 0$ implies $G^*(p) = 0$ for $p \leq 0$. This in particular means that in the above maximization problem one can suppose $p \geq 0$ (indeed, replacing p with p_+ does not change the G^* part, but improves the value of ϕ_0, considered as a function depending on p). The choice of using two variables (ϕ, p) connected by a PDE constraint instead of only ϕ is purely conventional, and it allows for a dual problem which has a sort of symmetry w.r.t. the primal one. Also the choice of the sign is conventional and due to the computation that we will perform later (in particular in Sect. 2.4).

Now, standard arguments in convex duality would allow to say that optimal pairs $(\rho, \mathbf{v})$ are obtained by looking at saddle points $((\rho, \mathbf{v}), (\phi, p))$, provided that there is no duality gap between the primal and the dual problems, and that both problems admit a solution. This would mean that, whenever $(\rho, \mathbf{v})$ minimizes $\mathcal{A}$, then there exists a pair (ϕ, p), solution of the dual problem, such that

- $\mathbf{v}$ minimizes $\frac{1}{2}\rho|\mathbf{v}|^2 + \nabla\phi \cdot \rho\mathbf{v}$, i.e. $\mathbf{v} = -\nabla\phi$ ρ-a.e. This gives (CE): $\partial_t\rho - \nabla \cdot (\rho\nabla\phi) = 0$.
- ρ minimizes $G(\rho) - p\rho$, i.e. $g(\rho) = p$ if $\rho > 0$ or $g(\rho) \geq p$ if $\rho = 0$ (in particular, when we have $g(0) = 0$, we can write $g(\rho) = p_+$); this gives (HJ): $-\partial_t\phi + \frac{1}{2}|\nabla\phi|^2 = V + g(\rho)$ on $\{\rho > 0\}$ (as the reader can see, there are some subtleties where the mass ρ vanishes;).
- ρ_T minimizes $(\Psi - \phi_T)\rho_T$ among $\rho_T \geq 0$. But this is not a condition on ρ_T, but rather on ϕ_T: we must have $\phi_T = \Psi$ on $\{\rho_T > 0\}$, otherwise there is no minimizer. This gives the final condition in (HJ).

This provides an informal justification for the equivalence between the equilibrium and the global optimization. It is only informal because we have not discussed whether we have or not duality gaps and whether or not the maximization in (ϕ, p) admits a solution. Moreover, even once these issues are clarified, what we will get will only be a very weak solution to the coupled system (CE)+(HJ). Nothing guaranteees that this solution actually encodes the individual minimization problem of each agent. This will be clarified in Sect. 2.2.3 where a Lagrangian point of view will be presented.

However, let us first give the duality result which can be obtained from a suitable application of Fenchel-Rockafellar's Theorem, and for which details are presented, in much wider generality, in [11].

Theorem 2.2.1 Set $\mathcal{D} = \{(\phi, p) \in C^1([0, T] \times \Omega) \times C^0([0, T] \times \Omega) \;:\; -\partial_t\phi + \frac{1}{2}|\nabla\phi|^2 = V + p, \phi_T \leq \Psi\}$ and $\mathcal{P} = \{(\rho, \mathbf{v}) \;:\; \partial_t\rho + \nabla \cdot (\rho\mathbf{v}) = 0, \rho_0 = \overline{\rho_0}\}$, where the continuity equation on $(\rho, \mathbf{v})$ is satisfied in the sense of (2.4) and $(\rho_t)_t$ is *a continuous curve of probability measures on Ω, with $\mathbf{v}_t \in L^2(\rho_t)$ for every t. We then have*

$$\min\{\mathcal{A}(\rho, \mathbf{v}) \,:\, (\rho, \mathbf{v}) \in \mathcal{P}\} = \sup\{-\mathcal{B}(\phi, p) \,:\, (\phi, p) \in \mathcal{D}\}.$$

Note that in the above theorem we called $\mathcal{D}$ and $\mathcal{P}$ the domains in the dual and primal problems respectively, with the standard confusion between dual and primal (officially it is the problem on measures which should be the dual of that on functions, and not viceversa) which is often done when we prefer to call "primal" the first problem that we meet and which is the main object of our analysis.

It is important to observe that the above theorem does not require the assumption on the growth rate of the Hamiltonian and of the congestion function G, which translate into this quadratic case into "$G(s) \leq C(s^q + 1)$ for an exponent $q < 1 + 2/d$", which is present in the paper [11]. This restriction was required in order to find a suitable relaxed solution to the dual problem, which has in general no solution in $\mathcal{D}$. This result is the object of the following theorem, where we omit this condition on q, since it has been later removed in more recent papers. Indeed, [12] was the first paper where this assumption disappears, for second-order MFG with possibly degenerate diffusion (which include the first-order case; also refer to [19], where duality was used for regularity purposes, which explicitly focuses on the first-order case).

Theorem 2.2.2 *Set $\tilde{\mathcal{D}} = \{(\phi, p) \in BV([0, T] \times \Omega) \times \mathcal{M}([0, T] \times \Omega) \,:\, -\partial_t\phi + \frac{1}{2}|\nabla\phi|^2 \leq V + p, \phi_T \leq \Psi\}$. We then have*

$$\min\{\mathcal{A}(\rho, \mathbf{v}) \,:\, (\rho, \mathbf{v}) \in \mathcal{P}\} = \max\{-\mathcal{B}(\phi, p) \,:\, (\phi, p) \in \tilde{\mathcal{D}}\}$$

and the max on the right hand side is attained.

A disambiguation is needed, when speaking of BV functions, about the final condition $\phi_T \leq \Psi$. Indeed, a BV function could have a jump exactly at $t = T$ and hence its pointwise value at the final time is not well-defined. The important point is that, if $\phi(T^-)$ does not satisfy the required inequality, but $\phi(T)$ is required to satisfy it, then a jump is needed, i.e. a singular part of the measure p concentrated at $\{t = T\}$, and this part will be considered in the dual cost (this is particularly important when the cost G^* in the dual problem has linear growth, and singular parts are allowed, which will be the case for the density-constrained case of Sect. 2.5).

However, most of these notes will not make use of this refined duality, both because we want to consider cases where the growth rate of G does not satisfy this inequality and because we will need to use (in Sect. 2.3) smooth test functions and apply the duality. For this sake, it will be more convenient to choose almost-maximizers $(\phi, p) \in C^1 \times C^0$ rather than maximizers with limited regularity.

We finish this section with a last variant, inspired by the crowd motion model of [31]. We would like to consider a variant where, instead of adding a penalization $g(\rho)$, we impose a capacity constraint $\rho \leq 1$. How to give a proper definition of equilibrium? A first, naive, idea, would be the following: when $(\rho_t)_t$ is given, every agent minimizes his own cost paying attention to the constraint $\rho_t(x(t)) \leq 1$. But if ρ already satisfies $\rho \leq 1$, then the choice of only one extra agent will not violate the constraint (since we have a non-atomic game), and the constraint becomes empty. As already pointed out in [38], this cannot be the correct definition.

In [38] an alternative model is formally proposed, based on the effect of the gradient of the pressure on the motion of the agents, but this model is not variational, and no solution has been proven to exist in general in its local and deterministic form.

A different approach to the question of density constraints in MFG was presented in [13]: the idea is to start from the the the variational problem

$$\min \left\{ \int_0^T \int_\Omega \left(\frac{1}{2}|\mathbf{v}_t|^2 + V \right) \, d\rho_t + \int_\Omega \Psi \, d\rho_T \ : \ \rho \leq 1 \right\}.$$

This means that we use $G = I_{[0,1]}$, i.e. $G(s) = 0$ for $s \in [0, 1]$ and $+\infty$ otherwise. The dual problem can be computed and we obtain

$$\sup \left\{ \int_\Omega \phi_0 \overline{\rho_0} - \int_0^T \int_\Omega p_+ \ : \ \phi_T \leq \Psi, \ -\partial_t \phi + \frac{1}{2}|\nabla \phi|^2 = V + p \right\}$$

(note that this problem is also obtained as the limit $m \to \infty$ of $g(\rho) = \rho^m$; indeed the functional $\frac{1}{m+1} \int \rho^{m+1}$ Γ-converges to the constraint $\rho \leq 1$ as $m \to \infty$).

By looking at the primal-dual optimality conditions, we get again $\mathbf{v} = -\nabla \phi$ and $\phi_T = \Psi$, but the optimality of ρ means

$$0 \leq \rho < 1 \Rightarrow p = 0, \quad \rho = 1 \Rightarrow p \geq 0.$$

This gives the following MFG system

$$\begin{cases} -\partial_t \varphi + \frac{|\nabla \varphi|^2}{2} = V + p, \\ \partial_t \rho - \nabla \cdot (\rho \nabla \varphi) = 0, \\ \varphi(T, x) = \Psi(x), \quad \rho(0, x) = \overline{\rho_0}(x), \\ p \geq 0, \ \rho \leq 1, \ p(1 - \rho) = 0. \end{cases} \tag{2.5}$$

It is important to understand that p is a priori just a measure on $[0, T] \times \Omega$, since it has a sign (and distributions with a sign are measures) and it is penalized in terms of its L^1 norm. Indeed, even if Theorem 2.2.2 is stated exactly by taking p in the space of measures, in general according to the function G^* it is possible to obtain extra summability (if G has growth.of order q then one obtains $p \in L^{q'}$). Here, instead, since G^* is linear we do not obtain more than measure bounds. This means that φ is not better than a BV function, and in particular it could have jumps. From the first equation in System (2.5) and the positivity of p we see that φ could have a jump at time $t = T$ in the sense that $\varphi(T^-) > \varphi(T) = \Psi$. Hence, an alternative way to write the same system is to remove the possible singular part of p concentrated on $t = T$ but consider as a final value for φ the value that it takes at T^-. In this way, we can re-write System (2.5) as

$$
\begin{cases}
-\partial_t \varphi + \frac{|\nabla \varphi|^2}{2} = V + p, \\
\partial_t \rho - \nabla \cdot (\rho \nabla \varphi) = 0, \\
\varphi(T, x) = \Psi(x) + P(x), \quad \rho(0, x) = \overline{\rho_0}(x), \\
p \geq 0, \ \rho \leq 1, \ p(1 - \rho) = 0, \\
P \geq 0, \ P(1 - \rho_T) = 0.
\end{cases}
\tag{2.6}
$$

Formally, by looking back at the relation between (HJ) and optimal trajectories, we can guess that each agent solves

$$
\min \int_0^T \left(\frac{|x'(t)|^2}{2} + h(t, x(t)) \right) dt + \tilde{\Psi}(x(T)),
\tag{2.7}
$$

where $h = p + V$ and $\tilde{\Psi} = \Psi + P$. Here p and P are pressures arising from the incompressibility constraint $\rho \leq 1$ and only present in the saturated zone $\{\rho = 1\}$, but they finally act as prices paid by the agents to travel through saturated regions. From the economical point of view this is meaningful: due to a capacity constraint, the most attractive regions develop a positive price to be paid to pass through them, and this price is such that, if the agents keep it into account in their choices, then their mass distribution will indeed satisfy the capacity constraints.

This problem of course presents the same difficulties of the case where congestion is penalized and not constrained: what does it mean to integrate p on a path if p is only a measure? We will see later on a technique to get rid of this difficulty, following an idea by Ambrosio and Figalli [2] for applications to the incompressible Euler equation, but this techniques requires at least that p is a sufficiently integrable function. In these notes, we will present Ambrosio and Figalli's ideas in the case where h is L^∞, insisting on he simplification that it brings, and in Sect. 2.5 we will provide indeed an L^∞ regularity result on both p and P. In the original paper on density-constrained MFG, [13], the L^∞ regularity result on p was not available, and suitable regularity results of the form $p \in L_t^2 BV_x$ were proven via a technique similar to that used in Sect. 2.3 of these notes. Since we have $L_t^2 BV_x \subset L_t^2 L_x^{d/(d-1)}$, this BV regularity result was enough to apply, at least to a certain extent, the theory developed in [2].

2.2.2 Lagrangian Formulation

We present now an alternative point of view for the overall minimization problem presented in the previous sections. As far as now, we only looked at an Eulerian point of view, where the motion of the population is described by means of its density ρ and of its velocity field $\mathbf{v}$. The Lagrangian point of view would be, instead, to describe the motion by describing the trajectory of each agent. Since the agents are supposed to be indistinguishable, then we just need to determine, for each possible trajectory, the number of agents following it (and not their names...); this means looking at a measure on the set of possible paths.

Set $C = H^1([0, T]; \Omega)$; this will be the space of possible paths that we use. In general, absolutely continuous paths would be the good choice, but we can restrict our attention to H^1 paths because of the kinetic energy term that we have in our minimization. We define the evaluation maps $e_t : C \to \Omega$, given for every $t \in [0, T]$ by $e_t(\omega) = \omega(t)$. Also, we define the kinetic energy functional $K : C \to \mathbb{R}$ given by

$$K(\omega) = \frac{1}{2} \int_0^T |\omega'|^2(t) \, dt.$$

We endow the space C with the uniform convergence (and not the strong H^1 convergence, so that we have compactness of the sublevel sets of K). For notational simplicity, we will also often write K_Ψ for the kinetic energy augmented by a final cost: $K_\Psi(\omega) := K(\omega) + \Psi(\omega(T))$; similarly, we will denote by $K_{\Psi,h}$ the same quantity when also a running cost is included: $K_{\Psi,h}(\omega) := K_\Psi(\omega) + \int_0^T h(t, \omega(t)) \, dt$.

Proposition 2.2.3 *Suppose* $(\rho, \mathbf{v})$ *satisfies the continuity equation* $\partial_t \rho + \nabla \cdot (\rho \mathbf{v}) = 0$ *and* $\int_0^T \int_\Omega \rho |\mathbf{v}|^2 < \infty$. *Then there exist a representative of* ρ *such that* $t \mapsto \rho_t$ *is weakly continuous, and a probability measure* $Q \in \mathcal{P}(C)$ *such that* $\rho_t = (e_t)_\# Q$ *and*

$$\int_C K(\omega) \, dQ(\omega) \leq \frac{1}{2} \int_0^T \int_\Omega \rho |\mathbf{v}|^2.$$

Conversely, if we have $\rho_t = (e_t)_\# Q$ *for a probability measure* $Q \in \mathcal{P}(C)$ *with* $\int_C K(\omega) \, dQ(\omega) < \infty$, *then* $t \mapsto \rho_t$ *is weakly continuous and there exists a time-dependent family of vector fields* $\mathbf{v}_t \in L^2(\rho_t)$ *such that* $\partial_t \rho + \nabla \cdot (\rho \mathbf{v}) = 0$ *and*

$$\frac{1}{2} \int_0^T \int_\Omega \rho |\mathbf{v}|^2 \leq \int_C K(\omega) \, dQ(\omega).$$

The above proposition comes from optimal transport theory and we will discuss a more refined version of it in Sect. 2.4. Its proof can be found combining, for instance, Theorems 5.14 and 5.31 in [39]. It allows to re-write the minimization problem

$$\min \{ \mathcal{A}(\rho, \mathbf{v}) \ : \ \partial_t \rho + \nabla \cdot (\rho \mathbf{v}) = 0 \},$$

in the following form:

$$
\min\left\{ J(Q) := \int_C K \, dQ + \int_0^T \mathcal{G}((e_t)_\# Q) \, dt + \int_\Omega \Psi \, d(e_T)_\# Q, \ \ Q \in \mathcal{P}(C), \ (e_0)_\# Q = \overline{\rho_0} \right\},
$$
(2.8)

where $\mathcal{G} : \mathcal{P}(\Omega) \to \overline{\mathbb{R}}$ is defined through

$$
\mathcal{G}(\rho) = \begin{cases} \int (V(x)\rho(x) + G(\rho(x))) \, dx & \text{if } \rho \ll \mathcal{L}^d, \\ +\infty & \text{otherwise.} \end{cases}
$$

The functional $\mathcal{G}$ is a typical local functional defined on measures (see [6]). It is lower-semicontinuous w.r.t. weak convergence of probability measures provided $\lim_{s \to \infty} G(s)/s = +\infty$ (which is the same as $\lim_{s \to \infty} g(s) = +\infty$), see, for instance, Proposition 7.7 in [39].

Under these assumptions, it is easy to prove, by standard semicontinuity arguments in the space $\mathcal{P}(C)$, that a minimizer of (2.8) exists. We summarize this fact, together with the corresponding optimality conditions, in the next proposition. The optimality conditions are obtained by standard convex perturbations: if $\bar{Q}$ is an optimizer and Q a competitor with finite energy, then one sets $Q_\varepsilon := (1-\varepsilon)\bar{Q} + \varepsilon Q$ and differentiates the cost w.r.t. ε at $\varepsilon = 0$. The idea is just that a point optimizes a convex functional on a convex set if and only if it optimizes its linearization around itself.

Proposition 2.2.4 *Suppose that Ω is compact, that G is a convex and superlinear function, and that V and Ψ are continuous functions on Ω. Then Problem (2.8) admits a solution $\bar{Q}$.*

Moreover, $\bar{Q}$ is a solution if and only if for any other competitor $Q \in \mathcal{P}(C)$ with $J(Q) < +\infty$ with $(e_0)_\# Q = \overline{\rho_0}$ we have

$$
J_h(Q) \geq J_h(\bar{Q}),
$$

where $J_{\Psi,h}$ is the linear functional

$$
J_{\Psi,h}(Q) = \int K \, dQ + \int_0^T \int_\Omega h(t, x) \, d(e_t)_\# Q + \int_\Omega \Psi \, d(e_T)_\# Q,
$$

the function h being defined through $\rho_t = (e_t)_\# \bar{Q}$ and $h(t, x) = V(x) + g(\rho_t(x))$.

Remark 2.1 The above optimality condition and its interpretation in terms of equilibria (see below), as well as the definition of the functional via an antiderivative, strongly recall the setting of *continuous Wardrop equilibria*, studied in [15] (see also [14] for a survey of the theory). Indeed, in [15] a traffic intensity i_Q (a positive measure on Ω) is associated with each measure Q on C, and we define a weighted length on curves ω using i_Q as a weighting factor. We then prove that the measure Q which minimizes a suitable functional minimizes its linearization, which in turn implies that the same Q is concentrated on curves which are geodesic for this weighted length, depending on Q itself. Besides some technical details about the precise mathematical form of the functionals, the main difference between Wardrop equilibria (which are traditionally studied in a discrete framework on networks, see [44]) and MFG is the fact that Wardrop's setting is static: in such a traffic notion we consider a continuous traffic flow, where some mass is constantly injected somewhere in the domain, and at the same time constantly absorbed somewhere else (see Chapter 4 of [39] for other models of this form).

We now consider the functional $J_{\Psi,h}$. Note that the function h is obtained from the densities ρ_t, which means that it is well-defined only a.e. However, the integral $\int_0^T \int_\Omega h(t,x) \, d(e_t)_\# Q$ is well-defined and does not depend on the representative of h, since $J(Q) < +\infty$ implies that the measures $(e_t)_\# Q$ are absolutely continuous for a.e. t. Hence, this functional is well-defined for $h \geq 0$ measurable.

Formally, we can also write

$$\int_0^T \int_\Omega h(t,x) \, d(e_t)_\# Q = \int_C dQ \int_0^T h(t, \omega(t)) \, dt$$

and hence we get

$$J_{\Psi,h}(Q) = \int_C dQ(\omega) \left(K(\omega) + \int_0^T h(t, \omega(t)) \, dt + \Psi(\omega(T)) \right) = \int_C K_{\Psi,h} \, dQ.$$

It is then tempting to interpret the optimality conditions on $\bar{Q}$ stated in Proposition 2.2.4 by considering that they can only be satisfied if $\bar{Q}$-a.e. curve ω satisfies

$$K(\omega) + \int_0^T h(t, \omega(t)) \, dt + \Psi(\omega(T)) \leq K(\tilde{\omega}) + \int_0^T h(t, \tilde{\omega}(t)) \, dt + \Psi(\tilde{\omega}(T)) \quad \text{for every } \tilde{\omega} \text{ s.t. } \tilde{\omega}(0) = \omega(0).$$

$$(2.9)$$

This would be exactly the equilibrium condition in the MFG. Indeed, the MFG equilibrium condition can be expressed in Lagrangian language in the following way: find Q such that, if we define $\rho_t = (e_t)_\# \bar{Q}$ and $h(t,x) = V(x) + g(\rho_t(x))$, then Q is concentrated on minimizers of $K_{\Psi,h}$ for fixed initial point.

Yet, there are two objections to this way of arguing. The first concerns the fact that the functional $K_{\Psi,h}$ does indeed depend on the representative of h that we choose and it looks suspicious that such an equilibrium statement could be true independently of the choice of the representative. Moreover, the idea behind the optimality in (2.9) would be to choose a measure Q concentrated on optimal, or almost optimal, curves starting from each point, and there is no guarantee that such a measure Q satisfies $J(Q) < +\infty$.

The approach that we present in Sect. 2.2.3 below, due to Ambrosio and Figalli [2] and first applied to MFG in [13] for the case of MFG with density constraints, is a way to rigorously overcome these difficulties. The goal is to find a suitably chosen representative $\hat{h}$ of h so that we can prove that if $\bar{Q}$ minimizes $J_{\Psi,h}$, then it is concentrated on curves minimizing $K_{\Psi,\hat{h}}$. We develop here this theory under the assumption $h \in L^\infty$, while the original proofs were more general, but required some technicalities that we will briefly address in a comment. We will explain in which points the L^∞ assumption allows to obtain cleaner and more powerful results. Moreover, we insist that we are allowed to stick to this more restrictive setting because we will see, in Sects 2.4 and 2.5, that we do have $h \in L^\infty$ in the cases of interest for us.

2.2.3 Optimality Conditions on the Level of Single Agent Trajectories

In this section we consider a measurable function $h : [0, T] \times \Omega \to \mathbb{R}$ and we suppose that h is upper-bounded by a constant H_0, i.e. $h \leq H_0$ a.e. As far as lower bounds are concerned, all the section is written supposing $h \in L^1([0, T] \times \Omega)$, but it is not difficult to adapt it to the case where h (or, rather, its negative part) is only a measure. Let us define then

$$h_r(t, x) := \fint_{B(x,r)} h(t, y)\, \mathrm{d}y \quad \text{if } B(x, r) \subset \Omega$$

and then

$$\hat{h}(t, x) := \begin{cases} \limsup_{r \to 0} h_r(t, x) & \text{if } x \notin \partial\Omega, \\ H_0 & \text{if } x \in \partial\Omega. \end{cases}$$

First, we observe that $\hat{h}$ is a representative of h, in the sense that we have $h = \hat{h}$ a.e. (in the case where h is a measure then h_r is defined as $h_t(B(x, r))/\mathcal{L}^d(B(x, r))$ and $\hat{h}$ is a representative of the absolutely continuous part of h). Indeed, a.e. point in Ω is a Lebesgue point for h, so that the above lim sup is indeed a limit and equal to h, and the boundary where the definition is not given as a limsup is supposed to be negligible. In some sense, we will obtain the desired result by writing estimates involving h_r and passing to the limit as $r \to 0$.

Proposition 2.2.5 *Suppose that $\bar{Q}$ minimizes $J_{\Psi,h}$ among measures with $J(Q) < +\infty$ and suppose that G is a convex function with polynomial growth, that Ψ is a continuous function and that Ω is a smooth domain. Define $\hat{h}$ as above and suppose that $(e_t)_{\#}Q$ is absolutely continuous for a.e. t. Then $\bar{Q}$ is concentrated on curves ω such that*

$$K_{\Psi,\hat{h}}(\omega) \leq K_{\Psi,\hat{h}}(\tilde{\omega}) \quad \text{for every } \tilde{\omega} \text{ s.t. } \tilde{\omega}(0) = \omega(0).$$

Proof The proof is an adaptation of those proposed in [2, 13].

Consider a countable set $D \subset H_\diamond^1([0, T])$, where $H_\diamond^1([0, T])$ is the Hilbert space of H^1 functions on $[0, T]$, valued in $\mathbb{R}^d$, and vanishing at $t = 0$ (but not necessarily at $t = T$), dense in $H_\diamond^1([0, T])$ for the H^1 norm. Also consider a curve $\gamma \in D$, a vector $y \in B(0, 1) \subset \mathbb{R}^d$, a number $r > 0$, and a cut-off function $\eta \in C^1([0, T])$, with $\eta(0) = 0$, and $\eta > 0$ on $(0, T]$, with $\eta(T) = 1$. Consider a Borel subset $E \subset C$, with $E \subset \{\omega : \omega(t) + \gamma(t) + B(0, r) \subset \Omega$ for every $t\}$ and define a map $S : C \to C$ as follows

$$S(\omega) = \begin{cases} \omega + \gamma + r\eta y & \text{if } \omega \in E \\ \omega & \text{if } \omega \notin E. \end{cases}$$

Defining $Q = S_\# \bar{Q}$ we can easily see that we have $J(Q) < +\infty$ (we use here the polynomial growth of G, since the density of $(e_t)_\# Q$ can be decomposed as the sum of two densities with finite value for G, and we need a bound for the sum, which is not available, for instance, for convex functions with exponential growth).

Comparing $J_{\Psi,h}(Q)$ to $J_{\Psi,h}(\bar{Q})$ and erasing, by linearity, the common terms (those coming from the integration on E^c) we get

$$\int_E K_\Psi(S(\omega)) \, d\bar{Q}(\omega) + \iint h(t, \cdot + \gamma(t) + r\eta(t)y) \, d(e_t)_\#(\bar{Q}\mathbb{1}_E) \geq \int_E K_\Psi(\omega) \, d\bar{Q}(\omega) + \iint h(t, \cdot) \, d(e_t)_\#(\bar{Q}\mathbb{1}_E).$$

The first parameter we get rid of is the parameter y, as we take the average among possible $y \in B(0, 1)$. It is important to note that we have

$$\fint_{B(0,1)} K(\omega + r\eta y) \, dy = K(\omega) + O(r^2)$$

(by symmetry, there is no first-order term in r, even if this is not important and we would only need terms tending to 0 as $r \to 0$, independently of their order) and to use the definition of h_r (and, by analogy, of Ψ_r) in order to obtain

$$\int_E \left(K_{\Psi_r}(\omega + \gamma) + O(r^2) \right) d\bar{Q}(\omega) + \iint h_{r\eta(t)}(t, \cdot + \gamma(t)) \, d(e_t)_\#(\bar{Q}\mathbb{1}_E) \geq$$

$$\geq \int_E K_\Psi(\omega) \, d\bar{Q}(\omega) + \iint h(t, \cdot) \, d(e_t)_\#(\bar{Q}\mathbb{1}_E).$$

Now, we observe that $h = \hat{h}$ a.e. together with $(e_t)_\# \bar{Q} \ll \mathcal{L}^d$ ($\mathcal{L}^d$ being the Lebesgue measure) allow to replace, in the right hand side, h with $\hat{h}$. Moreover, we rewrite some terms using the following equalities

$$\iint h_{r\eta(t)}(t, \cdot + \gamma(t)) \, d(e_t)_\#(\bar{Q}\mathbb{1}_E) = \int_E d\bar{Q}(\omega) \int h_{r\eta(t)}(t, \omega(t) + \gamma(t)) \, dt$$

$$\iint \hat{h}(t, \cdot) \, d(e_t)_\#(\bar{Q}\mathbb{1}_E) = \int_E d\bar{Q}(\omega) \int \hat{h}(t, \omega(t)) \, dt$$

We then use the arbitrariness of E, thus obtaining the following fact: for $\bar{Q}$ a.e. ω s.t. $\omega(t) + \gamma(t) + B(0, r) \subset \Omega$ for every t we have

$$K_{\Psi_r}(\omega + \gamma) + O(r^2) + \int_0^T h_{r\eta(t)}(t, \omega(t) + \gamma(t)) \, dt \geq K_{\Psi}(\omega) + \int_0^T \hat{h}(t, \omega(t)) \, dt.$$

This result is true for a.e. curve for fixed γ, while we would like to obtain inequalities which are valid on a full-measure set which is the same for every γ. This explains the use of the dense set D. On the same full-measure set this inequality is true for every $\gamma \in D$, since D is countable. Then, using the density of D and the continuity, for fixed $r > 0$, of all the quantities on the left-hand side, we obtain the following: for $\bar{Q}$ a.e. ω and for every $\gamma \in H_\diamond^1([0, T])$ s.t. $\omega(t) + \gamma(t) + B(0, 2r) \subset \Omega$ for every t we have

$$K_{\Psi_r}(\omega + \gamma) + O(r^2) + \int_0^T h_{r\eta(t)}(t, \omega(t) + \gamma(t)) \, dt \geq K_{\Psi}(\omega) + \int_0^T \hat{h}(t, \omega(t)) \, dt.$$

$$(2.10)$$

In order to obtain this result every $\gamma \in H_\diamond^1([0, T])$ is approximated by a sequence $\gamma_k \in D$, with both H^1 and uniform convergence, so that we have $|\gamma_k - \gamma| \leq r$ (which explains the condition $\omega(t) + \gamma(t) + B(0, 2r) \subset \Omega$ with a different radius now): the kinetic term $K(\omega + \gamma_k)$ passes to the limit because of strong H^1 convergence, while the integral term passes to the limit via dominated convergence (if h is bounded from above and below, otherwise we use a Fatou's lemma with a limsup, since we have at least $h \leq H_0$), since each function $h_{r\eta(t)}(t, \cdot)$ is continuous, and they are all bounded above by a same constant. This is a first point where we use the L^∞ upper bound $h \leq H_0$. Indeed, if we do not have a suitable bound on h, the L^∞ norm of $h_{r\eta(t)}$ could explose as $t \to 0$ since $\eta(t) \to 0$ (and, unfortunately, it is not possible in general to guarantee integrablity in time of this bound if we want $\eta \in H^1$).

Inequality (2.10) is true for fixed $r > 0$, but taking a countable sequence tending to 0 we can pass to the limit as $r \to 0$ on a full-measure set, thus obtaining the following: for $\bar{Q}$ a.e. ω and for every $\gamma \in H_\diamond^1([0, T])$ s.t. $\omega(t) + \gamma(t) \in \Omega$ for every t we have

$$K_{\Psi}(\omega + \gamma) + \int_0^T \hat{h}(t, \omega(t) + \gamma(t)) \, dt \geq K_{\Psi}(\omega) + \int_0^T \hat{h}(t, \omega(t)) \, dt.$$

Also this limit uses $h \leq H_0$ as an assumption to apply Fatou's lemma with limsup (we need to upper bound the terms with $h_{r\eta(t)}$). Of course some integrability on the curve of the maximal function of h would be enough, but this is a much trickier condition (see below in Remark 2.3). Note that on the left hand side we used the continuity of Ψ.

This shows optimality of a.e. ω compared to every curve lying in the interior of the domain Ω. In order to handle curves touching $\partial \Omega$, let us take a family of maps $\zeta_\delta : \Omega \to \mathring{\Omega}$ with the following properties: $\mathrm{Lip}(\zeta_\delta) \to 1$ as $\delta \to 0$, $|\zeta_\delta(x) - x| \leq C\delta$ for every $x \in \Omega$, and $\zeta_\delta(x) = x$ if $d(x, \partial \Omega) \geq \delta$. We just observe now that, for a given curve $\omega : [0, T] \to \Omega$, we have

$$K_{\Psi, \hat{h}}(\zeta_\delta \circ \omega) \leq \mathrm{Lip}(\zeta_\delta)^2 K(\omega) + \Psi(\omega(T)) + \int_0^T \hat{h}(t, \omega(t)) \, dt$$

$$+ |\Psi(\zeta_\delta(\omega(T)) - \Psi(\omega(T)| + H_0|\{t : 0 < d(\omega(t), \partial \Omega) < \delta\}| \to K_{\Psi, \hat{h}}(\omega).$$

Again we used $h \leq H_0 < \infty$. As a result, we obtain that $\bar{Q}$-a.e. curve ω optimizes $K_{\Psi, \hat{h}}$ in the class of H^1 curves staying in Ω and sharing the same starting point. $\square$

Remark 2.2 The proof can be easily adapted to the case where the function Ψ is not continuous but only bounded, but we need in this case to suppose that $(e_T)_\# \bar{Q}$ is absolutely continuous. It is then possible to treat Ψ exactly as h, replacing it with its representative $\hat{\Psi}$. This will be useful in the density-constrained case where Ψ is replaced by a new function $\Psi + P$.

Remark 2.3 Both in [2] and [13] h is not required to be bounded, but the statement is slightly different and makes use of the *Maximal function* $Mh := \sup_r h_r$. The result which is obtained is the optimality of $\bar{Q}$-a.e. curve in the class of curves $\tilde{\omega}$ with $\int Mh(t, \tilde{\omega}(t)) \, dt < +\infty$, and moreover the result is local in time (perturbations are only allowed to start from $t_0 > 0$). Besides this small technicality about locality in time, the optimality which is obtained is only useful if there are many curves $\tilde{\omega}$ satisfying this integrability condition on Mh. A typical statement is then "for Q-a.e. curve $\tilde{\omega}$ this is the case", but it is not straightforward for which measure Q should one require this. Again, the typical approach is to prove that this is the case for all measures Q with $J(Q) < +\infty$ (which are in some sense the relevant measures for this problem, and this corresponds to some integrability property of the densities $\rho_t := (e_t)_\# Q$). In this case, we can compute

$$\int \int Mh(t, \omega(t)) \, dt \, dQ(\omega) = \int dt \int_\Omega Mh(t, x) \, d(e_t)_\# Q.$$

We would like to guarantee that every Q with $J(Q) < +\infty$ is such that $\int \int Mh(t, \omega(t)) \, dt \, dQ(\omega) < \infty$. Since we know that $G((e_t)_\# Q)$ is integrable, it is enough to guarantee $G^*(Mh) \in L^1$. In the case where $G(s) \approx s^q$ (hence $g(s) \approx s^{q-1}$ we need $Mh \in L^{q'}$. Since in this case we know $\rho \in L^q$, then

$h \approx g(\rho) \in L^{q'}$ and this implies $Mh \in L^{q'}$ from standard theorems in harmonic analysis, as soon as $q' > 1$.

As we can see, the analysis of these equilibrium conditions motivates a deeper study of regularity issues, for several reasons. Indeed, in order to apply the previous considerations it would be important to obtain upper bounds on $h[\rho]$; when this is not possible, at least obtaining higher integrability (in particular when we only know $h \in L^1$, passing to $L^{1+\varepsilon}$ would be crucial) would be important in order to deal with the integrability of Mh. Higher integrability can sometimes be obtained via higher-order estimates (proving BV or Sobolev estimates). More generally, better regularity on ρ (or on the dual variable φ) could give "better" solutions to the (HJ) equation (instead of just a.e. solutions).

This is why in the next sections we will see some regularity techniques. In Sect. 2.3 we will prove Sobolev results on the optimal density ρ which are interesting in themselves, and also imply higher integrability. Then in Sect. 2.4 we will see how to directly obtain L^∞ results with a different technique. Finally, Sect. 2.5 is devoted to the density-constrained case: for this case, [13] presented a non-trivial variant of the technique used here in Sect. 2.3 and obtained BV estimates on the pressure, which implied that the pressure is a function belonging to a certain L^q space, $q > 1$: here, instead, we will present the approach of [27] which provides $p \in L^\infty$ (yet, we will choose an easier proof, not available in [27]).

2.3   Regularity via Duality

We present here a technique to prove Sobolev regularity results for the optimal density ρ. This technique, based on duality, is inspired from the work of [7], and has been applied to MFG in [13]. It is actually very general, and [42] shows how it can be used to prove (or re-prove) many regularity results in elliptic equations coming from convex variational problems.

We start from a lemma related to the duality results of Sect. 2.2.1.

Lemma 2.3.1 *For any $(\phi, p) \in \mathcal{D}$ and $(\rho, \mathbf{v}) \in \mathcal{P}$ we have*

$$\mathcal{B}(\phi, p) + \mathcal{A}(\rho, \mathbf{v}) = \int_\Omega (\Psi - \phi_T)\, \mathrm{d}\rho_T + \int_0^T \int_\Omega (G(\rho) + G^*(p) - \rho p)\, \mathrm{d}x\, \mathrm{d}t + \frac{1}{2} \int_0^T \int_\Omega \rho |\mathbf{v} + \nabla\phi|^2\, \mathrm{d}x\, \mathrm{d}t.$$

Proof We start from

$$\mathcal{B}(\phi, p) + \mathcal{A}(\rho, \mathbf{v}) = \int_0^T \int_\Omega \left(\frac{1}{2}\rho |\mathbf{v}|^2 + G(\rho) + G^*(p) + V\rho \right) \mathrm{d}x\, \mathrm{d}t + \int_\Omega \Psi\, \mathrm{d}\rho_T - \int_\Omega \phi_0\, \mathrm{d}\overline{\rho_0}. \tag{2.11}$$

Then we use

$$\int_\Omega \Psi \, d\rho_T - \int_\Omega \phi_0 \, d\overline{\rho_0} = \int_\Omega (\Psi - \phi_T) \, d\rho_T + \int_\Omega \phi_T \, d\rho_T - \int_\Omega \phi_0 \, d\overline{\rho_0}$$

and

$$\int_\Omega \phi_T \, d\rho_T - \int_\Omega \phi_0 \, d\overline{\rho_0} = \int_0^T \int_\Omega (-\phi \nabla \cdot (\rho \mathbf{v}) + \rho \partial_t \phi) \, dx \, dt$$

$$= \int_0^T \int_\Omega \left(\nabla\phi \cdot (\rho \mathbf{v}) + \rho \left(\frac{1}{2}|\nabla\phi|^2 - (p + V) \right) \right) dx \, dt.$$

If we insert this into (2.11) we get the desired result.   □

It is important to stress that we used the fact that ϕ is C^1 since $(\rho, \mathbf{v})$ only satisfies (CE) in a weak sense, i.e. tested against C^1 functions. The same computations above would not be possible for $(\phi, p) \in \tilde{\mathcal{D}}$.

The regularity proof will come from the previous computations applied to suitable translations in space and/or time.

In order to simplify the exposition, we will suppose that $\Omega = \mathbb{T}^d$ is the d-dimensional flat torus, which avoids boundary issues. To handle the case of a domain Ω with boundary, we refer to the computations in [42] which suggest how to adapt the method below. Finally, for simplicity, we will only prove in this paper local results in $(0, T)$, so that also the time boundary does not create difficulties.

Here is the intuition behind the proof in this spatially homogeneous case. First, we use Lemma 2.3.1 to deduce

$$\mathcal{B}(\phi, p) + \mathcal{A}(\rho, \mathbf{v}) \geq \int_0^T \int_\Omega \left(G(\rho) + G^*(p) - \rho p \right) dx \, dt$$

(since the other terms appearing in Lemma 2.3.1 are positive). Then, let us suppose that there exist two function $J, J_* : \mathbb{R} \to \mathbb{R}$ and a positive constant $c_0 > 0$ such that for all $a, b \in \mathbb{R}$ we have

$$G(a) + G^*(b) \geq ab + c_0 |J(a) - J_*(b)|^2. \tag{2.12}$$

Remark 2.4 Of course, this is always satisfied by taking $J = J_* = 0$, but there are less trivial cases. For instance, if $G(\rho) = \frac{1}{q}\rho^q$ for $q > 1$, then $G^*(p) = \frac{1}{q'}q^{r'}$, with $q' = q/(q-1)$ and

$$\frac{1}{q}|a|^q + \frac{1}{q'}|b|^{q'} \geq ab + \frac{1}{2\max\{q, q'\}}|a^{q/2} - b^{q'/2}|^2,$$

i.e. we can use $J(a) = a^{q/2}$ and $J_*(b) = b^{q'/2}$. Another easy case to consider is the one where $G'' \geq c_0 > 0$. In this case we can choose $J = \mathrm{Id}$ and $J^* = (G^*)'$.

We wish to show that if $(\rho, \mathbf{v})$ is a minimizer of $\mathcal{A}$ then $J(\rho) \in H^1_{\text{loc}}((0, T) \times \Omega)$. Should $\mathcal{B}$ admit a C^1 minimizer ϕ (more precisely, a pair (ϕ, p)), then by the Duality Theorem 2.2.1, we would have $\mathcal{B}(\phi, p) + \mathcal{A}(\rho, \mathbf{v}) = 0$. Using Lemma 2.3.1, we get $J(\rho) = J_*(p)$. If we manage to show that $\tilde{\rho}(t, x) := \rho(t + \eta, x + \delta)$ with a corresponding velocity field $\tilde{\mathbf{v}}$ satisfies

$$\mathcal{A}(\tilde{\rho}, \tilde{\mathbf{v}}) \leq \mathcal{A}(\rho, \mathbf{v}) + C(|\eta|^2 + |\delta|^2) \tag{2.13}$$

for small $\eta \in \mathbb{R}$, $\delta \in \mathbb{R}^d$, then we would have

$$C(|\eta|^2 + |\delta|^2) \geq \mathcal{A}(\tilde{\rho}, \tilde{\mathbf{v}}) + \mathcal{B}(\phi, p) \geq c\|J(\tilde{\rho}) - J_*(p)\|^2_{L^2}.$$

Using then $J_*(p) = J(\rho)$, we would get

$$C(|\eta|^2 + |\delta|^2) \geq c\|J(\tilde{\rho}) - J(\rho)\|^2_{L^2},$$

which would mean that $J(\rho)$ is H^1 as we have estimated the squared L^2 norm of the difference between $J(\rho)$ and its translation by the squared length of the translation. Of course, there are some technical issues that need to be taken care of, for instance $\tilde{\rho}$ is not even well-defined (as we could need the value of ρ outside $[0, T] \times \Omega$), does not satisfy the initial condition $\tilde{\rho}(0) = \overline{\rho_0}$, we do not know if $\mathcal{B}$ admits a minimizer, and we do not know whether (2.13) holds.

To perform our analysis, let us fix $t_0 < t_1$ and a cut-off function $\zeta \in C^\infty_c(]0, T[)$ with $\zeta \equiv 1$ on $[t_0, t_1]$. Let us define

$$\begin{cases} \rho^{\eta,\delta}(t, x) := \rho(t + \zeta(t)\eta, x + \zeta(t)\delta), \\ \mathbf{v}^{\eta,\delta}(t, x) := \mathbf{v}(t + \zeta(t)\eta, x + \zeta(t)\delta)(1 + \zeta'(t)\eta) - \zeta'(t)\delta. \end{cases} \tag{2.14}$$

It is easy to check that the pair $(\rho^{\eta,\delta}, \mathbf{v}^{\eta,\delta})$ satisfies the continuity equation together with the initial condition $\rho^{\eta,\delta}(0) = \overline{\rho_0}$. Therefore it is an admissible competitor in $\mathcal{A}$ for any choice of (η, δ). We may then consider the function

$$M : \mathbb{R} \times \mathbb{R}^d \to \mathbb{R}, \quad M(\eta, \delta) := \mathcal{A}(\rho^{\eta,\delta}, \mathbf{v}^{\eta,\delta}).$$

The key point here is to show that M is $C^{1,1}$.

Lemma 2.3.2 *Suppose* $V \in C^{1,1}$. *Then, the function* $(\eta, \delta) \mapsto M(\eta, \delta)$ *defined above is also* $C^{1,1}$.

Proof We have

$$\mathcal{A}(\rho^{\eta,\delta}, \mathbf{v}^{\eta,\delta}) = \int_0^T \int_{\mathbb{T}^d} \frac{1}{2}\rho^{\eta,\delta}|\mathbf{v}^{\eta,\delta}|^2 \, dx \, dt + \int_0^T \int_{\mathbb{T}^d} V \, d\rho^{\eta,\delta} + \int_0^T \int_{\mathbb{T}^d} G(\rho^{\eta,\delta}) \, dx \, dt + \int_{\mathbb{T}^d} \Psi(x) \, d\rho^{\eta,\delta}_T.$$

Since $\rho^{\eta,\delta}(T, x) = \rho(T, x)$, the last term does not depend on (η, δ). For the other terms, we use the change-of-variable

$$(s, y) = (t + \zeta(t)\eta, x + \zeta(t)\delta)$$

which is a C^∞ diffeomorphism for small η. Then we can write

$$\int_0^T \int_{\mathbb{T}^d} \left(G(\rho^{\eta,\delta}(x, t)) + V(x)\rho^{\eta,\delta}(x, t) \right) dx \, dt = \int_0^T \int_{\mathbb{T}^d} \left(G(\rho(y, s)) + V(y - \zeta(t)\delta)\rho(y, s) \right) dx \, dt$$

$$= \int_0^T \int_{\mathbb{T}^d} \left(G(\rho(y, s)) V(y - \zeta(t)\delta)\rho(y, s) \right) K(\eta, \delta, s) \, dy \, ds,$$

where $K(\eta, \delta, s)$ is a smooth Jacobian factor (which does not depend on y since the change of variable is only a translation in space). Hence, this term depends smoothly on (η, δ), with the same regularity as that of V.

We also have

$$\int_0^T \int_{\mathbb{T}^d} \rho^{\eta,\delta} |\mathbf{v}^{\eta,\delta}|^2 \, dx \, dt = \int_0^T \int_{\mathbb{T}^d} \rho(s, y)|(1 + \eta\zeta'(t))\mathbf{v}(s, y) - \delta\zeta'(t)|^2 \, dx \, dt$$

$$= \int_0^T \int_{\mathbb{T}^d} \rho(s, y)|(1 + \eta\zeta'(t(\eta, s)))\mathbf{v}(s, y) - \delta\zeta'(t(\eta, s))|^2 K(\eta, \delta, s) \, dy \, ds,$$

where $K(\eta, \delta, s)$ is the same Jacobian factor as before, and $t(\eta, s)$ is obtained by inversing, for fixed $\eta > 0$, the relation $s = t + \eta\zeta'(t)$, and is also a smooth map. Hence, this term is also smooth. $\square$

We can now apply the previous lemma to the estimate we need.

Proposition 2.3.3 *There exists a constant C, independent of (η, δ), such that for $|\eta|, |\delta| \leq 1$, we have*

$$|M(\eta, \delta) - M(0, 0)| = |\mathcal{A}(\rho^{\eta,\delta}, \mathbf{v}^{\eta,\delta}) - \mathcal{A}(\rho, \mathbf{v})| \leq C(|\eta|^2 + |\delta|^2).$$

Proof We just need to use Lemma 2.3.2 and the optimality of $(\rho, \mathbf{v})$. This means that M achieves its minimum at $(\eta, \delta) = (0, 0)$, therefore its first derivative must vanish at $(0, 0)$ and we may conclude by a Taylor expansion, using boundedness of the second derivatives (as a consequence of the $C^{1,1}$ regularity). $\square$

With this result in mind, we may easily prove the following

Theorem 2.3.4 *If $(\rho, \mathbf{v})$ is a solution to the primal problem $\min \mathcal{A}$, if $\Omega = \mathbb{T}^d$ and if J satisfies (2.12), then $J(\rho)$ satisfies, for every $t_0 < t_1$,*

$$\|J(\rho(\cdot + \eta, \cdot + \delta)) - J(\rho)\|_{L^2([t_0, t_1] \times \mathbb{T}^d)}^2 \leq C(|\eta|^2 + |\delta|^2)$$

(where the constant C depends on t_0, t_1 and on the data), and hence is of class $H^1_{loc}(]0, T[\times\mathbb{T}^d))$.

Proof Let us take a minimizing sequence (ϕ_n, p_n) for the dual problem, i.e. $\phi_n \in C^1$, $p_n + V = -\partial_t\phi_n + \frac{1}{2}|\nabla\phi_n|^2$ and

$$\mathcal{B}(\phi_n, p_n) \leq \inf_{(\phi,p)\in\mathcal{F}} \mathcal{B}(\phi, p) + \frac{1}{n}.$$

We use $\tilde{\rho} = \rho^{\eta,\delta}$ and $\tilde{\mathbf{v}} = \mathbf{v}^{\eta,\delta}$ as in the previous discussion. Using first the triangle inequality and then Lemma 2.3.1 we have (where the L^2 norme denotes the norm in $L^2((0, T) \times \mathbb{T}^d)$)

$$c_0\|J(\rho^{\eta,\delta}) - J(\rho)\|^2_{L^2} \leq 2c_0(\|J(\rho^{\eta,\delta}) - J_*(p_n)\|^2_{L^2} + \|J(\rho) - J_*(p_n)\|^2_{L^2})$$

$$\leq 2(\mathcal{B}(\phi_n, p_n) + \mathcal{A}(\rho^{\eta,\delta}, \mathbf{v}^{\eta,\delta}) + \mathcal{B}(\phi_n, p_n) + \mathcal{A}(\rho, \mathbf{v})),$$

hence

$$\|J(\rho^{\eta,\delta}) - J(\rho)\|^2_{L^2} \leq C(\mathcal{B}(\phi_n, p_n) + \mathcal{A}(\rho, \mathbf{v})) + C(|\eta|^2 + |\delta|^2) \leq \frac{C}{n} + C(|\eta|^2 + |\delta|^2).$$

Letting n go to infinity and restricting the L^2 norm to $[t_0, t_1] \times \mathbb{T}^d$ gives the claim.
□

Remark 2.5 If one restricts to the case $\eta = 0$, then it is also possible to use a cut-off function $\zeta \in C^\infty_c(]0, T])$ with $\zeta(T) = 1$, as we only perform space translations. In this case, however, the final cost $\int_{\mathbb{T}^d} \Psi(x) \, d\rho^{\eta,\delta}_T$ depends on δ, and one needs to assume $\Psi \in C^{1,1}$ to prove $M \in C^{1,1}$. This allows to deduce H^1 regularity in space, local in time far from $t = 0$, i.e. $J(\rho) \in L^2_{loc}(]0, T]; H^1(\mathbb{T}^d))$.

A finer analysis of the behavior at $t = T$ also allows to extend the above H^1 regularity result in space time till $t = T$, but needs extra tools (in particular defining a suitable extension of ρ for $t > T$). This is developed in [37]. Moreover, it is also possible to obtain regularity results till $t = 0$, under additional assumptions on $\overline{\rho_0}$ and at the price of some extra technical work, as it is done in [19].

Remark 2.6 From $J(\rho) = J_*(p)$, the above regularity result on ρ can be translated into a corresponding regularity result on p whenever an optimal pair (ϕ, p) exists (even if the dual problem is stated in $\tilde{D}$: we could indeed prove that there exists a maximizing sequence composed of smooth functions, satisfying suitable H^1 bounds, which would imply the same regularity for the maximizer of the relaxed dual problem).

Remark 2.7 When $G(\rho) = \rho^q$, $q > 1$, the above H^1 result can be applied to $\rho^{q/2}$ and combined with the Sobolev injection $H^1 \subset L^{2^*}$. This shows that we have $\rho \in L^{\tilde{q}}_{loc}((0, T) \times \Omega)$ for an exponent $\tilde{q} > q$, given by $q(d+1)/(d-1)$ in dimension $d > 1$ (and any exponent $\tilde{q} < \infty$ if $d = 1$). This is a better integrability than just L^q, which came from the finiteness of the functional. The exponent has been computed using the Sobolev injection in dimension $d + 1$, the dimension of $(0, T) \times \Omega$. If we distinguish the behavior in time and space, just using $J(\rho) \in L_t^2 H_x^1$, we get $\rho \in L_t^2 L_x^{qd/(d-2)}$ for $d > 2$, $\rho \in L_t^2 L_x^{\tilde{q}}$ for any $\tilde{q} < \infty$ in dimension $d = 2$, and $L_t^2 L_x^\infty$ in dimension $d = 1$.

Finally, we finish this section by underlining the regularity results in the density-constrained case [13]: the same kind of strategy, but with many more technical issues, which follow the same scheme as in [7] and [1], and the result is much weaker. Indeed, it is only possible to prove in this case $p \in L_{loc}^2((0, T); BV(\mathbb{T}^d))$ (exactly as in [1]). Even if very weak, this result is very important in what it gives higher integrability on p, which was a priory only supposed to be a measure and this allows to get the necessary summability of the maximal function that we briefly mentioned in Sect. 2.2.3.

2.4 Regularity via OT, Time Discretization, and Flow Interchange

In this section we will interpret the Eulerian variational formulation as the search for an optimal curve in the Wasserstein space, i.e. the space of probability measures endowed with a particular distance coming from optimal transport. This will lead to a very efficient time discretization on which we are able to perform suitable computations providing strong bounds.

2.4.1 Tools from Optimal Transport and Wasserstein Spaces

The space $\mathcal{P}(\Omega)$ of probability measures on Ω can be endowed with the Wasserstein distance: if μ and ν are two elements of $\mathcal{P}(\Omega)$, the 2-Wasserstein distance $W_2(\mu, \nu)$ between μ and ν is defined via

$$W_2(\mu, \nu) := \sqrt{\min \left\{ \iint_{\Omega \times \Omega} |x - y|^2 \, d\gamma(x, y) \; : \; \gamma \in \mathcal{P}(\Omega \times \Omega) \text{ and } (\pi_x)_\# \gamma = \mu, \; (\pi_y)_\# \gamma = \nu \right\}}.$$
(2.15)

In the formula above, π_x and $\pi_y : \Omega \times \Omega \to \Omega$ stand for the projections on respectively the first and second component of $\Omega \times \Omega$. If $T : X \to Y$ is a measurable application and μ is a measure on X, then the image measure of μ by T, denoted by $T_\# \mu$, is the measure defined on Y by $(T_\# \mu)(B) = \mu(T^{-1}(B))$ for any measurable set $B \subset Y$. For general results about optimal transport, the reader might refer to [3, 43], or [39].

The Wasserstein distance admits a dual formulation, the dual variables being the so-called Kantorovich potentials. The main properties of these potentials, in the case which is of interest to us, are summarized in the proposition below. We restrict to the cases where the measures have a strictly positive density a.e., as in this particular case the potentials are unique (up to a global additive constant). The proof of these results can be found, for instance, in [39, Chapters 1 and 7].

Proposition 2.4.1 Let $\mu, \nu \in \mathcal{P}(\Omega)$ be two absolutely continuous probability measures with strictly positive density. Then there exists a unique (up to adding a constant to φ and subtracting it from ψ) pair (φ, ψ) of Kantorovich potentials satisfying the following properties.

1. The squared Wasserstein distance $W_2^2(\mu, \nu)$ can be expressed as

$$\frac{1}{2} W_2^2(\mu, \nu) = \int_\Omega \varphi \mu + \int_\Omega \psi \nu.$$

2. The "vertical" derivative of $W_2^2(\cdot, \nu)$ at μ is φ: if $\tilde{\mu} \in \mathcal{P}(\Omega)$ is any probability measure, then

$$\lim_{\varepsilon \to 0} \frac{\frac{1}{2} W_2^2((1-\varepsilon)\mu + \varepsilon\tilde{\mu}, \nu) - \frac{1}{2} W_2^2(\mu, \nu)}{\varepsilon} = \int_\Omega \varphi(\tilde{\mu} - \mu).$$

3. The potentials φ and ψ are one the c-transform of the other, meaning that we have

$$\begin{cases} \varphi(x) &= \inf_{y \in \Omega} \frac{|x-y|^2}{2} - \psi(y) \\ \psi(y) &= \inf_{x \in \Omega} \frac{|x-y|^2}{2} - \varphi(x). \end{cases}$$

4. There holds $(\mathrm{Id} - \nabla\varphi)_\# \mu = \nu$ and the transport plan $\gamma := (\mathrm{Id}, \mathrm{Id} - \nabla\varphi)_\# \mu$ is optimal in the problem (2.15). We also say that the map $x \mapsto x - \nabla\varphi(x)$ is the optimal transport map from μ to ν.

The function φ (resp. ψ) is called the Kantorovich potential from μ to ν (resp. from ν to μ).

We will denote by Γ the space of absolutely continuous curves from $[0, 1]$ to $\mathcal{P}(\Omega)$ endowed with the Wasserstein distance W_2.

Definition 2.4.2 We say that a curve ρ is absolutely continuous if there exists a function $a \in L^1([0, 1])$ such that, for every $0 \leqslant t \leqslant s \leqslant 1$,

$$W_2(\rho_t, \rho_s) \leqslant \int_t^s a(r) \, dr.$$

We say that ρ is 2$-$absolutely continuous if the function a above can be taken in $L^2([0, 1])$

This space will be equipped with the distance d_Γ of the uniform convergence, i.e.

$$d_\Gamma(\rho^1, \rho^2) := \max_{t \in [0,1]} W_2(\rho_t^1, \rho_t^2).$$

The main interest of the notion of absolute continuity for curves in the Wasserstein space lies in the following theorem, which we recall without proof (but we refer to [3] or to Chapter 5 in [39]).

Theorem 2.4.3 *For $\rho \in \Gamma$ the quantity*

$$|\dot{\rho}_t| := \lim_{h \to 0} \frac{W_2(\rho_{t+h}, \rho_t)}{h}$$

exists and is finite for a.e. t. Moreover, we have the following

- *if $\rho \in \Gamma$ is a 2-absolutely continuous curve, there exists for a.e. t a vector field $\mathbf{v}_t \in L^2(\rho_t)$ such that $||\mathbf{v}_t||_{L^2(\rho_t)} \leq |\dot{\rho}_t|$ and such that the continuity equation $\partial_t \rho + \nabla \cdot (\rho \mathbf{v}) = 0$ holds in distributional sense;*
- *if $\rho \in \Gamma$ is such that there exists a family of vector fields $\mathbf{v}_t \in L^2(\rho_t)$ satisfying $\int_0^T \int_\Omega |\mathbf{v}_t|^2 \, d\rho_t \, dt < +\infty$ and $\partial_t \rho + \nabla \cdot (\rho \mathbf{v}) = 0$, then ρ is a 2-absolutely continuous curve and $||\mathbf{v}_t||_{L^2(\rho_t)} \geq |\dot{\rho}_t|$ for a.e. t.*

Finally, we can represent $\int_0^1 |\dot{\rho}_t|^2 \, dt$ in various ways such as

$$\int_0^1 |\dot{\rho}_t|^2 \, dt = \sup_{N \geq 2} \sup_{0 \leq t_1 < t_2 < \ldots < t_N \leq 1} \sum_{k=2}^N \frac{W_2^2(\rho_{t_{k-1}}, \rho_{t_k})}{t_k - t_{k-1}} \qquad (2.16)$$

$$= \min \left\{ \int_0^1 \int_\Omega |\mathbf{v}_t|^2 \, d\rho_t \, dt \ : \ \partial_t \rho + \nabla \cdot (\rho \mathbf{v}) = 0 \right\}. \qquad (2.17)$$

Observe that the kinetic energy in (2.16) is exactly the same quantity appearing in Sect. 2.2.3.

2.4.2 Discretization in Time of Variational MFG and Optimality Conditions

We first start from the observation that the above tools from optimal transport theory allow to re-write the variational problem defining MFG equilibria into the following form

$$\min \left\{ \int_0^T \frac{1}{2} |\dot{\rho}_t|^2 \, dt + \int_0^T \mathcal{G}(\rho_t) \, dt + \int_\Omega \Psi \, d\rho_T \ : \ \rho : [0, T] \to \mathcal{P}(\Omega), \ \rho_0 = \overline{\rho_0} \right\}.$$

A useful approximation can be obtained via time-discretization: we fix a time step $\tau = T/N$ and we look for a sequence $\rho_0 = \overline{\rho_0}, \rho_1, \ldots, \rho_N$ solving

$$\min \left\{ \sum_{k=0}^{N-1} \left(\frac{W_2^2(\rho_k, \rho_{k+1})}{2\tau} + \tau G(\rho_k) \right) + \int_\Omega \Psi \, d\rho_N \right\}.$$

If $\overline{\rho_0}, \rho_1, \ldots, \rho_N$ solves the above minimization problem then, for each $0 < k < N$, the measure ρ_k solves

$$\min \left\{ \frac{W_2^2(\rho, \rho_{k-1})}{2\tau} + \frac{W_2^2(\rho, \rho_{k+1})}{2\tau} + \tau G(\rho) \quad : \quad \rho \in \mathcal{P}(\Omega) \right\},$$

i.e. it solves a minimization problem similar to what we see in the JKO scheme for gradient flows (see [3, 22, 40]), which would be of the form

$$\min \left\{ \frac{W_2^2(\rho, \rho_{k-1})}{2\tau} + G(\rho) \right\}.$$

By the way, for $k = N$, we have a true JKO-style problem with one only Wasserstein distance.

From this similarity with the JKO scheme, we are lead to apply techniques which have been previously applied to this other setting, and in particular the notion of *flow interchange*, developed in [30].

Consider the functional $\mathcal{F}_m(\rho) := \int F_m(\rho(x)) \, dx$, where $F_m(s) := s^m$. The important point about this functional, if we suppose Ω to be convex, is that it is a geodesically convex functional on the W_2 Wasserstein space (see [32]). This means that it is convex along constant-speed geodesic interpolations in $\mathbb{W}_2(\Omega)$. Consider now $(\rho_s)_s$ be the gradient flow of $\mathcal{F}_m(\rho)$, i.e. a solution of $\partial_s \rho - \nabla \cdot (\rho \nabla (F_m'(\rho))) = 0$, with initial datum at $s = 0$ equal to the optimal ρ at step k. From the EVI definition of gradient flows [3] and the geodesic convexity of F_m we obtain the following inequality, valid for every v

$$\frac{d}{ds} \frac{W_2^2(\rho_s, v)}{2} \leq \mathcal{F}_m(v) - \mathcal{F}_m(\rho_s).$$

We can also compute

$$\frac{d}{ds} G(\rho_s) = - \int \nabla (g(\rho_s) + V) \cdot \nabla (F_m'(\rho_s)) \, d\rho_s.$$

On the other hand, the optimality of ρ_k implies that the derivative of the sum of the Wasserstein terms and of the G term should be non-negative, which provides

$$\int \nabla(g(\rho_k) + V) \cdot \nabla(F'_m(\rho_k)) \, \mathrm{d}\rho_k \leq \frac{\mathcal{F}_m(\rho_{k+1}) - 2\mathcal{F}_m(\rho_k) + \mathcal{F}_m(\rho_{k-1})}{\tau^2}.$$

Let us start from the easier case $V = 0$: in this case we get

$$0 \leq \int g'(\rho_k) F''_m(\rho_k) \rho_k |\nabla \rho_k|^2 \leq \frac{\mathcal{F}_m(\rho_{k+1}) - 2\mathcal{F}_m(\rho_k) + \mathcal{F}_m(\rho_{k-1})}{\tau^2}.$$

This shows that $k \mapsto \mathcal{F}_m(\rho_k)$ is (discretely) convex. If $\overline{\rho_0} \in L^m$, and if for some reason we suppose $\rho_T \in L^m$, then, after passing to the limit $\tau \to 0$, we deduce a uniform bound on $||\rho_t||_{L^m}$. This also works for $m = \infty$. This was essentially a result proven by P.-L. Lions in his course ([28], lecture of November 27, 2009), in a more general setting (still with no x-dependence, but with more general Hamiltonians than the quadratic one).

Note that the case where ρ_T is prescribed is known under the name of *planning problem* (see, for instance, [20, 34, 35]) but is out of the scopes of these notes. When, instead, we have a final penalization, the same flow interchange technique provides

$$\int \nabla \Psi \cdot \nabla(F'_m(\rho_N)) \, \mathrm{d}\rho_N \leq \frac{\mathcal{F}_m(\rho_{N-1}) - \mathcal{F}_m(\rho_N)}{\tau}.$$

After an integration by parts, using $\nabla(F'_m(\rho_N))\rho_N = m(m-1)\rho_N^{m-1}\nabla\rho_N = (m-1)\nabla(F_m(\rho_N))$, and assuming $\Psi \in C^{1,1}$ and $\partial\Psi/\partial n \geq 0$ on $\partial\Omega$, we obtain

$$\mathcal{F}_m(\rho_N) \leq \mathcal{F}_m(\rho_{N-1}) + \tau(m-1)\int F_m(\rho_N)\Delta\Psi,$$

i.e.

$$(1 - C\tau)\mathcal{F}_m(\rho_N) \leq \mathcal{F}_m(\rho_{N-1}), \quad \text{for } C = (m-1)||(\Delta\Psi)_+||_{L^\infty}. \tag{2.18}$$

This shows that not only $k \mapsto \mathcal{F}_m(\rho_k)$ is convex, but that we control its final derivative. From a continuous point of view, it is as if we had a function $u \geq 0$, with $u'' \geq 0$ and $u'(T) \leq Cu(T)$. This is not enough to provide a bound on $u(T)$ as, for instance, all functions of the form $u(t) = \lambda(1 - C(T - t))_+$ satisfy these assumptions (note by the way that, in case $CT > 1$, we also have $u(0) = 0$, which shows that adding an assumption on the initial data would not be enough). Yet, we can obtain $u(T) \leq 2C \int_0^T u$. This can be, for instance, applied to the case where the two functionals G and F_m have the same order of growth: $G \approx F_m$. From the finiteness of the integral of $\mathcal{F}_m$ we would deduce in this case a uniform bound for $\mathcal{F}_m(\rho_T)$ and, if $\mathcal{F}_m(\overline{\rho_0}) < \infty$, a uniform bound in time.

However, we are able, following the non-trivial computations in [26], to obtain much more.

To give an idea of the method, let us stick to the case $V = 0$ and let us impose a very stringent assumption on the congestion function g. We will suppose $g'(s) \geq cs^{-1}$ an assumption which is satisfied in the entropy case $G(s) = s \log s$. We will see that the important assumption is indeed the inequality $g'(s) \geq cs^\alpha$ for $\alpha \geq -1$. The idea is to exploit the positive term $\int g'(\rho_k) F_m''(\rho_k) \rho_k |\nabla \rho_k|^2$. In this case we have

$$\int g'(\rho_k) F_m''(\rho_k) \rho_k |\nabla \rho_k|^2 \geq c \int \rho_k^{m-2} |\nabla \rho_k|^2 = c||\nabla(\rho_k^{m/2})||_{L^2}^2.$$

We then apply the Sobolev injection of H^1 into L^β, for an exponent $2\beta > 2$. This allows, for instance, to write

$$||(\rho_k^{m/2})||_{L^{2\beta}}^2 \leq C||\nabla(\rho_k^{m/2})||_{L^2}^2 + C \int \rho_k^m$$

for a suitable constant C. As the last term in the right hand side is just $\mathcal{F}_m(\rho_k)$, we obtain a bound on $\mathcal{F}_{m\beta}(\rho_k)$ in terms of $\mathcal{F}_m(\rho_k)$ and of its second variation in k. The idea is then to apply Moser's iteration on exponents $m_j \approx \beta^j$. This is delicate, since in order to take care of the second derivative in time (even if it is discrete) we need to integrate in time, and the integral (sum over k in the discrete setting) in time of the $L^{2\beta}$ norms raised to the power 2 is not the $L^{2\beta}$ norm in time-space. This can be dealt with using the fact that all the functionals $\mathcal{F}_m$ are convex in time, which allows to obtain reversed Jensen inequalities: if a function $u \geq 0$ is convex, indeed, we have

$$\left(\int_{T_1}^{T_2} u(t) \, dt\right)^{1/\beta} \leq \frac{(T_2 - T_1)^{1\beta}}{\varepsilon} \int_{T_1-\varepsilon}^{T_2+\varepsilon} u^{1/\beta}(t) \, dt.$$

This allows hence to obtain an estimate of the form

$$\left(\int \int_{T_1}^{T_2} F_{m\beta}(\rho(t)) \, dt\right)^{1/\beta} \leq C(m, \varepsilon) \int_{T_1-\varepsilon}^{T_2+\varepsilon} F_m(\rho(t)) \, dt$$

and, choosing suitable values of $\varepsilon = \varepsilon_m$ and exploiting the polynomial behaviour of $C(m, \varepsilon)$ in m and ε^{-1}, it is possible to iterate this estimate in the spirit of the work of Moser [33] for elliptic regularity, thus obtaining an estimate on $||\rho||_{L^\infty([T_1,T_2]\times\Omega)}$ in terms of $\int_{[T_1-\varepsilon,T_2+\varepsilon]\times\Omega} G(\rho) \, dx \, dt$.

Even if the computations are less straightforward it is not difficult to see that the assumption $g'(s) \geq cs^{-1}$ can be replaced by a more general one where we use $g'(s) \geq cs^\alpha$ for an exponent $\alpha \geq -1$, and that it is enough, in order to obtain L^∞ bounds, that this inequality is satisfied for $s \geq s_0$ (see [26]).

The situation is trickier when there is an exterior potential V. In this case we have

$$\int g'(\rho_k)F_m''(\rho_k)\rho_k|\nabla\rho_k|^2 \leq \frac{\mathcal{F}_m(\rho_{k+1}) - 2\mathcal{F}_m(\rho_k) + \mathcal{F}_m(\rho_{k-1})}{\tau^2} - \int (\nabla V \cdot \nabla\rho_k)F_m''(\rho_k)\rho_k.$$

The new term needs to be estimated in terms of V and $\mathcal{F}_m$, which can be done in two possible ways. Either we integrate by parts, as we did for the final cost Ψ, and suppose $V \in C^{1,1}$ and $\partial V/\partial n \geq 0$, in which case we use $\nabla\rho_k F_m''(\rho_k)\rho_k = (m-1)\nabla(F_m(\rho))$ and we get

$$-\int (\nabla V \cdot \nabla\rho_k)F_m''(\rho_k)\rho_k \leq (m-1)\int (\Delta V)F_m(\rho_k),$$

or we use a Young inequality:

$$-\int (\nabla V \cdot \nabla\rho_k)F_m''(\rho_k)\rho_k \leq \frac{1}{2}\int |\nabla V|^2 F_m''(\rho_k)\rho_k^2 + \frac{1}{2}\int |\nabla\rho_k|^2 F_m''(\rho_k).$$

The first term in the right-hand side can be bounded by $Cm^2\mathcal{F}_m(\rho_k)$ as soon as V is Lipschitz continuous, and the second can be bounded in terms of $\int g'(\rho_k)F_m''(\rho_k)\rho_k|\nabla\rho_k|^2$ as soon as $g'(\rho) \geq \rho^\alpha$ with $\alpha \geq -1$. We will see in the statement of Theorem 2.4.4 that this computation (only assuming V to be Lipschitz) can only be exploited for L^∞ regularity under some very restrictive assumptions.

However, a difficulty arising in this case is that $k \mapsto \mathcal{F}_m(\rho_k)$ is no more convex. From a continuous point of view, we do not have anymore a time-dependent function u with $u'' \geq 0$, but rather a solution of $u'' + \omega^2 u \geq 0$, for a constant ω depending on m. Differently from convexity, in general this inequality cannot provide bounds, if we think that functions of the form $u(t) = \lambda \sin \omega t$ solve the equality case for any λ, on intervals of the form $[0, T]$, $T = k\pi/\omega$. Hence, this inequality can only provide bounds on short intervals of time, smaller than π/ω. In particular, when doing Moser's iterations, we need to divide every interval into smaller ones; since the reverse Jensen inequality requires to enlarge these intervals, there will be many new integrals on overlapping intervals. As a result, this will bring to a larger multiplicative constant depending on m (since ω also depends on m, and the parameter ε_m in the enlargement of the intervals also depends on m) in the estimates. This is not a problem as soon as the dependence is polynomial.

A final remark about the case where $(g's) \geq s^\alpha$ but $\alpha < -1$. This case is called in [26] "weak congestion". In this case, we only have a control of $\mathcal{F}_m$ in terms of $\mathcal{F}_{\beta(m+1+\alpha)}$. Thus we must start the iterative procedure with a value m such that $m < \beta(m+1+\alpha)$, i.e. we must impose a priori some L^m regularity on ρ (with an exponent m which depends on α and β, the latter depending itself only on the dimension of the ambient space). Such a regularity can be obtained, for instance, by assuming that $\overline{\rho_0}$ (which is fixed) is in $L^m(\Omega)$ and that T is small enough. Indeed, if this is the case, the boundary condition (2.18) combined with the interior estimate

$u'' + \omega^2 u \geq 0$ show that if T is small enough (given the potentials and the congestion function f), the L^m norm of ρ on $[0, T] \times \Omega$ must be bounded.

We do not develop all the details, which are very technical, here but we summarize here below the L^∞ results which can be found in [26]. The results are based on the above estimates obtained in the time-discrete setting, together with a suitable use of the limit $\tau \to 0$.

Theorem 2.4.4 *Consider a running cost of the form* $h[\rho] = V(x) + g(\rho)$. *Suppose that the inequality* $g'(s) \geq s^\alpha$ *is satisfied for every* $s \geq s_0$. *Then, we have:*

- *If V is Lipschitz, $\alpha \geq -1$, and $s_0 = 0$ then $\rho \in L^\infty_{loc}((0, T) \times \overline{\Omega})$.*
- *The same result holds if $s_0 > 0$ but $V \in C^{1,1}$ and $\partial V / \partial n \geq 0$.*
- *These results extend to $(0, T]$ if $\Psi \in C^{1,1}$ and $\partial \Psi / \partial n \geq 0$.*
- *If $\alpha < -1$, then the same results, for $V, \Psi \in C^{1,1}$, $\partial V / \partial n \geq 0$ and $\partial \Psi / \partial n \geq 0$, are true if we already know $\rho \in L^{m_0}((0, T) \times \overline{\Omega})$ for $m_0 > d|\alpha + 1|/2$. This is true in particular if $\overline{\rho_0} \in L^{m_0}$ and T is small enough.*

It is now straightforward to apply the L^∞ bounds on ρ to obtain boundedness from above of $h[\rho]$, and then apply the content of Sect. 2.2.3 in order to transform the optimality into a the equilibrium condition characterizing optimal tranjectories in MFG.

2.5   Density-Constrained Mean Field Games

In this section we are concerned with the model presentd in [13] (but, compared to such a paper, we will restrict to the case where the cost is quadratic in the velocity): the variational problem to be considered is

$$\min\left\{ \int_0^T \int_\Omega \left(\frac{1}{2}|\mathbf{v}_t|^2 + V \right) d\rho_t \, dt + \int_\Omega \Psi \, d\rho_T \; : \; \rho \leq 1 \right\}.$$

This can be translated into

$$\min\left\{ \int_0^T \frac{1}{2}|\dot{\rho}_t|^2 \, dt + \int_0^T \mathcal{G}(\rho_t) \, dt + \int_\Omega \Psi \, d\rho_T \; : \; \rho : [0, T] \to \mathcal{P}(\Omega), \; \rho_0 = \overline{\rho_0} \right\},$$

where $\mathcal{G}$ is a very degenerate functional:

$$\mathcal{G}(\rho) := \begin{cases} \int V \, d\rho & \text{if } \rho \leq 1, \\ +\infty & \text{if not.} \end{cases}$$

We already discussed that this provides the following MFG system

$$
\begin{cases}
-\partial_t \varphi + \frac{|\nabla \varphi|^2}{2} = V + p, \\
\partial_t \rho - \nabla \cdot (\rho \nabla \varphi) = 0, \\
\varphi(T, x) = \Psi(x) + P(x), \quad \rho(0, x) = \overline{\rho_0}(x), \\
p \geq 0, \ \rho \leq 1, \ p(1 - \rho) = 0, \\
P \geq 0, \ P(1 - \rho_T) = 0.
\end{cases}
\tag{2.19}
$$

and that the running cost of every agent is in the end $V + p$ (note that this is coherent with the general formula $V + G'(\rho)$, where the derivative $G' = g$ should be replaced here by a generic element of the subdifferntial ∂G). Note that in this case we also have an effect on the final cost, where Ψ is replaced by $\Psi + P$. This can be interpreted in two ways. In general, we did not put any density penalization at final time (i.e. the final cost is not of the form $\Psi + g(\rho_T)$ but only of the form Ψ), but here the constraint $\rho_T \leq 1$ is also present on the final density, and lets its subdifferential appear. On the other hand, we can consider that the constraint $\rho_t \leq 1$ for all $t < T$ is enough to impose the same (by continuity in the Wasserstein space of the curve $t \mapsto \rho_t$) for $t = T$, so that in the final cost functional we can omit the constraint part. If we interpret this in this way, how can we justify the presence of a final cost P? The answer comes from the fact that the natural regularity for the pressure p, which is supposed to be positive, is being a positive measure (since distributions with a sign are measures, and also because in the dual problem p is penalized in a L^1 sense). Hence, the extra cost P represents the singular part of p concentrated on $t = T$. What we will prove in this section is that we have $p \in L^\infty([0, T] \times \Omega)$ and $P \in L^\infty(\Omega)$, thus decomposing the pressure into a bounded density in time-space and a bounded density at the final time.

This problem can also be discredited in the same way as in Sect. 2.4, and this discretization technique will be the one which will rigorously provide the estimates we look for. Yet, before looking at the details, we prefer first to give an heuristic derivation of the main idea in continuous time. The key point will consist in proving $\Delta(V + p) \geq 0$ on $\{p > 0\}$. To do this, we consider System (2.19), and denote by $D_t := \partial_t - \nabla \varphi \cdot \nabla$ the convective derivative. The idea is to look at the quantity $-D_{tt}(\log \rho)$. Indeed, the continuity equation in (2.19) can be rewritten $D_t(\log \rho) = \Delta \varphi$. On the other hand, taking the Laplacian of the Hamilton–Jacobi equation, it is easy to get, dropping a positive term, $-D_t(\Delta \varphi) \leq \Delta(p + V)$. Hence,

$$
- D_{tt}(\log \rho) \leq \Delta(p + V).
\tag{2.20}
$$

Then, we observe that $\log \rho$ is maximal where $\rho = 1$, hence we have $-D_{tt}(\log \rho) \geq 0$. This implies $\Delta p \geq -\Delta V$ on $\{p > 0\}$.

Let us say that the strategy of looking at the convective derivative of $\log \rho$ was already used by Loeper [29] to study a similar problem (related to the reconstruction of the early universe). Moreover, also in [29] the rigorous proof was done by time-discretization.

As the tools which are required to study the L^∞ regularity are much less technical than for the density-penalized case, we will develop here more details. In particular, we will write here the optimality conditions for the discrete problems and see that quantities acting like a pressure appear. For the convergence of these quantities to the true pressures p and P, we refer to [27], whose results are also recalled in Sect. 2.5.3.

Some regularity will be needed in order to be able to correctly perform our analysis. In particular, we will assume that $\bar\rho_0$ is smooth and strictly positive and that V and Ψ are C^2 function. We will also add a small entropy penalization to the term $\mathcal{G}$, thus considering

$$\mathcal{G}_\lambda(\rho) =:= \begin{cases} \int V \, d\rho + \lambda \int \rho \log \rho & \text{if } \rho \leq 1, \\ +\infty & \text{if not} \end{cases}$$

and we will also add the same entropy penalization to the final cost, thus solving

$$\min \left\{ \sum_{k=0}^{N-1} \left(\frac{W_2^2(\rho_k, \rho_{k+1})}{2\tau} + \tau \mathcal{G}_\lambda(\rho_k) \right) + \int_\Omega \Psi \, d\rho_N + \lambda \int_\Omega \rho_N \log \rho_n \, dx \right\}.$$

Yet, all the estimates that we establish will not depend on the smoothness of $\bar\rho_0$, V and Ψ or on the value of λ.

2.5.1   Optimality Conditions and Regularity of p

In this subsection, we fix $N \geq 1$ and $k \in \{1, 2, \ldots, N-1\}$ a given instant of time. We will fix an optimal sequence $(\rho_0 = \bar\rho_0, \rho_1, \ldots \rho_N$ and set $\bar\rho := \rho_k$; we also denote $\mu := \rho_{k-1}$ and $\nu := \rho_{k+1}$. From the same consideration of the previous section, we know that $\bar\rho$ is a minimizer, among all probability measures with density bounded by 1, of

$$\rho \mapsto \frac{W_2^2(\mu, \rho) + W_2^2(\rho, \nu)}{2\tau} + \tau \mathcal{G}_\lambda(\rho)$$

Lemma 2.5.1 *The density $\bar\rho$ is strictly positive a.e.*

Proof The proof is based on the fact that the derivative of the function $s \mapsto s \log s$ at $s = 0$ is $-\infty$, so that minimizers avoid the value $\rho = 0$. It can be obtained following the procedure in [39, Lemma 8.6], or of [26, Lemma 3.1], as the construction done in these proofs preserves the constraint of having a density smaller than 1.   □

Proposition 2.5.2 *Let us denote by φ_μ and φ_ν the Kantorovich potentials for the transport from $\bar\rho$ to μ and ν respectively (this potentials are unique up to additive constants because $\bar\rho > 0$). There exists $p \in L^1(\Omega)$, positive, such that $\{p > 0\} \subset \{\bar\rho = 1\}$ and a constant C such that*

$$\frac{\varphi_\mu + \varphi_\nu}{\tau^2} + V + p + \lambda\log(\bar\rho) = C \quad a.e. \tag{2.21}$$

Moreover p and $\log(\bar\rho)$ are Lipschitz continuous.

Proof Let $\tilde\rho \in \mathcal{P}(\Omega)$ such that $\tilde\rho \le 1$. We define $\rho_\varepsilon := (1 - \varepsilon)\bar\rho + \varepsilon\tilde\rho$ and use it as a competitor. Clearly $\rho_\varepsilon \le 1$, i.e. it is an admissible competitor. We will obtain the desired optimality conditions comparing the cost of ρ_ε to that of ρ. Using Proposition 2.4.1, as $\bar\rho > 0$, the Kantorovich potentials φ_μ and φ_ν are unique (up to a constant) and

$$\lim_{\varepsilon \to 0} \frac{W_2^2(\mu, \rho_\varepsilon) - W_2^2(\mu, \bar\rho) + W_2^2(\rho_\varepsilon, \nu) - W_2^2(\bar\rho, \nu)}{2\tau^2} = \int_\Omega \frac{\varphi_\mu + \varphi_\nu}{\tau}(\tilde\rho - \bar\rho).$$

The term involving V is straightforward to handle as it is linear. The only remaining term is the one involving the entropy. For this term (following, for instance, the reasoning in [26, Proposition 3.2]), we can obtain the inequality

$$\limsup_{\varepsilon \to 0} \frac{\int \rho_\varepsilon \log \rho_\varepsilon - \int \bar\rho \log \bar\rho}{\varepsilon} \le \int_\Omega \log(\bar\rho)(\tilde\rho - \bar\rho).$$

Putting the pieces together, we see that $\int_\Omega u\,(\tilde\rho - \bar\rho) \ge 0$ for any $\tilde\rho \in \mathcal{P}(\Omega)$ with $\tilde\rho \le 1$, provided that u is defined by

$$u := \frac{\varphi_\mu + \varphi_\nu}{\tau^2} + V + \lambda\log(\bar\rho)$$

It is known, analogously to [31, Lemma 3.3], that this leads to the existence of a constant C such that

$$\begin{cases} \bar\rho = 1 & \text{on } \{u < C\} \\ \bar\rho \le 1 & \text{on } \{u = C\} \\ \bar\rho = 0 & \text{on } \{u > C\} \end{cases} \tag{2.22}$$

Specifically, C is defined as the smallest real $\tilde C$ such that $\mathcal{L}^d(\{u \le \tilde C\}) \ge 1$, and it is quite straightforward to check that this choice works. Note that the case $\{u > C\}$ can be excluded by Lemma 2.5.1. We then define the pressure p as $p = (C - u)_+$, thus (2.21) holds. It satisfies $p \ge 0$, and $\bar\rho < 1$ implies $p = 0$.

It remains to answer the question of the Lipschitz regularity of p and $\log(\bar\rho)$. Notice that p is positive, and non zero only on $\{\bar\rho = 1\}$. On the other hand, $\log(\bar\rho) \le$

0 and it is non zero only on $\{\bar{\rho} < 1\}$. Hence, one can write

$$p = \left(C - \frac{\varphi_\mu + \varphi_\nu}{\tau^2} + V\right)_+ \quad \text{and} \quad \log(\bar{\rho}) = -\frac{1}{\lambda}\left(C - \frac{\varphi_\mu + \varphi_\nu}{\tau^2} + V\right)_-.$$

(2.23)

Given that the Kantorovich potentials and V are Lipschitz, it implies the Lipschitz regularity for p and $\log(\bar{\rho})$.   □

Let us note that φ_μ and φ_ν have additional regularity properties, even though this regularity heavily depends on τ.

Lemma 2.5.3 *The Kantorovich potentials φ_μ and φ_ν belong to $C^{2,\alpha}(\mathring{\Omega}) \cap C^{1,\alpha}(\Omega)$ and $p \in C^{2,\alpha}(\{p > 0\})$.*

Proof If $k \in \{2, \ldots, N\}$, thanks to Proposition 2.5.2 (applied in $k - 1$ and $k + 1$), we know that μ and ν have a Lipschitz density and are bounded from below. Using the regularity theory for the Monge Ampère-equation [43, Theorem 4.14], we can conclude that φ_μ and φ_ν belong to $C^{2,\alpha}(\mathring{\Omega}) \cap C^{1,\alpha}(\Omega)$.

   Once we have the regularity of $\varphi_\mu + \varphi_\nu$, as we were supposing $V \in C^2$, we get $C^{2,\alpha}$ regularity for $p + \lambda \log \bar{\rho}$, which in turns implies the same regularity for $p = (p + \lambda \log \bar{\rho})_+$ in the open set $\{p > 0\}$.   □

Theorem 2.5.4 *We have the following L^∞ estimate:*

$$p \leq \max V - \min V.$$

Proof First we will prove that, on the open set $\{p > 0\}$, we have $\Delta(p + V) \geq 0$.

   In order to do this, we consider the (optimal) transport map from $\bar{\rho}$ to μ given by $\mathrm{Id} - \nabla\varphi_\mu$, and similarly for ν. Let us define the following quantity:

$$D(x) := -\frac{\log(\mu(x - \nabla\varphi_\mu(x))) + \log(\nu(x - \nabla\varphi_\nu(x))) - 2\log(\bar{\rho}(x))}{\tau^2}.$$

Notice that if $\bar{\rho}(x) = 1$, then by the constraint $\mu(x - \nabla\varphi_\mu(x)) \leq 1$ and $\nu(x - \nabla\varphi_\nu(x)) \leq 1$ the quantity $D(x)$ is positive. On the other hand, using $(\mathrm{Id} - \nabla\varphi_\mu)_\#\bar{\rho} = \mu$ and the Monge-Ampére equation, for all $x \in \mathring{\Omega}$ there holds

$$\mu(x - \nabla_\mu\varphi_\mu(x)) = \frac{\bar{\rho}(x)}{\det(I - D^2\varphi_\mu(x))},$$

and a similar identity holds for φ_ν. Hence the quantity $D(x)$ is equal, for all $x \in \mathring{\Omega}$, to

$$D(x) = \frac{\log(\det(I - D^2\varphi_\mu(x))) + \log(\det(I - D^2\varphi_\nu(x)))}{\tau^2}.$$

Diagonalizing the matrices $D^2\varphi_\mu$, $D^2\varphi_\nu$ and using the convexity inequality $\log(1 - y) \leqslant -y$, we end up with

$$D(x) \leqslant -\frac{\Delta(\varphi_\mu(x) + \varphi_\nu(x))}{\tau^2}.$$

This shows that, on the region $\{p > 0\}$, we have the desired inequality $\Delta(p + V) \geq 0$, thanks to (2.21).

We want now to determine where does $p + V$ attain its maximum. Because of subharmonicity this should be on the boundary of $\{p > 0\}$. This boundary is composed by points on $\partial\Omega$ and by points where $p = 0$.

To handle the boundary $\partial\Omega$, recall that $\nabla\varphi_\mu$ is continuous up to the boundary and that $x - \nabla\varphi_\mu(x) \in \Omega$ for every $x \in \Omega$ as $(\mathrm{Id} - \nabla\varphi_\mu)_\#\bar\rho = \mu$. Given the convexity of Ω, it implies $\nabla\varphi_\mu(x) \cdot \mathbf{n}_\Omega(x) \geqslant 0$ for every point $x \in \partial\Omega$, where $\mathbf{n}_\Omega(x)$ is the corresponding outward normal vector. This translates, applying this first to φ_μ and then to φ_ν, into $\nabla(p + V)(x) \cdot \mathbf{n}_\Omega(x) \leq 0$. We are then in this situation: a certain function u satisfies $\Delta u \geq 0$ in the interior of a domain (which is here $\{p > 0\}$) and $\partial u/\partial n \leq 0$ on a part of the boundary. By applying an easy maximum principle to $u_\varepsilon := u + \varepsilon v$ where v is a fixed harmonic function with $\partial v/\partial n < 0$ on the same part of the boundary shows that the maximum of u is attained on the other part of the boundary (we prefer not to evoke Hopf's lemma as we do not want to discuss the regularity of $\partial\Omega$, and we do not need the strong maximum principle). We then deduce that the maximum of $p + V$ is attained on $\{p = 0\}$.

This easily implies

$$\max_{\{p>0\}} p + \min_{\{p>0\}} V \leq \max_{\{p>0\}} (p + V) \leq \max_{\{p=0\}} V,$$

which gives the claim.   □

Remark 2.8 The same proof actually shows the stronger inequality $p + V \leq \max V$.

2.5.2 Optimality Conditions and Regularity of P

We look now at the optimality conditions satisfied by ρ_N. The situation is even simple than the one in Sect. 2.5.1. Set $\bar\rho := \rho_N$ and $\mu := \rho_{N-1}$. We can see that $\bar\rho$ is a minimizer, among all probability measures with density bounded by 1, of

$$\rho \mapsto \frac{W_2^2(\mu, \rho)}{2\tau} + \int_\Omega \Psi \, d\rho + \lambda \int_\Omega \rho \log(\rho) \, dx.$$

This time, we will assume that Ψ is smooth, but the estimates on P will not depend on its smoothness. As most of the arguments are the same as in Sect. 2.5.1 we resume the results in just two statements.

Proposition 2.5.5 *The optimal $\bar{\rho}$ is strictly positive a.e.. Denoting by φ_μ the Kantorovich potential for the transport from $\bar{\rho}$ to μ (which is unique up to additive constants), there exists $P \in L^1(\Omega)$, positive, such that $\{p > 0\} \subset \{\bar{\rho} = 1\}$ and a constant C such that*

$$\frac{\varphi_\mu}{\tau} + \Psi + P + \lambda \log(\bar{\rho}) = C \quad a.e. \tag{2.24}$$

Moreover $\varphi_\mu \in C^{2,\alpha}(\overset{\circ}{\Omega}) \cap C^{1,\alpha}(\Omega)$, P and $\log(\bar{\rho})$ are Lipschitz continuous, and $P \in C^{2,\alpha}(\{P > 0\})$.

Proof The proof is just an adaptation of those of Lemma 2.5.1, Proposition 2.5.2, and Lemma 2.5.3.   □

Theorem 2.5.6 *We have the following L^∞ estimate:*

$$P \leq \max \Psi - \min \Psi.$$

Proof The proof is just an adaptation of that of Theorem 2.5.4, defining now

$$D(x) := -\frac{\log(\mu(x - \nabla\varphi_\mu(x))) - \log(\bar{\rho}(x))}{\tau}.$$

□

Another useful result concerns the H^1 regularity of P. This results could have also be obtained in the case of p, and improves the result of [13] (since it consists in $L_t^\infty H_x^1$ regularity under the only assumption $V \in H^1$ compared to $L_t^2 BV_x$ for $V \in C^{1,1}$, but in [13] more general cost functions (with non-quadratic Hamiltonians) were also considered. Anyway, it is only for P that we will use it.

Theorem 2.5.7 *Suppose $\Psi \in H^1(\Omega)$. We then have $P \in H^1(\Omega)$ and*

$$\int_\Omega |\nabla P|^2 \leq \int_\Omega |\nabla\Psi|^2.$$

Proof In the proof of Theorem 2.5.6, which is based on that of Theorem 2.5.4, we also obtained $\Delta(\Psi + P) \geq 0$ on $\{P > 0\}$. By multiplying times P and integrating by parts, we obtain

$$\int_\Omega |\nabla P|^2 \leq -\int_\Omega \nabla\Psi \cdot \nabla P, \tag{2.25}$$

from which the claim follows.   □

Remark 2.9 From the inequality (2.25) wa can also obtain $\int |\nabla P|^2 + \int |\nabla(P + \Psi)|^2 \leq \int |\nabla\Psi|^2$, which is a stronger result.

2.5.3 Approximation and Conclusions

We now want to explain how to deduce results on the continuous-time pressure p from the estimates that we detailed in the discrete case. We fix a, integer number $N > 1$ and take $\tau = T/N$ as a time step. We will build an approximate value function ϕ^N together with an approximate pressure p^N which will converge, as $N \to +\infty$, to a pair which solves the (continuous) dual problem.

Let us start from the solution of the discrete problem $\bar\rho^N := (\bar\rho_0^N, \bar\rho_1^N, \ldots, \bar\rho_N^N)$. For any $k \in \{0, 1, \ldots, N-1\}$, we choose (φ_k^N, ψ_k^N) a pair of Kantorovich potential between $\bar\rho_k^N$ and $\bar\rho_{k+1}^N$, such choice being unique up to an additive constant. We then know that there exist a pressure p_k^N and P^N, positive and Lipschitz, and constants C_k^N and C^N such that

$$\begin{cases} \frac{\psi_{k-1}^N+\varphi_k^N}{\tau^2} + V_N + p_k^N + \lambda_N \log(\bar\rho_k^N) = C_k^N & k \in \{1, 2, \ldots, N-1\}, \\ \frac{\psi_{k-1}^N}{\tau} + \Psi_N + P^N + \lambda_N \log(\bar\rho_k^N) = C^N & k = N. \end{cases} \quad (2.26)$$

We define the following value function, defined on the whole interval $[0, T]$ which can be thought as a function which looks like a solution of what could be called a discrete dual problem.

Definition 2.5.8 Let ϕ^N the function defined as follows. The "final" value is given by

$$\phi^N(T^-, \cdot) := \Psi + P^N. \quad (2.27)$$

Provided that the value $\phi^N((k\tau)^-, \cdot)$ is defined for some $k \in \{1, 2, \ldots, N\}$, the value of ϕ^N on $((k-1)\tau, k\tau) \times \Omega$ is defined by

$$\phi^N(t, x) := \inf_{y\in\Omega} \left(\frac{|x-y|^2}{2(k\tau-t)} + \phi^N((k\tau)^-, y) \right). \quad (2.28)$$

If $k \in \{1, 2, \ldots, N-1\}$, the function ϕ^N has a temporal jump at $t = k\tau$ defined by

$$\phi^N((k\tau)^-, x) := \phi^N((k\tau)^+, x) + \tau \left(V_N + p_k^N \right)(x) \quad (2.29)$$

We now also define a measure $\pi \in \mathcal{M}([0, T] \times \Omega)$ which will play the role of the continuous pressure.

Definition 2.5.9 Let π^N be the positive measures on $[0, T] \times \Omega$ defined in the following way: for any test function $a \in C([0, 1] \times \Omega)$, we set

$$\int_{[0,1]\times\Omega} a \, d\pi^N := \tau \sum_{k=1}^{N-1} \int_\Omega a(k\tau, \cdot)p_k^N + \int_\Omega a(T, \cdot)P^N.$$

In other words, π^N is a sum of singular measures corresponding to the jumps of the value function ϕ^N.

Provided that we set $\phi^N(0^-, \cdot) = \phi^N(0^+, \cdot)$ and $\phi^N(T^+, \cdot) = \Psi_N$, the following equation holds in the sense of distributions on $[0, 1] \times \Omega$:

$$-\partial_t \phi^N + \frac{1}{2}|\nabla \phi^N|^2 \leq \pi^N + V \tag{2.30}$$

It is then possible to prove the following (see Section 4 in [27]).

Theorem 2.5.10 *The sequence* (ϕ^N, π^N) *is bounded in* $\left(BV([0, T] \times \Omega) \cap L^2([0, T]; H^1(\Omega))\right) \times \mathcal{M}([0, T] \times \Omega)$ *and converges, up to subsequences, to a pair* $(\bar\phi, \bar\pi) \in \tilde{\mathcal{D}}$, *the convergence being in the sense of distributions. This limit pair* $(\bar\phi, \bar\pi) \in \tilde{\mathcal{D}}$ *is optimal in the relaxed dual problem. When the functions* p_k^N *and* P^N *are uniformly bounded in* L^∞ *then the measure* $\bar\pi$ *is the sum of an* L^∞ *density (w.r.t. the space-time Lebesgue measure* $\mathcal{L}^{d+1}$) p *on* $[0, T] \times \Omega$ *and of a singular part on* $t = T$ *with an* L^∞ *density (with respect to the space Lebesgue measure* $\mathcal{L}^d$) P, *and we can write System (2.19). Moreover,* $\bar\phi$ *is the value function of the value function of an optimization problem of the form (2.1) for a running cost given by* $V + p$ *and a final cost given by* $\Psi + P$.

Remark 2.10 The reader can observe that we obtain here the existence of an optimal pair $(\phi, \pi) \in \tilde{\mathcal{D}}$, as in Theorem 2.2.2. This was already proven in [13] without passing through the discrete approximation.

It remains to be convinced that the optimal measure Q in the Lagrangian problem, in the present case of density constraints, optimizes a functional of the form $J_{\Psi,h}$. This was obtained in the density-penalized case by differentiating along perturbations Q_ε but here the additional term in h is not obtained as a derivative of $G(\rho)$ but comes from the constraint and is in some sense a Lagrange multiplier (and a similar term appears at $t = T$). This makes the proof more difficult, but we can obtain the desired result by using the duality.

Theorem 2.5.11 *Suppose that* (ϕ, π) *is an optimal pair in the relaxed dual problem and that* π *decomposes into a density* $p \in L^1([0, T] \times \Omega)$ *and a singular measure on* $\{t = T\}$ *with a density* $P \in L^1(\Omega)$. *Then we have*

- *for every measure* $Q \in \mathcal{P}(C)$ *such that* $(e_t)_\# Q$ *is uniformly* L^∞ *and* $(e_0)_\# Q = \bar\rho_0$, *we have* $J_{\Psi+P,V+p}(Q) \geq \int \phi(0^+)d\bar\rho_0$,
- *if* $\bar Q$ *is optimal in (2.8) for the density-constrained problem (i.e. when* $G = I_{[0,1]}$), *then we have* $J_{\Psi+P,V+p}(\bar Q) = \int \phi(0^+)d\bar\rho_0$,

In particular $\bar Q$ *optimizes* $J_{\Psi+P,V+p}$ *among measures on curves such that* $(e_t)_\# Q$ *is uniformly* L^∞ *and, when* $\Psi + P$ *and* $V + p$ *are* L^∞, *it is concentrated on curves optimizing* $K_{\widehat{\Psi+P},\widehat{V+p}}$.

Proof In order to prove the first statement, we consider a pairs of functions $\phi \in C^1([0, T] \times \Omega)$ and $h \in C^0([0, T] \times \Omega)$ such that $-\partial_t \phi + \frac{1}{2}|\nabla \phi|^2 \leq h$. We then have

$$J_{\phi(T),h}(Q) = \int dQ(\gamma) \left(\int_0^T \left(\frac{1}{2}|\gamma'(t)|^2 + h(t, \gamma(t)) \right) dt + \phi(T, \gamma(T)) \right)$$

and for every curve γ, using $-\partial_t \phi + \frac{1}{2}|\nabla|^2 \leq h$ and $\frac{1}{2}|\gamma'(t)|^2 + \frac{1}{2}|\nabla\phi(t, \gamma(t))|^2 \geq -\nabla\phi(t, \gamma(t)) \cdot \gamma'(t)$, we have

$$\int_0^T \left(\frac{1}{2}|\gamma'(t)|^2 + h(t, \gamma(t)) \right) dt + \phi(T, \gamma(T)) \geq \int_0^T \frac{d}{dt}\phi(t, \gamma(t)) \, dt + \phi(T, \gamma(T)) = \phi(0, \gamma(0)).$$

This would be sufficient to prove the desired inequality if we had enough regularity. The same inequality in the case of the optimal relaxed function ϕ together with $h = V + p$ can be obtained if we regularize by space-time convolution. Let us consider a convolution kernel η supported in $[0, 1] \times B_1$, and use convolutions with rescaled versions of this kernel $\eta_\delta(t, x) = \delta^{-(d+1)}\eta(t/\delta, x/\delta)$, so that we do not need to look at times $t < 0$. On the other hand, this requires first to extend ϕ for $t > T$, and it can be done by taking $\phi(t, x) = \Psi(x) + P(x)$ for every $t > T$. As a consequence, one should also extend $h := V + p$, and in this case we use $h(t, x) := \frac{1}{2}|\nabla(\Psi + P)|^2$, which belongs to L^1 thanks to Theorem 2.5.7 (this explains why we prefer to do an asymmetric convolution looking at the future and not at the past, since we do not know whether $\phi_0 \in H^1$ or not). It is then necessary to extend ϕ and h outside Ω as well, for space convolution. As we assumed that the boundary of Ω is smooth, there exists a C^1 map R, defined on a neighborhood of Ω and valued into Ω, such that its jacobian $DR(x)$ has a determinant bounded from below and from above close to $\partial\Omega$ and its operator norm $||DR(x)||$ tends to 1 as $d(x, \partial\Omega) \to 0$ (a typical example is the reflection map when $\Omega = \{x_1 > 0\}$, possibly composed with a diffeomorphism which rectifies the boundary). Then, It is enough to define $\tilde{\phi}_\varepsilon(t, x) := \phi((1 + \varepsilon)t, R(x))$ and $\tilde{h}_\varepsilon(t, x) := (1 + \varepsilon)h((1 + \varepsilon)t, R(x))$ and take $\phi_\varepsilon := \eta_\delta * \tilde{\phi}_\varepsilon$ and $h_\varepsilon := \eta_\delta * \tilde{h}_\varepsilon$, for a suitable choice $\delta = \delta_\varepsilon$, provided δ_ε is such that $||DR(x)|| \leq \sqrt{1 + \varepsilon}$ for x such that $d(x, \partial\Omega) \leq \delta_\varepsilon$. In this way we obtain smooth functions $(\phi_\varepsilon, h_\varepsilon)$ such that $-\partial_t\phi_\varepsilon + \frac{1}{2}|\nabla\phi_\varepsilon|^2 \leq h_\varepsilon$. This allows to write

$$J_{\phi_\varepsilon(T),h_\varepsilon}(Q) \geq \int \phi_\varepsilon(0) \, d\overline{\rho_0}.$$

We then need to pass to the limit as $\varepsilon \to 0$. We have $h_\varepsilon \to h$ in L^1 which, together with the L^∞ bound on $(e_t)_\# Q$, allows to deal with the h-term. The kinetic term does not depend on h, and we are only left to consider the terms with $\phi_\varepsilon(T)$ and $\phi_\varepsilon(0)$: since ϕ is a BV function, these functions converge in $L^1(\Omega)$ to $\phi(0^+)$ and $\phi(T^+) = \Psi + P$, respectively, which provides the desired inequality.

We are now left to prove that we have equality if we choose $Q = \bar{Q}$, the optimal measure on curves. For this, we use the equality between the primal and the dual problem (knowing that the value of the primal can be expressed either in its Eulerian formulation or in its Lagrangian one). We then have

$$\int_C K_{\Psi,V}\,\mathrm{d}\bar{Q} = \int \phi_\varepsilon(0)\,\mathrm{d}\overline{\rho_0} - \int_{[0,T]\times\Omega}\,\mathrm{d}\pi = \int \phi_\varepsilon(0)d\overline{\rho_0} - \int_{[0,T]\times\Omega} p - \int_\Omega P.$$

We then use the fact that we have, by primal-dual optimality conditions (which can also be seen in System (2.19)), $p_t(1 - \rho_t) = 0$ and $P(1 - \rho_T) = 0$, where $\rho_t = (e_t)_\# \bar{Q}$. Then we obtain

$$\int_C K_{\Psi,V}\,\mathrm{d}\bar{Q} = \int \phi_\varepsilon(0)\,\mathrm{d}\overline{\rho_0}\int_{[0,T]\times\Omega} p\,\mathrm{d}(e_t)_\#\bar{Q} - \int_\Omega P\,\mathrm{d}(e_T)_\#\bar{Q},$$

which can be re-written in terms of $J_{\Psi+P,V+p}$ and gives the claim. $\qquad\square$

References

1. L. Ambrosio, A. Figalli, On the regularity of the pressure field of Brenier's weak solutions to incompressible Euler equations. Calc. Var. PDE **31**(4), 497–509 (2008)
2. L. Ambrosio, A. Figalli, Geodesics in the space of measure-preserving maps and plans. Arch. Rational Mech. Anal. **194**, 421–462 (2009)
3. L. Ambrosio, N. Gigli, G. Savaré, *Gradient Flows in Metric Spaces and in the Space of Probability Measures.* Lectures in Mathematics (Birkhäuser, Basel, 2005)
4. J.-D. Benamou, Y. Brenier, A computational fluid mechanics solution to the Monge-Kantorovich mass transfer problem. Numer. Math. **84**, 375–393 (2000)
5. J.D. Benamou, G. Carlier, F. Santambrogio, Variational mean field games, in *Active Particles, Volume 1: Theory, Models, Applications*, ed. by N. Bellomo, P. Degond, E. Tadmor (Springer, Berlin, 2017), pp. 141–171
6. G. Bouchitté, G. Buttazzo, New lower semicontinuity results for nonconvex functionals defined on measures. Nonlinear Anal. **15**, 679–692 (1990)
7. Y. Brenier, Minimal geodesics on groups of volume-preserving maps and generalized solutions of the Euler equations. Commun. Pure Appl. Math. **52**(4), 411–452 (1999)
8. G. Buttazzo, C. Jimenez, E. Oudet, An optimization problem for mass transportation with congested dynamics. SIAM J. Control Optim. **48**, 1961–1976 (2010)
9. P. Cardaliaguet, Notes on mean field games. https://www.ceremade.dauphine.fr/~cardalia/MFG20130420.pdf
10. P. Cardaliaguet, Weak solutions for first order mean field games with local coupling (2013). http://arxiv.org/abs/1305.7015
11. P. Cardaliaguet, J. Graber, Mean field games systems of first order. ESAIM: Control Optim. Calc. Var. **21**(3), 690–722 (2015)
12. P. Cardaliaguet, J. Graber, A. Porretta, D. Tonon, Second order mean field games with degenerate diffusion and local coupling. Nonlinear Differ. Equ. Appl. **22**, 1287–1317 (2015)
13. P. Cardaliaguet, A.R. Mészáros, F. Santambrogio, First order mean field games with density constraints: pressure equals price. SIAM J. Control Optim. **54**(5), 2672–2709 (2016)
14. G. Carlier, F. Santambrogio, A continuous theory of traffic congestion and Wardrop equilibria, in *Proceedings of Optimization and Stochastic Methods for Spatially Distributed Information*, St Petersburg (2010). J. Math. Sci. **181**(6), 792–804 (2012)
15. G. Carlier, C. Jimenez, F. Santambrogio, Optimal transportation with traffic congestion and Wardrop equilibria. SIAM J. Control Optim. **47**, 1330–1350 (2008)

16. R.J. DiPerna, P.-L. Lions, Ordinary differential equations, transport theory and Sobolev spaces. Invent. Math. **98**, 511–547 (1989)
17. I. Ekeland, R. Temam, Convex Analysis and Variational Problems. Classics in Mathematics (Society for Industrial and Applied Mathematics, Philadelphia, 1999)
18. P.J. Graber, Optimal control of first-order Hamilton-Jacobi equations with linearly bounded Hamiltonian. Appl. Math. Optim. **70**(2), 185–224 (2014)
19. P.J. Graber, A. Mészáros Sobolev regularity for first order Mean Field Games. Annales de l'Institut Henri Poincaré (C) Analyse Non Linéaire **35**(6), 1557–1576 (2018)
20. P.J. Graber, A.R. Mészáros, F. Silva, D. Tonon, The planning problem in Mean Field Games as regularized mass transport. Calc. Var. Partial Differ. Equ. **58**, 115 (2019)
21. M. Huang, R.P. Malhamé, P.E. Caines, Large population stochastic dynamic games: closed-loop McKean-Vlasov systems and the Nash certainty equivalence principle. Commun. Inf. Syst. **6**(3), 221–252 (2006)
22. R. Jordan, D. Kinderlehrer, F. Otto, The variational formulation of the Fokker-Planck equation. SIAM J. Math. Ann. **20**, 1–17 (1998)
23. J.-M. Lasry, P.-L. Lions, Jeux à champ moyen. I. Le cas stationnaire. C. R. Math. Acad. Sci. Paris **343**(9), 619–625 (2006)
24. J.-M. Lasry, P.-L. Lions, Jeux à champ moyen. II. Horizon fini et contrôle optimal. C. R. Math. Acad. Sci. Paris **343**(10), 679–684 (2006)
25. J.-M. Lasry, P.-L. Lions, Mean field games. Jpn. J. Math. **2**(1), 229–260 (2007)
26. H. Lavenant, F. Santambrogio, Optimal density evolution with congestion: L^∞ bounds via flow interchange techniques and applications to variational Mean Field Games. Commun. Partial Differ. Equ. **43**(12), 1761–1802 (2018)
27. H. Lavenant, F. Santambrogio, New estimates on the regularity of the pressure in density-constrained mean field games. J. Lond. Math. Soc. **100**(2), 644–667 (2019)
28. P.-L. Lions, Cours au Collège de France. www.college-de-france.fr
29. G. Loeper The reconstruction problem for the Euler-Poisson system in cosmology. Arch. Rational Mech. Anal. **179**(2), 153–216 (2006)
30. D. Matthes, R.J. McCann, G. Savaré, A family of nonlinear fourth order equations of gradient flow type. Commun. Partial Differ. Equ. **34**, 1352–1397 (2009)
31. B. Maury, A. Roudneff-Chupin, F. Santambrogio, A macroscopic crowd motion model of gradient flow type. Math. Models Methods Appl. Sci. **20**(10), 1787–1821 (2010)
32. R.J. McCann, A convexity principle for interacting gases. Adv. Math. **128**, 153–179 (1997)
33. J. Moser, A new proof of De Giorgi's theorem concerning the regularity problem for elliptic differential equations. Commun. Pure Appl. Math. **13**(3), 457–468 (1960)
34. C. Orrieri, A. Porretta, G. Savaré A variational approach to the mean field planning problem. J. Funct. An. **277**(6), 1868–1957 (2019)
35. A. Porretta, On the planning problem for the mean field games system. Dyn. Games Appl. **4**, 231–256 (2014)
36. A. Porretta, Weak Solutions to Fokker-Planck equations and mean field games. Arch. Ration. Mech. Anal. **216**, 1–62 (2015)
37. A. Prosinski, F. Santambrogio, Global-in-time regularity via duality for congestion-penalized Mean Field Games. Stochastics **89**, (6–7) (2017). Proceedings of the Hammamet Conference, 19–23 October 2015, 923–942
38. F. Santambrogio, A modest proposal for MFG with density constraints. Netw. Heterog. Media **7**(2), 337–347 (2012)
39. F. Santambrogio, Optimal transport for applied mathematicians, in *Progress in Nonlinear Differential Equations and Their Applications*, vol. 87 (Birkhäuser, Basel, 2015)
40. F. Santambrogio, {Euclidean, metric, and wasserstein} gradient flows: an overview. Bull. Math. Sci. **7**(1), 87–154 (2017)
41. F. Santambrogio, Crowd motion and population dynamics under density constraints. ESAIM: Proceedings SMAI 2017 - 8e Biennale Française des Mathématiques Appliquées et Industrielles, vol. 64, (2018), pp. 137–157

42. F. Santambrogio, Regularity via duality in calculus of variations and degenerate elliptic PDEs. J. Math. Anal. Appl. **457**(2), 1649–1674 (2018)
43. C. Villani, *Topics in Optimal Transportation*. Graduate Studies in Mathematics (AMS, Providence, 2003)
44. J.G. Wardrop, Some theoretical aspects of road traffic research. Proc. Inst. Civ. Eng. **2**, 325–378 (1952)

Chapter 3
Master Equation for Finite State Mean Field Games with Additive Common Noise

François Delarue

Abstract The goal of these notes is to address the solvability of the master equation for mean field games with a common noise. Whilst the methodology is mostly inspired from earlier works in the field on continuous state mean field games, see in particular the monograph (P. Cardaliaguet et al., The Master Equation and the Convergence Problem in Mean Field Games. Annals of Mathematics Studies, vol. 201. Princeton University Press, Princeton, NJ, 2019), the text focuses on a specific type of finite state mean field games subjected to a common noise. Although the rationale for switching from continuous to discrete state spaces is mostly dictated by pedagogical reasons, it turns out that not only the results in their own but also the structure of the underpinning common noise are new in the literature on mean field games.

3.1 Introduction

3.1.1 Mean Field Games with a Common Noise

In the theory of Mean Field Games (MFGs), the master equation was first introduced by Lions in his seminal lectures at *Collège de France* on MFGs. Although we are here at an early stage of the notes, we feel fair to refer the reader to [35] for the whole collection of videos and also to [36] for a more specific seminar that Lions gave on the solvability of this equation.

One of Lions' motivation for reformulating MFGs through the master equation was precisely to address games subjected to a common noise. This may be easily understood: The usual characterization of MFGs comes through a forward-backward system of two Partial Differential Equations (PDEs), one forward Fokker–Planck (FP) equation describing the evolution of the statistical state of the popula-

F. Delarue (✉)
Université Côte d'Azur, CNRS, Laboratoire J.A. Dieudonné, Nice, France
e-mail: delarue@unice.fr

© The Editor(s) (if applicable) and The Author(s), under exclusive license
to Springer Nature Switzerland AG 2020

P. Cardaliaguet, A. Porretta (eds.), *Mean Field Games*,
Lecture Notes in Mathematics 2281, https://doi.org/10.1007/978-3-030-59837-2_3

tion at equilibrium and one backward Hamilton–Jacobi–Bellman (HJB) equation describing the evolution of the cost to one tagged player in the population when all the others follow the equilibrium policy given by the forward equation. Without any common noise, this system has been widely studied in the literature and in various settings: For deterministic or stochastic games, for a finite or an infinite time horizon, for a continuous or a finite state space... However, the story becomes somewhat different whenever the game is subjected to a common (or systemic) noise.

From a modelling prospect, MFGs with a common noise are designed to handle large stochastic games in which all the players are subjected to a common source of randomness. Obviously, such games with hence correlated noises come in contrast to most of the stochastic models that have been treated in the MFG literature. Most of the time, players, whenever there are finitely many of them, are indeed subjected to idiosyncratic noises only—i.e., to noises that are independent—. However, as demonstrated in earlier texts on MFGs, see for instance [9, 12, 24, 26, 30] for a tiny example, MFGs with a common noise are of a great interest from the practical point of view: Should we have to give one illustration, say that, in economics or in finance, incorporating common sources of randomness might be especially important in systemic risk analysis (see again [12]). On a more elaborated level, which we however do not discuss in the rest of the paper, MFGs with major and minor players form also an important class of examples where all the players are correlated, see for instance [31, 39–41] for pioneering results in this direction together with the more recent contributions [7, 10, 11]. One key fact with all the models featuring some correlation is that equilibria—i.e., solutions of the corresponding MFG—become random themselves under the action of the common noise. This feature comes in fact as a by-product of earlier results on the asymptotic behavior of large (but uncontrolled) systems of particles with mean field interaction that are subjected to a common noise, see among others [17, 33, 34, 47]. Roughly speaking, the standard property of propagation of chaos, which is at the roots of the mean field formulation, then remains true, but at the price of conditioning upon the realization of the common noise. Equivalently, propagation of chaos becomes conditional and, accordingly, the mean field limit is random and reads (at least in simpler cases) as a measurable function of the common noise. Recast in the framework of MFGs, this says that equilibria can no longer be described by a standard FP equation. Because of the common noise, the latter becomes stochastic and, in turn, the HJB equation also becomes stochastic. At the end of the day, the whole MFG system is stochastic, which makes it of a more intricate nature than in the standard case without common noise. To wit, because it is set backwards in time, the stochastic HJB equation popping up in the MFG system cannot be a mere randomization of the standard deterministic HJB equation addressed in MFGs without common noise. As we explain later on in the text, it must be understood as a Backward Stochastic Differential Equation (BSDE) using the terminology from the seminal work of Pardoux and Peng [43], see also on this subject the two recent monographs [44, 48] together with Sect. 3.3 for the application to our framework.

To the best of our knowledge, the first description of the MFG system for games with a common noise is due to [6]. The framework therein is quite general and addresses games with a continuous state space. Accordingly, both the FP and HJB equations read in the form of two Stochastic Partial Differential Equations (SPDEs), the HJB equation being in fact a Backward SPDE, namely a BSDE set on a space of infinite dimension. Subsequently, the fact that the MFG system is both stochastic and infinite dimensional makes the overall analysis rather difficult. Our aim below is mostly to revisit the arguments developed in [6] but in the simpler case when the MFG system is finite dimensional. To do so, we just work here on state spaces that are finite. Similar to the content of [6], our first result in this framework is to prove existence and uniqueness of a solution to the MFG system under the so-called Lasry–Lions monotonicity condition. Although it is of course of a somewhat limited scope from a practical point of view, the monotonous setting is in fact pretty convenient for addressing the solvability of the related master equation (importantly, this remark also applies when there is no common noise, see [6, 9, 16] together with [2, 14] for finite state MFGs). Moreover, the reader must be also aware of the fact that, as demonstrated in [13] and [9, Chapter 3], solutions to MFGs with a common noise may be of a rather subtle nature: Similar to weak solutions of Stochastic Differential Equations (SDEs), they may not be adapted to the filtration generated by the common noise. Fortunately, there exists a version of the Yamada–Watanabe theorem states that, whenever a strong form of uniqueness holds true, the hence unique solution is necessarily strong, namely it must be an adapted function of the common noise. In short, this is exactly what happens under the monotonicity condition and this makes the overall analysis much simpler, which is another strong case for restricting the entire analysis here to the monotonous setting. Even more, it is fair to say that the monotonous setting is actually so robust that, in the end, there is no need to invoke any Yamada–Watanabe argument in order to guarantee that the solutions are indeed strong. Indeed, as we explain later on, existence and uniqueness are shown to hold true by means of a continuation argument, which is the same as the one used in [6] and which is in fact quite standard in the literature on forward-backward SDEs (at least whenever the latter are in finite-dimension), see among others [45]. Again, part of our objective here is to explain how this continuation argument works in the presence of a common noise but in a simpler setting than the one addressed in [6].

3.1.2 Master Equation

We already alluded to the master equation in the previous paragraph. Generally speaking, the master equation is a nonlinear PDE set on an enlarged state space, the latter being obtained by tensorizing the physical state space carrying a tagged player with the probability space carrying the distributions of all the other players. In particular, whenever the MFG is over a continuous state space, the master equation is a PDE set on an infinite dimensional space. Obviously, the latter fact

makes it rather difficult to study. This is all the more true that the underlying
infinite dimensional space is not flat, which requires some care when defining a
relevant form of derivatives. In this respect, several approaches are conceivable:
We may think of embedding the space of probability measures into the space of
(signed) measures, in which case the derivative is said to be flat; We may also
follow earlier works from optimal transportation theory and equip the so-called
Wasserstein space of probability measures with a finite second moment with a kind
of Riemannian structure (see for instance [1, 37, 42]); Lastly, we may also lift the
space of probability measures onto an L^2 space of random variables and then use
Fréchet derivatives on this Hilbert space (see [5, 35] together with [8, Chapter 5]).
At the end of the day, all these notions of derivatives lead to the same form of master
equation: Compare for instance [6], in which the flat derivative is used, with [16] and
[9, Chapter 5], in which the L^2 approach is preferred. In contrast, we just focus here
on a finite dimensional version of the master equation, which is consistent with the
fact that we restrict ourselves to a finite state space. Fortunately, there is no need in
this framework to introduce derivatives on the Wasserstein space, since probability
measures on the state space are then identified with elements of a finite-dimensional
simplex.

Whether the state space is finite or continuous, a common possible approach
to the master equation is to regard the underpinning MFG system as a system of
characteristics. This fits exactly the approach used in [6]. Unfortunately, the latter
remains rather complicated due to the fact that the state space therein is continuous,
hence our choice here to switch to a finite state space. Anyway, such an approach—
as it based upon characteristics—requires the MFG system to be uniquely solvable,
which is indeed the case whenever the Lasry–Lions monotonicity condition is in
force. In this regard, it is worth saying that, in the MFG folklore, very few is known
about the master equation when the MFG system is not uniquely solvable: We refer
to [15, 20] for two very specific instances when the master equation reduces to a
scalar conservation law and can hence be studied even though the characteristics are
not unique. Actually, one the main result of [6] is to show that, in the monotonous
setting and under sufficiently strong regularity assumptions on the coefficients, the
master equation has a unique classical solution. For sure, this requires much more
than proving that the MFG system is uniquely solvable. In a shell, the idea developed
in [6] (see also [25] but in short time and without common noise) is to prove that the
MFG system defines a flow that is differentiable with respect to the initial condition,
the latter being understood as the initial state of the population. The derivative of the
flow is then shown to solve a linearized version of the MFG system, which turns out
to be uniquely solvable in the monotonous setting. In this respect, it is fair to say
that the monotonous setting actually permits to kill two birds with one stone: Not
only this allows us to solve for the MFG system and for the linearized version of
it, but it also supplies us with sufficiently strong stability estimates to prove that the
linearized version is indeed the derivative of the flow and that it is itself regular with
respect to the underlying initial condition. A probabilistic variant of this approach
is used in [16] and [9, Chapter 5]: Therein, the point is to directly differentiate the
L^2-valued flow generated by the (random) state of a tagged player in the population

whenever the latter is at equilibrium (this is obviously different from the approach used [6] which consists in differentiating the measure-valued flow generated by the state of the population.) We here reproduce the approach used in [6], but again we hope that the finite state framework makes it easier for the reader. For another approach to the solvability of the master equation, we refer to the recent contribution [7].

Generally speaking, one of the interest of the master equation is that it permits to address the convergence of the equilibria of the corresponding N-player game towards the solution of the MFG. This point was proved first in [6] and then extended in different ways in other contributions: See [2, 14] for the analogue for finite state MFGs, including as well the analysis of the fluctuations and of the large deviations; See [21, 22] for the fluctuations and the large deviations for continuous state MFGs; See [9] for MFGs with Hamiltonians of different growths. Our feeling is that this aspect of the theory of MFGs has been well understood (to wit, the increasing number of publications). Also, since the argument is pretty much the same whether there is or not a common noise, we feel useless to address it in these notes, even though this is certainly an important piece of the field. Instead, we feel better to focus here on the construction of a classical solution to the master equation with a common noise, which remains, to our point of view, relatively little addressed in the literature (except in the aforementioned references).

Lastly, it is worth mentioning that there is a subtle difference between the master equation for MFGs without common noise and the master equation for MFGs with a common noise. In short, the master equation for MFGs with a common noise features additional second-order terms in the direction of the measure. This is exemplified in [6] for continuous state MFGs, but at the price of a relevant notion of second-order derivatives for functions defined on the space of probability measures. In the discrete setting addressed below, things are simpler since the common noise is shown to manifest in the master equation through an additional Laplace operator, see Eq. (3.25). However, we feel important to stress that, most of the time and whatever the cardinality of the state space, the second-order structure induced by the common noise is degenerate. This is the case in [6]: Therein, the common noise is finite-dimensional only whilst the space of probability measures is of infinite dimension. This is also the case in these notes: The additional Laplace operator that is here generated by our choice of common noise just acts on the elements of the state space and not on the weights of those elements under the distribution of the equilibrium. To make the latter point clear, it might be welcome to give a flavor of the form of the common noise used below. In a shell, we let the common noise act additively onto the state space, namely the elements of the state space are shifted, with time, along the realization of some Brownian motion. Our choice for an additive common noise is hence completely consistent with the framework addressed in [6], which is also additive. On a more prospective level, investigating more complex types of common noise for which the master equation may be non-degenerate turns out to be a very interesting question: We refer to [46] for a first example for linear-quadratic MFGs, to [19] for a more general example with an infinite dimensional common noise, and finally to [3] for another example but over a finite state space.

3.1.3 Finite State MFGs with a Common Noise

MFGs with a finite state space were introduced in [27–29]. We also refer to [8, Chapter 7] and to the references therein for more examples.

Generally speaking, and at least when there is no common noise, MFGs with a finite state space are easier to handle than MFGs with a continuous state space. To wit, the MFG system in that case becomes a mere forward-backward system of two ODEs. Accordingly, the shape of the master equation becomes simpler since, as we already alluded to, there is then no real need for a differential calculus on the infinite dimensional space of probability measures. In short, it is indeed sufficient to use mere derivatives on the finite dimensional simplex in order to formulate the master equation, the latter reading in the end as a nonlinear system of first-order PDEs. Existence of a classical solution to the finite state variant of the master equation is addressed in the aforementioned two references [2, 14] provided that a convenient form of monotonicity holds true.

Finite state MFGs with a common noise are more subtle. In [4], the authors addressed a first systematic method to generate a common noise on a finite state space: Back to the finite player approximation of MFGs, the idea is to force a macroscopic fraction of the players to jump simultaneously from time to time. For instance, all the players may switch, according to some deterministic transformation, from one state to another whenever a common exponential clock rings. This approach was revisited in the more recent contribution [3]. Therein, the shape of the common noise is inspired from the so-called Wright-Fischer model used in population genetics: When the common noise rings, all the players sample their new state according to the current empirical distribution of the system.[1] One of the thrust of [3] is that the common noise forces the master equation to be non-degenerate and hence to admit a classical solution even though the coefficients do not satisfy the standard Lasry–Lions monotonicity conditions.

In the sequel of the notes, we do not address similar smoothing properties of common noises. Instead, we provide a new form of common noise for finite state MFGs. The main advantage of this new model is that it accommodates quite well the method developed in [6] (as we already explained, things become even easier, which is one of the reasons why we focus on this example below). The downside (or, at least, the limitation) is that the common noise does not leave the state space invariant. To make it clear, the state space has to be thought of as a collection of d reals $\varsigma^1, \cdots, \varsigma^d$, but those reals are allowed to depend on time, which means that the state space evolves with time and hence has to be written in the form $\varsigma_t^1, \cdots, \varsigma_t^d$, for t denoting the time variable. We then postulate that the common noise acts additively, meaning that ς_t^i expands in the form $\varsigma_0^i + \eta W_t$, where $(W_t)_{t \geq 0}$ is a standard Brownian motion and η accounts for the intensity of the common noise. Besides the pedagogical interest of this model, we feel that it might be useful

---

[1] Actually, this picture is only true at equilibrium. For deviating players, it is no longer true and another interpretation is needed. We refer to [3] for the details.

in practice: Obviously, we may think of applications with a time-dependent state space. To wit, this model is consistent with the model used in [6], where the common noise also manifests in the form of a mere additive white noise.

We give a more detailed presentation of the model in Sect. 3.2. The stochastic MFG system is formulated and then studied in Sect. 3.3 under a suitable version of the Lasry–Lions monotonicity condition. In Sect. 3.4, we define the notion of master field and then derive, at least informally, the shape of the master equation. The existence of a classical solution is addressed in Sect. 3.5. As we already explained, we feel better not to discuss the convergence problem in these notes. This would increase the length significantly but the interest would remain rather limited. Indeed, the approach developed in [6] and based upon the master equation has been revisited in several articles, including two for finite state MFGs, see [2, 14]. Even though the latter two contributions do not include a common noise, they make clear how the underlying machinery works in the discrete setting. We refer the interested reader to both of them.

3.2 Formulation of the Finite State MFG with a Common Noise

Throughout the notes, we consider a state space indexed by the elements of $E = \{1, \cdots, d\}$, for an integer $d \geq 1$. Accordingly, we write $\mathcal{S}_d$ for the simplex $\{(p_1, \cdots, p_d) \in [0, 1]^d : p_1 + \cdots + p_d = 1\}$; obviously, $\mathcal{S}_d$ identifies with the set of probability measures on E. The space of probability measures on $\mathbb{R}$ is denoted by $\mathcal{P}(\mathbb{R})$. Also, we denote by T the finite time horizon on which the MFG is defined and by Leb_1 the Lebesgue measure on $\mathbb{R}$.

3.2.1 Finite State MFG Without Common Noise

Without common noise, the MFG addressed in the notes has a pretty simple form, which is directly taken from [28] and [8, Chapter 7].

3.2.1.1 Optimal Control Problem

In words, a player may control the instantaneous rates $\boldsymbol{\alpha} = ((\alpha_t^{i,j})_{i,j \in E})_{0 \leq t \leq T}$ at which she may jump from one state to another. Obviously, $\boldsymbol{\alpha}$ is required to take values in the set $\mathcal{A}$ defined as

$$\mathcal{A} = \left\{ (\alpha_{i,j})_{i,j \in E} : \begin{array}{ll} \alpha_{i,j} \geq 0, & j \neq i, \quad i, j \in E \\ \alpha_{i,i} = -\sum_{j \in E: j \neq i} \alpha_{i,j}, & i \in E \end{array} \right\}.$$

Below, we ask any control $\boldsymbol{\alpha}$ to be a square-integrable (measurable) mapping from $[0, T]$ to $\mathcal{A}$, in which case $\boldsymbol{\alpha}$ is said to be admissible. For such an admissible control and for a given initial statistical state $p^{\text{init}} \in \mathcal{S}_d$, the marginal laws $((q_t^i)_{i \in E})_{0 \leq t \leq T}$ of the player then obey the following discrete FP equation, which is nothing but an Ordinary Differential Equation (ODE):

$$\dot{q}_t^i = \sum_{j \in E} q_t^j \alpha_t^{j,i} = \sum_{j \in E: j \neq i} q_t^j \alpha_t^{j,i} - \sum_{j \in E: j \neq i} q_t^i \alpha_t^{i,j}, \quad t \in [0, T] \quad ; \quad q_0 = p^{\text{init}}.$$

(3.1)

Whilst we may use (3.1) to describe the statistical state of the given tagged player playing her own control $\boldsymbol{\alpha}$, we need another continuous path from $[0, T]$ into $\mathcal{S}_d$, say $\boldsymbol{p} = (p_t)_{0 \leq t \leq T}$, to account for the statistical state of the population within which the tagged player evolves. With a control $\boldsymbol{\alpha}$ taking values in $\mathcal{A}$ and the related solution $\boldsymbol{q} = (q_t)_{0 \leq t \leq T}$ to (3.1), we hence associate the following cost functional in environment $\boldsymbol{p}$:

$$\mathcal{J}(\boldsymbol{\alpha}; \boldsymbol{p}) = \sum_{i \in E} q_T^i G^i(p_T) + \sum_{i \in E} \int_0^T q_t^i \Big(F^i(t, p_t) + \tfrac{1}{2} \sum_{j \in E: j \neq i} |\alpha_t^{i,j} - \gamma|^2 \Big) dt,$$

(3.2)

where γ is a non-negative constant (the role of which is detailed later on) and where the coefficients

$$F : E \times [0, T] \times \mathbb{R} \to \mathbb{R}, \quad G : E \times \mathbb{R} \to \mathbb{R},$$

are respectively called *running* and *terminal* costs of (3.2), both being obviously required to be Borel measurable. Below, we take as typical instance for F and G (the rationale for choosing such a form is justified in the sequel of the text):

$$F_t^i(p) = f\big(t, \varsigma_i, \mu^\varsigma[p]\big), \quad G^i(p) = g\big(\varsigma_i, \mu^\varsigma[p]\big), \quad t \in [0, T], \ i \in E, \ p \in \mathcal{S}_d,$$

(3.3)

where $(\varsigma_1, \cdots, \varsigma_d)$ are d fixed elements of $\mathbb{R}$ (we could replace $\mathbb{R}$ by $\mathbb{R}^m$, for some integer $m \geq 1$, in the analysis below), $\mu^\varsigma[p]$ denotes the finitely supported probability measure

$$\mu^\varsigma[p] = \sum_{i \in E} p_i \delta_{\varsigma_i},$$

(3.4)

and f and g are two Borel-measurable functions

$$f : [0, T] \times \mathbb{R} \times \mathcal{P}(\mathbb{R}) \to \mathbb{R}, \quad g : \mathbb{R} \times \mathcal{P}(\mathbb{R}) \to \mathbb{R},$$

(3.5)

the space $\mathcal{P}(\mathbb{R})$ being here equipped with the vague topology. From a practical point of view, (3.3) should be understood as follows: Under (3.3), the elements of E are regarded as mere labels while the true values that really make sense for computing the costs are not those labels but the reals $\varsigma_1, \cdots, \varsigma_d$; equivalently, the true finite state space is $\mathfrak{S} = \{\varsigma_i, \ i \in E\}$.

Remark 3.1 The role of the coefficient γ in (3.2) is to mollify the Hamiltonian associated with the cost functional $\mathcal{J}(\cdot \, ; \boldsymbol{p})$. Even though we have said nothing so far on the methodology that may be used to minimize $\mathcal{J}(\cdot \, ; \boldsymbol{p})$, it should not come as a surprise for the reader that, in the end, part of the analysis relies on the structure of the following Hamiltonian:

$$\forall w \in \mathbb{R}, \quad H(w) = \inf_{a \geq 0} \left(aw + \tfrac{1}{2} a^2 - a\gamma \right) + \tfrac{1}{2} \gamma^2. \tag{3.6}$$

The minimizer in the definition of H is given by $a^\star(w) = (\gamma - w)_+$ and, accordingly, $H(w) = -\tfrac{1}{2}(\gamma - w)_+^2 + \tfrac{1}{2}\gamma^2$. In particular, $a^\star$ is smooth (and in fact affine) on $(-\infty, \gamma]$: Later on in the text, we choose γ large enough with respect to the coefficients F and G so that, in all our computations, the variable w is restricted to this interval. Obviously, this is a way to force the Hamiltonian H to be smooth, which plays a key role in the subsequent analysis.

3.2.1.2 Definition of an MFG Equilibrium and Monotonicity Condition for Uniqueness

In this framework, the definition of an MFG equilibrium is given by:

Definition 3.1 For a given initial condition $p^{\mathrm{init}} \in \mathcal{S}_d$ as before, we call an MFG equilibrium a continuous function $\boldsymbol{p} = (p_t)_{0 \leq t \leq T}$ from $[0, T]$ to $\mathcal{S}_d$ satisfying the following two features:

- There exists an admissible control $\boldsymbol{\alpha}$ such that $\boldsymbol{p}$ solves the FP equation (3.1) with p^{init} as initial condition;
- For any other admissible control $\boldsymbol{\beta}$, it holds that

$$\mathcal{J}(\boldsymbol{\alpha}; \boldsymbol{p}) \leq \mathcal{J}(\boldsymbol{\beta}; \boldsymbol{p}).$$

This definition is absolutely standard. We refer to the aforementioned references on MFGs without common noise for more details if needed.

Before we introduce a common noise, we feel useful to recall the definition of the Lasry–Lions monotonicity condition, which is the standard assumption to ensure uniqueness (see for instance [8, Chapter 7]):

Definition 3.2 The two running and terminal costs F and G are said to be monotonous if, for any $t \in [0, T]$ and $p, q \in \mathcal{S}_d$,

$$\sum_{i \in E} (p_i - q_i)\big(F^i(t, p) - F^i(t, q)\big) \geq 0 ; \quad \sum_{i \in E} (p_i - q_i)\big(G^i(p) - G^i(q)\big) \geq 0.$$

(3.7)

In this respect, it is useful to reformulate (3.7) whenever F and G are given by (3.3). For instance, under (3.3), F is monotonous if, for any $t \in [0, T]$ and $p, q \in \mathcal{S}_d$,

$$\sum_{i \in E} (p_i - q_i)\big(f(t, \varsigma_i, \mu^{\varsigma}[p]) - f(t, \varsigma_i, \mu^{\varsigma}[q])\big) \geq 0.$$

The monotonicity condition for F rewrites

$$\int_{\mathbb{R}} \Big(f\big(t, x, \mu^{\varsigma}[p]\big) - f\big(t, x, \mu^{\varsigma}[q]\big)\Big)\big(\mu^{\varsigma}[p] - \mu^{\varsigma}[q]\big)(dx) \geq 0, \qquad (3.8)$$

and similarly for G. Above, we felt better to write $(\mu^{\varsigma}[p] - \mu^{\varsigma}[q])(dx)$ instead of $d(\mu^{\varsigma}[p] - \mu^{\varsigma}[q])(x)$ to denote the measure underpinning the integration.

We thus understand that F and G are monotonous if f and g are monotonous in the following sense, which fits in fact the definition of the standard Lasry–Lions monotonicity condition for MFGs with a continuous state space:

Definition 3.3 The coefficients f and g lying above F and G are said to satisfy the Lasry–Lions monotonicity condition if, for any $t \in [0, T]$ and $\mu, \nu \in \mathcal{P}(\mathbb{R})$,

$$\int_{\mathbb{R}} (f(t, x, \mu) - f(t, x, \nu))d(\mu - \nu)(x) \geq 0 ; \quad \int_{\mathbb{R}} (g(x, \mu) - g(x, \nu))d(\mu - \nu)(x) \geq 0,$$

(3.9)

at least whenever the above integrals make sense.

Obviously, the two integrals right above always make sense whenever μ and ν have a finite support, which is the case of $\mu^{\varsigma}[p]$ and $\mu^{\varsigma}[q]$ in (3.8). We refer to [8, Chapter 3] for various instances of coefficients f and g that satisfy (3.9).

3.2.2 Common Noise

As we explained in introduction, there might be several ways to produce a common noise in a finite state MFG. In [3] and [4], part of the difficulty in the construction of the common noise is that the state space is required to be fixed, independently of the choice of the common noise. We here proceed differently. Inspired by the additive form of the common noise used in [6], we indeed design a common noise that directly acts on the state space itself. Assuming that the common noise manifests (up to a multiplicative constant) in the form of a one-dimensional Brownian motion $W = (W_t)_{0 \leq t \leq T}$ and recalling that E, as introduced in the previous section, denotes the labels of the elements $\varsigma_1, \cdots, \varsigma_d$ that show up in the cost coefficients (3.3), we postulate that W acts additively on those elements $\varsigma_1, \cdots, \varsigma_d$. Equivalently, for some initial reals $\varsigma_1^{\text{init}}, \cdots, \varsigma_d^{\text{init}}$, we call *true* state space at initial time the set $\mathfrak{S} = \{\varsigma_i^{\text{init}}, \ i \in E\}$; at any time $t \in [0, T]$, we then define the *true* state space at time t as being the set

$$\mathfrak{S}_t = \{\varsigma_i^{\text{init}} + \eta W_t, \ i \in E\} = \mathfrak{S} + \eta W_t,$$

where $\eta > 0$ denotes the intensity of the common noise. Below, we often use the notation $\varsigma_t^i = \varsigma_i^{\text{init}} + \eta W_t$, for $i \in E$ and $t \in [0, T]$. Moreover, we merely write $(\varsigma_1, \cdots, \varsigma_d)$ for $(\varsigma_1^{\text{init}}, \cdots, \varsigma_d^{\text{init}})$.

With these definitions in hand, we may revisit the definitions of the state dynamics (3.1) and of the cost functional (3.2). In this respect, the first point to clarify is the notion of admissible controls. Indeed, in order to accommodate the stochasticity of W, we must allow controls to be random, which leads us to set:

Definition 3.4 Denoting by $(\Omega, \mathcal{F}, \mathbb{P})$ the probability space carrying W and by $\mathbb{F} = (\mathcal{F}_t)_{0 \leq t \leq T}$ the completion of the natural filtration generated by W, we call an admissible control an $\mathbb{F}$-progressively measurable process α with values in $\mathcal{A}$ such that

$$\sum_{i,j \in E: i \neq j} \mathbb{E} \int_0^T |\alpha_t^{i,j}|^2 dt < \infty.$$

The state equation associated with α is still (3.1), but the latter is now an ODE with random coefficients. In particular, for a given (deterministic) initial condition p^{init} as before, the solution q is a continuous $\mathbb{F}$-adapted process.

Accordingly, for a continuous and $\mathbb{F}$-adapted process p with values in S_d, we may define the following variant of $\mathcal{J}$ in (3.2):

$$\mathcal{J}^\eta(\alpha; p) = \mathbb{E}\left[\sum_{i \in E} q_T^i g\left(\varsigma_T^i, \mu^{\varsigma_T}[p_T]\right) \right.$$
$$\left. + \sum_{i \in E} \int_0^T q_t^i \left(f\left(t, \varsigma_t^i, \mu^{\varsigma_t}[p_t]\right) + \tfrac{1}{2} \sum_{j \in E: j \neq i} |\alpha_t^{i,j} - \gamma|^2 \right) dt \right],$$

$$(3.10)$$

where f and g are exactly as in (3.5). Of course, it must be clear for the reader that the above cost functional depends on the common noise through the process $(\varsigma_t)_{0 \leq t \leq T}$.

We then have the analogue of Definition 3.1:

Definition 3.5 For a given initial condition $p^{\text{init}} \in S_d$ as before, we call an MFG equilibrium (of the MFG with common noise associated with (3.1) and (3.10)) a continuous $\mathbb{F}$-adapted process $p = (p_t)_{0 \leq t \leq T}$ with values in S_d satisfying the following two features:

- There exists an admissible control α such that p solves, almost surely, the FP equation (3.1) with p^{init} as initial condition;
- For any other admissible control β, it holds that

$$\mathcal{J}^\eta(\alpha; p) \leq \mathcal{J}^\eta(\beta; p).$$

3.2.3 Assumption

Throughout the paper, we assume that f and g satisfy the Lasry–Lions monotonicity condition, as given by Definition 3.3. Also, we require them to be smooth enough in the x and μ variables (x standing for the space variable and μ for the measure argument) as specified below.

3.2.3.1 Differentiability in μ

While smoothness with respect to x may be defined in a pretty standard fashion, regularity in the variable μ is more subtle. We here borrow the following material from [6] and [8, Chapter 5].

Definition 3.6 A function h from $\mathcal{P}(\mathbb{R})$ to $\mathbb{R}$ is said to be flat continuously differentiable if there exists a continuous function (the first factor being equipped with the vague topology)

$$\frac{\delta h}{\delta m} : \mathcal{P}(\mathbb{R}) \times \mathbb{R} \ni (\mu, v) \mapsto \frac{\delta h}{\delta m}(\mu)(v)$$

such that:

- There exists $\Lambda > 0$ such that, for any $\mu \in \mathcal{P}(\mathbb{R})$,

$$\left| \frac{\delta h}{\delta m}(\mu)(v) \right| \leq \Lambda, \quad v \in \mathbb{R};$$

- For all $\mu, v \in \mathcal{P}(\mathbb{R})$, for any $t \in [0, 1]$,

$$h\big(\mu + t(v - \mu)\big) - h(\mu) = \int_0^t \left(\int_{\mathbb{R}} \frac{\delta h}{\delta m}(\mu + s(v - \mu))(v)d(v - \mu)(v) \right) ds.$$

Since the second condition remains true if we add a constant to the flat derivative, we require the latter to satisfy

$$\int_{\mathbb{R}} \frac{\delta h}{\delta m}(\mu)(v)d\mu(v) = 0.$$

Importantly, by choosing $t = 1$ in the second equation of Definition 3.6, we get:

$$\left| h(v) - h(\mu) \right| \leq \Lambda \sup_{\ell:\|\ell\|_\infty \leq 1} \int_{\mathbb{R}} \ell(v)d(v - \mu)(v), \qquad (3.11)$$

the supremum in the right-hand side being taken over all the Borel functions ℓ from $\mathbb{R}$ into itself with a supremum norm less than 1. In fact, the former supremum is nothing but (up to a constant 2) the total variation distance between μ and v, which is defined by

$$d_{\mathrm{TV}}(\mu, v) = \sup_{A \in \mathcal{B}(\mathbb{R})} \left| (\mu - v)(A) \right|,$$

where $\mathcal{B}(\mathbb{R})$ denotes the Borel σ-field on $\mathbb{R}$. In others words, requiring $\delta h/\delta m$ to be bounded implies that h is Lipschitz continuous with respect to d_{TV}. Unfortunately, assuming the various functions in hand to be Lipschitz continuous with respect to d_{TV} is not enough for our purpose. Instead, we will focus on functions h for which $\delta h/\delta m$ is not only bounded but is also (say Λ) Lipschitz continuous with respect to the variable v, in which case the bound (3.11) becomes

$$\left| h(v) - h(\mu) \right| \leq \Lambda \sup_{\ell:\|\ell\|_{1,\infty} \leq 1} \int_{\mathbb{R}} \ell(v)d(v - \mu)(v), \qquad (3.12)$$

the supremum in the right-hand side being taken over all the Borel function ℓ from $\mathbb{R}$ into itself with a supremum norm less than 1 and with a Lipschitz constant that is also less than 1. This prompts us to introduce the following distance on $\mathcal{P}(\mathbb{R})$:

$$d_{\mathrm{BL}}(\mu, v) = \sup_{\ell:\|\ell\|_{1,\infty} \leq 1} \int_{\mathbb{R}} \ell(v)d(v - \mu)(v).$$

The interested reader will observe that d_{BL} is not only bounded by d_{TV} but also by the 1-Wasserstein metric. In fact, it is show in [23] that d_{BL} metricizes the vague topology on $\mathcal{P}(\mathbb{R})$. As far as we are concerned, we will make use of the following lemma:

Lemma 3.1 *Let* $p = (p_i)_{i \in E}$ *and* $p' = (p'_i)_{i \in E}$ *be two elements of* S_d *and* $x = (x_i)_{i \in E}$ *and* $x' = (x'_i)_{i \in E}$ *be two elements of* $\mathbb{R}$. *Then, using the same notation as in* (3.4),

$$d_{\mathrm{BL}}\left(\mu^x[p], \mu^{x'}[p'] \right) \leq \sum_{i \in E} \left(|x_i - x'_i| + |p_i - p'_i| \right).$$

Proof Take a function ℓ that is bounded by 1 and that is 1-Lipschitz continuous on $\mathbb{R}$. Then,

$$\int_{\mathbb{R}} \ell(v) d\big[\mu^x[p] - \mu^{x'}[p']\big](v) = \sum_{i \in E} p_i \ell(x_i) - p_i' \ell(x_i') \leq \sum_{i \in E} \big(|p_i - p_i'| + |x_i - x_i'|\big),$$

which is the claim. $\square$

3.2.3.2   Detailed Regularity Assumptions

Throughout the notes, the space of probability measures $\mathcal{P}(\mathbb{R})$ is equipped with the distance d_{BL} and we require:

1. The function f is bounded; it is Lipschitz continuous in (x, μ), uniformly in time; it is Hölder continuous in time, uniformly in (x, μ), for some given Hölder exponent in $(0, 1]$.
2. For any $(t, x) \in [0, T] \times \mathbb{R}^d$, the function $\mathcal{P}(\mathbb{R}) \ni m \mapsto f(t, x, m)$ is flat continuously differentiable. The function $\delta f / \delta m$ is bounded; it is Lipschitz continuous in (x, v, μ), uniformly in time.
3. The function g is twice differentiable with respect to x. The functions g, $\partial_x g$ and $\partial_x^2 g$ are bounded and Lipschitz continuous in (x, μ).
4. The function g is flat continuously differentiable with respect to m. The function $\delta g / \delta m$ is bounded; it is Lipschitz continuous in (x, v, μ).

3.3   Stochastic MFG System

The purpose of this section is to characterize the equilibria of the MFG with common noise through a suitable version of the MFG system and then to address the unique solvability of the latter.

3.3.1   Stochastic HJB Equation

Our first step towards a convenient formulation of the MFG system is to characterize the minimizers of the cost functional $\mathcal{J}^\eta(\,\cdot\,; p)$ for a continuous $\mathbb{F}$-adapted process p with values in S_d. Recalling the definition of H from (3.6), we have the following statement:

Proposition 3.1 *For a given continuous $\mathbb{F}$-adapted process p with values in S_d, the cost functional $\mathcal{J}^\eta(\,\cdot\,; p)$ has a unique minimizer α. It is given by*

$$\alpha_t^{i,j} = \big(u_t^i - u_t^j + \gamma\big)_+, \quad i, j \in E : i \neq j, \quad t \in [0, T], \tag{3.13}$$

where $((u_t^i)_{i \in E})_{0 \leq t \leq T}$ is the unique solution of the BSDE:

$$du_t^i = -\left[\sum_{j \in E} H(u_t^j - u_t^i) + f(t, \varsigma_t^i, \mu^{\varsigma_t}[p_t])\right]dt + v_t^i dW_t, \quad t \in [0, T],$$

$$u_T^i = g(\varsigma_T^i, \mu^{\varsigma_T}[p_T]), \quad i \in E.$$

$$\tag{3.14}$$

Moreover, u_0^i coincides with the optimal cost $\inf_\beta \mathcal{J}^\eta(\boldsymbol{\beta}; \boldsymbol{p})$ when $\boldsymbol{q}$ in (3.1) is required to start from the Dirac mass at point i at time 0.

Remark 3.2 In the absence of common noise, the term v_t^i in (3.14) should be understood as 0. We then recover the standard MFG system for games without common noise.

In this regard, the statement of Proposition 3.1 may look rather sloppy as we said nothing about the process $((v_t^i)_{i \in E})_{0 \leq t \leq T}$. In fact, $((v_t^i)_{i \in E})_{0 \leq t \leq T}$ is part of the solution itself, which means in particular that there are two unknowns in Eq. (3.14). This might seem rather strange at first sight but this feature is actually at the basic roots of the theory of BSDEs. Indeed, the reader must remember that solutions to (3.14) must be non-anticipative, as otherwise our candidate (3.13) for solving the optimization problem $\mathcal{J}^\eta(\cdot; \boldsymbol{p})$ would not be progressively-measurable. As a result, (3.14) should not be read as a single equation, but as an equation plus a constraint on the measurability properties of $((u_t^i)_{i \in E})_{0 \leq t \leq T}$. This makes two conditions for two unknowns, which sounds fair in the end.

To make the case even stronger, we feel useful to emphasize that the process $((u_t^i)_{i \in E})_{0 \leq t \leq T}$ is required to be a continuous $\mathbb{F}$-adapted process satisfying the integrability condition[2]

$$\sup_{i \in E} \mathbb{E}\left[\sup_{t \in [0,T]} |u_t^i|^2\right] < \infty,$$

and that the process $((v_t^i)_{i \in E})_{0 \leq t \leq T}$ is required to be an $\mathbb{F}$-progressively measurable process satisfying

$$\sup_{i \in E} \mathbb{E} \int_0^T |v_t^i|^2 dt < \infty.$$

Taking conditional expectation in the first line of (3.14), we then get that

$$u_t^i = \mathbb{E}\left[g(\varsigma_T^i, \mu^{\varsigma_T}[p_T]) + \int_t^T \left(\sum_{j \in E} H(u_s^j - u_s^i) + f(s, \varsigma_s^i, \mu^{\varsigma_s}[p_s])\right)ds \mid \mathcal{F}_t\right],$$

$$\tag{3.15}$$

---

[2] In fact, we will see in the next Remark 3.3 that solutions must be bounded.

or equivalently, the process

$$\left(u_t^i + \int_0^t \left(\sum_{j \in E} H\left(u_s^j - u_s^i\right) + f\left(s, \varsigma_s^i, \mu^{\varsigma_s}[p_s]\right) \right) ds \right)_{0 \le t \le T}$$

is, for each $i \in E$, an $\mathbb{F}$-martingale. Then, the construction of the stochastic integral in (3.14) follows from the representation theorem for martingales with respect to Brownian filtrations, see [32, Chapter 3, Theorem 4.15 and Problem 4.16].

Remark 3.3 As a corollary of the proof, we get that the process $((u_t^i)_{i \in E})_{0 \le t \le T}$ is bounded by $M = \|g\|_\infty + T\|f\|_\infty$. The latter constant plays a crucial role in the sequel of the paper. In particular, we notice that the positive part in H and $a^\star$, see (3.6), can be hence removed if $\gamma \ge 2M$. This observation is consistent with the discussion in Remark 3.1.

Remark 3.4 Imitating (3.3), we use the convenient notation:

$$\widetilde{F}_t^i(p) = f\left(t, \varsigma_t^i, \mu^{\varsigma_t}[p]\right),$$
$$\widetilde{G}^i(p) = g(\varsigma_T^i, \mu^{\varsigma_T}[p]), \quad t \in [0, T], \ i \in E, \ p \in \mathcal{S}_d.$$

It must be paid attention that both $\widetilde{F}$ and $\widetilde{G}$ are hence random.

Remark 3.5 BSDE (3.14) should be regarded as a discrete Stochastic Hamilton-Jacobi-Bellman (SHJB) equation. It is the discrete analogue of the SHJB equation addressed in [6, Chapter 4].

Remark 3.6 In the literature on BSDEs, the term $\sum_{j \in E} H(u_t^j - u_t^i) + f(t, \varsigma_t^i, \mu^{\varsigma_t}[p_t])$ in (3.14) (which is nothing but the dt term, up to the sign minus in front of the whole) is called the driver of the equation. We use quite often this terminology in the sequel of the text.

Proof of Proposition 3.1 Our first step is to replace the Hamiltonian H in the driver of the BSDE by the following truncated version:

$$H_c(w) = \inf_{0 \le a \le c} \left(aw + \tfrac{1}{2}a^2 - a\gamma\right) + \tfrac{1}{2}\gamma^2$$
$$= -\tfrac{1}{2}(\gamma - w)_+^2 \mathbf{1}_{\{\gamma - w \le c\}} - \left(c(\gamma - w) - \tfrac{1}{2}c^2\right)\mathbf{1}_{\{\gamma - w > c\}} + \tfrac{1}{2}\gamma^2, \quad w \in \mathbb{R},$$

where c is a positive constant, the value of which is fixed later on. It is clear that H_c is Lipschitz continuous and coincides with H on $[\gamma - c, +\infty)$. Importantly, the minimizer in the definition of $H_c(w)$ is $a_c^\star(w) = \min((\gamma - w)_+, c)$. By [43], (3.14) with H replaced by H_c therein is uniquely solvable. We denote by $((u_t^i, v_t^i)_{i \in E})_{0 \le t \le T}$ its solution.

We then observe that, for any admissible control β whose off-diagonal coefficients are bounded by c and for $q = (q_t)_{0 \le t \le T}$ the solution to (3.1) with β as control therein, we have

$$d\left[\sum_{i\in E} q_t^i u_t^i + \int_0^t \sum_{i\in E} q_s^i\left(\widetilde{F}_s^i(p_s) + \tfrac{1}{2}\sum_{j\in E:j\neq i}|\beta_t^{i,j} - \gamma|^2\right)ds\right]$$

$$= \sum_{i\in E} q_t^i\left[\sum_{j\in E:j\neq i}\left(\beta_t^{i,j}(u_t^j - u_t^i) + \tfrac{1}{2}|\beta_t^{i,j} - \gamma|^2\right) - H_c(u_t^j - u_t^i)\right]dt + \sum_{i\in E} q_t^i v_t^i dW_t.$$

Recalling the definition of H_c, integrating between 0 and T and then taking expectation, we deduce that

$$\mathbb{E}\left[\sum_{i\in E} q_T^i \widetilde{G}^i(p_T) + \int_0^T \sum_{i\in E} q_s^i\left(\widetilde{F}_s^i(p_s) + \tfrac{1}{2}\sum_{j\in E:j\neq i}|\beta_s^{i,j} - \gamma|^2\right)ds\right] \geq \sum_{i\in E} q_0^i u_0^i.$$

$$(3.16)$$

Equality is achieved by choosing $\boldsymbol{\beta}$ as $((\beta_t^{i,j} = a_c^\star(u_t^j - u_t^i))_{i,j\in E:i\neq j})_{0\leq t\leq T}$, from which we deduce that, whenever H is replaced by H_c in (3.14), u_0^i is in fact the minimum of $\mathcal{J}^\eta(\cdot\,;\boldsymbol{p})$ over admissible processes with off-diagonal entries that are bounded by c and over processes $\boldsymbol{q}$ that start from the Dirac mass at point i at time 0. Using the non-negativity of the Lagrangian, we get as trivial bound $u_0^i \geq -M$, see Remark 3.3 for the definition of M. Choosing $\boldsymbol{\beta} \equiv 0$ as control, we get the opposite bound, namely $u_0^i \leq M$. In fact, by initializing the process $\boldsymbol{q}$ from the Dirac mass at point i at any time $t \in [0, T]$ and then by working with conditional expectation given $\mathcal{F}_t$ in (3.16), we get in a similar way that

$$\mathbb{P}\big(\forall i \in E, \ \forall t \in [0, T], \ |u_t^i| \leq M\big) = 1. \qquad (3.17)$$

Since M is independent of c, we can choose c large enough so that $H_c(u_t^j - u_t^i) = H(u_t^j - u_t^i)$ in the BSDE satisfied by $((u_t^i)_{i\in E})_{0\leq t\leq T}$. This shows the existence of a solution to (3.14).

By similar computations, we deduce that the hence constructed solution to (3.14) satisfies (3.16) for any admissible control $\boldsymbol{\beta}$ (no need to assume the latter to be bounded), with equality when $\boldsymbol{\beta}$ is chosen as in (3.13). In fact, this holds true for any other solution to (3.14), hence proving that all the solutions to (3.14) satisfy (3.17). As a result, there exists $c > 0$ such that all the solutions to (3.14) satisfy (3.14) but with respect to the Hamiltonian H_c. We deduce that (3.14) is uniquely solvable. $\qquad\square$

3.3.2 Formulation of the MFG System

We now have all the ingredients to characterize the equilibria of the MFG under study through a suitable form of the MFG system.

As a direct application of Proposition 3.1, we get:

Proposition 3.2 *For a given initial condition $p^{\mathrm{init}} \in S_d$, a continuous adapted process $\boldsymbol{p} = (p_t)_{0 \le t \le T}$ with values in the simplex is an equilibrium, as defined in Definition 3.5, if and only if there exists a pair $(((u_t^i)_{i \in E})_{0 \le t \le T}, ((v_t^i)_{i \in E})_{0 \le t \le T})$ satisfying*

$$\forall i \in E, \quad \mathbb{E}\left[\sup_{t \in [0,T]} |u_t^i|^2 + \int_0^T |v_t^i|^2 dt \right] < \infty, \tag{3.18}$$

together with the forward-backward system:

$$dp_t^i = \left[\sum_{j \in E: j \ne i} p_t^j \left(\gamma + u_t^j - u_t^i \right)_+ - p_t^i \sum_{j \in E: j \ne i} \left(\gamma + u_t^i - u_t^j \right)_+ \right] dt,$$

$$du_t^i = -\left[\sum_{j \in E} H\left(u_t^j - u_t^i \right) + \widetilde{F}_t^i(p_t) \right] dt + v_t^i dW_t, \quad t \in [0, T], \tag{3.19}$$

$$u_T^i = \widetilde{G}^i(p_T), \quad i \in E.$$

Remark 3.7 In probability literature, the system (3.19) is referred to as a forward-backward SDE.

In comparison with [6], (3.19) is the analogue of (4.7) therein: Using the fact that the common noise acts in an additive manner, we are indeed able to write the forward equation as a random ODE and not as an SDE (which would be more complicated). This observation is at the roots of the analysis provided in [6] and we here duplicate it in the discrete setting.

Here is now our first main result:

Theorem 3.1 *Under the assumption of Sect. 3.2.3 and under the condition $\gamma \ge 2M$ (see Remark 3.3), for any initial condition $p^{\mathrm{init}} \in S_d$, there exists a unique solution to the MFG system (3.19).*

In particular, the MFG with common noise has a unique equilibrium for any given initial condition.

We prove Theorem 3.1 in the next section by means of a continuation argument. This argument is similar to the one used in [6, Chapter 4], but, in fact, continuation method for forward-backward SDEs goes back to the paper [45].

Remark 3.8 In the proofs below, we often remove, for simplicity, the superscript *init* in the initial condition p^{init}.

3.3.3 Proof of the Solvability Result

This section is devoted to the proof of Theorem 3.1.

In order to proceed, we introduce the following notation. For any $\lambda \in [0, 1]$, for any bounded $\mathbb{F}$-progressively measurable process $\tilde{f} = ((\tilde{f}_t^i)_{i \in E})_{0 \le t \le T}$ and any bounded $\mathcal{F}_T$-measurable random variable $\tilde{g} = (\tilde{g}_T^i)_{i \in E}$, we denote by $\mathcal{E}(\lambda, \tilde{f}, \tilde{g})$ the forward-backward system:

$$dp_t^i = \left[\sum_{j \in E: j \ne i} p_t^j \left(\gamma + u_t^j - u_t^i\right)_+ - p_t^i \sum_{j \in E: j \ne i} \left(\gamma + u_t^i - u_t^j\right)_+ \right] dt,$$

$$du_t^i = -\left[\sum_{j \in E} H\left(u_t^j - u_t^i\right) + \lambda \tilde{F}_t^i(p_t) + \tilde{f}_t^i \right] dt + v_t^i dW_t, \quad t \in [0, T], \qquad (3.20)$$

$$u_T^i = \lambda \tilde{G}^i(p_T) + \tilde{g}_T^i, \quad i \in E,$$

with $p_0 = p^{\text{init}}$ as initial condition, for a fixed $p^{\text{init}} \in \mathcal{S}_d$. Whenever $\mathcal{E}(\lambda, \tilde{f}, \tilde{g})$ has a solution, we often denote it in the shorten forms (p, u, v) or (p, u).

With those notations in hand, we formulate the following lemma, which makes clear the basic mechanism of the continuation method:

Lemma 3.2 *There exists* $\epsilon > 0$ *with the following property: If, for a given* $\lambda \in [0, 1)$ *and for any* $((\tilde{f}_t^i)_{i \in E})_{0 \le t \le T}$ *and* $(\tilde{g}_T^i)_{i \in E}$ *as above with*

$$\forall i \in E, \quad \begin{cases} \mathbb{P} \otimes \text{Leb}_1\left(\left\{(\omega, t) : |\tilde{f}_t^i(\omega)| > (1 - \lambda)\|f\|_\infty\right\}\right) = 0, \\ \mathbb{P}\left(\left\{\omega : |\tilde{g}_T^i(\omega)| > (1 - \lambda)\|g\|_\infty\right\}\right) = 0, \end{cases} \qquad (3.21)$$

the system $\mathcal{E}(\lambda, \tilde{f}, \tilde{g})$ *has a unique solution (satisfying the integrability conditions (3.18)), then, for any* $\lambda' \in [\lambda, \min(1, \lambda + \epsilon)]$ *and for any* $((\tilde{f}_t^i)_{i \in E})_{0 \le t \le T}$ *and* $(\tilde{g}_T^i)_{i \in E}$ *as above with*

$$\forall i \in E, \quad \begin{cases} \mathbb{P} \otimes \text{Leb}_1\left(\left\{(\omega, t) : |\tilde{f}_t^i(\omega)| > (1 - \lambda')\|f\|_\infty\right\}\right) = 0, \\ \mathbb{P}\left(\left\{\omega : |\tilde{g}_T^i(\omega)| > (1 - \lambda')\|g\|_\infty\right\}\right) = 0, \end{cases} \qquad (3.22)$$

the system $\mathcal{E}(\lambda', \tilde{f}, \tilde{g})$ *is also uniquely solvable.*

Remark 3.9 Condition (3.21) is really important. It says that the process $((\lambda \tilde{F}_t^i(p_t) + \tilde{f}_t^i)_{i \in E})_{0 \le t \le T}$ is bounded by $\|f\|_\infty$ and that the variable $(\lambda \tilde{G}^i(p_t) + \tilde{f}_T^i)_{i \in E}$ is bounded by $\|g\|_\infty$. Duplicating the proof of Proposition 3.1, this implies that any solution (p, u, v) to $\mathcal{E}(\lambda, \tilde{f}, \tilde{g})$ satisfies the bound

$$\forall i \in E, \quad \mathbb{P}\left(\forall t \in [0, T], |u_t^i| \le M\right) = 1.$$

In particular, since $\gamma \ge 2M$, the effective values that are inserted in the Hamiltonian H in (3.20) are restricted to the interval where $H(w)$ is equal to $-\frac{1}{2}(\gamma - w)^2 + \frac{1}{2}\gamma^2$.

Take Lemma 3.2 for granted. Then, observing that, whenever $\lambda = 0$ in (3.20), the system $\mathcal{E}(0, \tilde{f}, \tilde{g})$ is decoupled, meaning that the backward equation may be solved first and then the forward equation may be solved next by inserting therein the solution of the backward equation, we deduce that the solvability of $\mathcal{E}(0, \tilde{f}, \tilde{g})$ can be treated on the same model as the solvability of (3.14). Existence and uniqueness of a solution to $\mathcal{E}(0, \tilde{f}, \tilde{g})$ easily follows. By iterating in Lemma 3.2 on the value of λ (choosing $\lambda = \epsilon$, and then $\lambda = 2\epsilon, \ldots$, up to $\lambda = n\epsilon$ for $n = \lfloor 1/\epsilon \rfloor$ and then $\lambda = 1$), we deduce that $\mathcal{E}(1, 0, 0)$ is uniquely solvable, which is Theorem 3.1.

In fact, the proof of Lemma 3.2 is based upon the following stability lemma:

Lemma 3.3 *For any* $\lambda \in [0, 1)$, *consider* $\tilde{f} = ((\tilde{f}_t^i)_{i \in E})_{0 \leq t \leq T}$ *and* $\tilde{f}' = ((\tilde{f}_t^{',i})_{i \in E})_{0 \leq t \leq T}$ *two* $\mathbb{F}$-*progressively measurable processes that are bounded by* $(1 - \lambda)\|f\|_\infty$ *and* $\tilde{g} = (\tilde{g}_T^{',i})_{i \in E}$ *and* $\tilde{g}' = (\tilde{g}_T^i)_{i \in E}$ *two* $\mathcal{F}_T$-*measurable random variables that are bounded by* $(1 - \lambda)\|g\|_\infty$, *see (3.21). Assume that* (p, u) *and* (p', u') *solve respectively* $\mathcal{E}(\lambda, \tilde{f}, \tilde{g})$ *and* $\mathcal{E}(\lambda, \tilde{f}', \tilde{g}')$. *Then, there exists a constant* C *only depending on the underlying constants in the assumption stated in Sect. 3.2.3 (in particular,* C *is independent of* λ) *such that*

$$\mathbb{E}\left[\sum_{i \in E} \sup_{t \in [0,T]} \left(|p_t^i - p_t^{',i}|^2 + |u_t^i - u_t^{',i}|^2 \right) + \sum_{i \in E} \int_0^T |v_t^i - v_t^{',i}|^2 dt \right]$$

$$\leq C\mathbb{E}\left[\sum_{i \in E} |\tilde{g}_T^i - \tilde{g}_T^{',i}|^2 + \sum_{i \in E} \int_0^T |\tilde{f}_t^i - \tilde{f}_t^{',i}|^2 dt \right].$$

Remark 3.10 Interestingly enough, the reader may double-check that the proof easily extends to the case when the two processes p and p' start from different initial conditions at time 0, in which case we need to include $\sum_{i \in E} |p_0^i - p_0^{',i}|^2$ in the right-hand side.

Proof of Lemma 3.3 We call (p, u, v) and (p', u', v') the two solutions considered in the statement. The proof is then quite classical in MFG theory and consists in expanding $(\sum_{i \in E}(p_t^i - p_t^{',i})(u_t^i - u_t^{',i}))_{0 \leq t \leq T}$ by Itô's formula. We get

$$d\left(\sum_{i \in E}(p_t^i - p_t^{',i})(u_t^i - u_t^{',i}) \right)$$

$$= - \sum_{i,j \in E}(p_t^i - p_t^{',i})\left(H(u_t^j - u_t^i) - H(u_t^{',j} - u_t^{',i}) \right)dt$$

$$- \sum_{i \in E}(p_t^i - p_t^{',i})\left(\lambda \tilde{F}_t^i(p_t) - \lambda \tilde{F}_t^i(p_t') + \tilde{f}_t^i - \tilde{f}_t^{',i} \right)dt + \sum_{i \in E}(p_t^i - p_t^{',i})(v_t^i - v_t^{',i})dW_t$$

$$+ \sum_{i,j \in E} \left(p_t^j(\gamma + u_t^j - u_t^i)_+ - p_t^{',j}(\gamma + u_t^{',j} - u_t^{',i})_+ \right)(u_t^i - u_t^j - (u_t^{',i} - u_t^{',j}))dt.$$

By monotonicity of f, the first term on the penultimate line is non-positive. As for the first and last terms on the right-hand side, we have

$$- \sum_{i,j \in E} \left[p_t^i \left(H(u_t^j - u_t^i) - H(u_t'^{,j} - u_t'^{,i}) \right) - p_t^i \left(\gamma + u_t^j - u_t^i \right)_+ \left(u_t^i - u_t^j - (u_t'^{,i} - u_t'^{,j}) \right) \right]$$

$$= \sum_{i,j \in E} p_t^i \left(H(u_t'^{,j} - u_t'^{,i}) - H(u_t^j - u_t^i) - \partial_w H(u_t^j - u_t^i)(u_t'^{,j} - u_t'^{,i} - (u_t^j - u_t^i)) \right).$$

Since $\gamma \geq 2M$, we are here on the part where H is strictly concave, see Remark 3.9. We deduce that the last term right above is upper bounded by $-\frac{1}{2} \sum_{i,j \in E} p_t^i [u_t'^{,j} - u_t'^{,i} - (u_t^j - u_t^i)]^2$. Exchanging the primes and the non-primes in the above identity and then taking expectation in the expansion of $(\sum_{i \in E} (p_t^i - p_t'^{,i})(u_t^i - u_t'^{,i}))_{0 \leq t \leq T}$, we deduce that

$$\mathbb{E} \left[\sum_{i \in E} (p_T^i - p_T'^{,i})(u_T^i - u_T'^{,i}) + \frac{1}{2} \sum_{i,j \in E} (p_t^i + p_t'^{,i})[u_t'^{,j} - u_t'^{,i} - (u_t^j - u_t^i)]^2 \right]$$

$$\leq \mathbb{E} \int_0^T \sum_{i \in E} |p_t^i - p_t'^{,i}| |\tilde{f}_t^i - \tilde{f}_t'^{,i}| dt.$$

Replacing u_T^i and $u_T'^{,i}$ by their expressions in the left-hand side and then using the monotonicity of g, we finally obtain:

$$\frac{1}{2} \mathbb{E} \sum_{i,j \in E} (p_t^i + p_t'^{,i})[u_t'^{,j} - u_t'^{,i} - (u_t^j - u_t^i)]^2$$

$$\leq \mathbb{E} \sum_{i \in E} |p_T^i - p_T'^{,i}| |\tilde{g}_T^i - \tilde{g}_T'^{,i}| + \mathbb{E} \int_0^T \sum_{i \in E} |p_t^i - p_t'^{,i}| |\tilde{f}_t^i - \tilde{f}_t'^{,i}| dt. \tag{3.23}$$

Back to the forward equation, we write, for any $i \in E$ and $t \in [0, T]$,

$$|p_t^i - p_t'^{,i}| \leq (2M + \gamma) \sum_{j \in E} \int_0^t |p_s^j - p_s'^{,j}| + \sum_{j \in E} \int_0^t p_s^j |u_s'^{,j} - u_s'^{,i} - (u_s^j - u_s^i)| ds.$$

Summing over $i \in E$, applying Gronwall's lemma and then taking the square, we deduce that there exists a constant C, only depending on d, M and γ, such that

$$\sum_{i \in E} \sup_{t \in [0,T]} |p_t^i - p_t'^{,i}|^2 \leq C \sum_{i,j \in E} \int_0^T p_s^j |u_s'^{,j} - u_s'^{,i} - (u_s^j - u_s^i)|^2 ds.$$

Inserting the above bound in (3.23), we get the announced estimate for the term $\mathbb{E} \sum_{i \in E} \sup_{t \in [0,T]} |p_t^i - p_t'^{,i}|^2$. In order to complete the proof, we inject in turn this estimate in the backward equation. Using the fact that the argument in the Hamiltonian H is bounded, we can easily apply standard estimates for BSDEs with Lipschitz continuous coefficients, see [48, Chapter 3]. □

Proof of Lemma 3.2 We prove Lemma 3.2 by a contraction argument. For $\lambda \in (0, 1], \lambda' \in (\lambda, 1]$ and $((\tilde{f}_t^i)_{i \in E})_{0 \leq t \leq T}$ and $(\tilde{g}_T^i)_{i \in E}$ as in (3.22), we create a mapping Φ that maps the set of continuous $\mathbb{F}$-adapted and $\mathcal{S}_d$-valued processes into itself. For $\boldsymbol{p}$ a continuous and adapted process with values in $\mathcal{S}_d$, we define $\Phi(\boldsymbol{p})$ as follows: We solve for $\mathcal{E}(\lambda, \tilde{f} + \delta \tilde{F}(\boldsymbol{p}), \tilde{g} + \delta \tilde{G}(\boldsymbol{p}))$ with $\delta = \lambda' - \lambda$, where $\tilde{F}(\boldsymbol{p})$ is a shorten notation for $((\tilde{F}_t^i(p_t))_{i \in E})_{0 \leq t \leq T}$ and $\tilde{G}(\boldsymbol{p})$ is a shorten notation for $(\tilde{G}^i(p_T))_{i \in E}$. We call the solution $((\Phi_t^i(\boldsymbol{p}))_{i \in E}, (u_t^i)_{i \in E}, (v_t^i)_{i \in E})_{0 \leq t \leq T}$.

Now, for two inputs $\boldsymbol{p}$ and $\boldsymbol{p}'$, Lemma 3.3 yields (for C independent of λ and δ)

$$
\mathbb{E}\left[\sum_{i \in E} \sup_{t \in [0,T]} \left(|\Phi_t^i(\boldsymbol{p}) - \Phi_t^i(\boldsymbol{p}')|^2 \right) \right]
$$

$$
\leq C\delta^2 \mathbb{E}\left[\sum_{i \in E} |\tilde{G}^i(p_T) - \tilde{G}^i(p_T')|^2 + \sum_{i \in E} \int_0^T |\tilde{F}_t^i(p_t) - \tilde{F}_t^i(p_t')|^2 dt \right].
$$

Since $\tilde{G}$ and $\tilde{F}$ are Lipschitz continuous in the variable p, see the assumption in Sect. 3.2.3 together with Lemma 3.1 therein, we easily deduce that Φ is a contraction for δ small enough, independently of the value of λ. The proof ends up by Banach fixed point theorem. □

3.4 Master Equation

We now address the form and the solvability of the master equation. In the whole section, we assume that $\gamma \geq 2M$.

3.4.1 Master Field

Our first step towards the master equation is to introduce the notion of master field. In comparison with [6], it is here tailored-made to our choice of common noise. In this regard, it is worth recalling that, in the MFG literature, the construction of the master field has been almost exclusively addressed when equilibria are unique (as quoted in introduction, see [15, 20] for a few examples when uniqueness does not hold). Even more, the standard approach to the master field is in fact intrinsically connected with a Markov property that the equilibria should satisfy whenever uniqueness indeed holds true. Obviously, this Markov property especially makes sense when the game is subjected to a common noise, since, as we already mentioned several times, equilibria are then random.

However, it is pretty easy to see that, in our case, equilibria, as defined by the sole process $\boldsymbol{p}$ solving the forward equation in the MFG system (3.19), cannot be Markovian on their own. This is here a subtlety that is due to the way the common noise is assumed to act on the game. To make it clear, assume that the state of the population at a given time $t \in [0, T]$ is given by some $\mathcal{F}_t$-measurable random variable with values in $\mathcal{S}_d$. Obviously, this information is not enough to recompute the future evolution of the equilibrium since the latter also depends on the new form that the state space takes at time t. Recall indeed that the effective state space at time t is given by $\mathfrak{S}_t = \mathfrak{S} + \eta W_t$. So, here, the right candidate for being a Markov process is the pair $(\eta W, \boldsymbol{p})$. For sure, the reader might find it rather strange, but in fact it is completely consistent with the approach taken in [6]: Therein the process $(\tilde{m}_t)_{0 \le t \le T}$ that solves the forward equation in [6, (4.7)] is not a Markov process with values in $\mathcal{P}(\mathbb{R}^d)$ (d being the dimension in [6]); the Markov process therein is made of the image of $\tilde{m}_t$ by the (random) translation $\mathbb{R}^d \ni x \mapsto x + \eta W_t$. The situation is absolutely similar here: Our *true* Markov process is in fact the $\mathcal{P}(\mathbb{R})$-valued process:

$$\left(\mu^{\varsigma_t}[p_t] = \sum_{i \in E} p_t^i \delta_{\varsigma_i + \eta W_t} \right)_{0 \le t \le T}.$$

However, to keep the whole discussion at a reasonable level, we feel better not to consider general $\mathcal{P}(\mathbb{R})$-valued processes (as otherwise we would face the same kind of complexity as in [6]). Instead, we only focus on the locations of the Dirac masses in $\mu^{\varsigma_t}[p_t]$ through the position of ηW_t and on the weights of each of those Dirac masses through the element p_t of $\mathcal{S}_d$. As a by-product, we are able to write below the master equation as a finite-dimensional PDE.

Clearly, our willingness to regard the pair $(\eta W, \boldsymbol{p})$ as a Markov process (with values in $\mathbb{R} \times \mathcal{S}_d$) should prompt us to allow, as initial condition for the game, any triplet $(t, x, p) \in [0, T] \times \mathbb{R} \times \mathcal{S}_d$, meaning that ηW is then required to start from x at time t and $\boldsymbol{p}$ to start from p at the same time t. It is absolutely obvious that Theorem 3.1 extends to this new case. This allows us to let:

Definition 3.7 We call master field the function $U : [0, T] \times \mathbb{R} \times \mathcal{S}_d \to \mathbb{R}^d$ that associates, with any $(t, x, p) \in [0, T] \times \mathbb{R} \times \mathcal{S}_d$, the d-tuple $U(t, x, p) = (U^i(t, x, p))_{i \in E}$, where $U^i(t, x, p)$ is equal to u_t^i in the SHJB equation in (3.19) when $(p_s)_{t \le s \le T}$ therein starts from p (at time t) and $(\varsigma_s)_{t \le s \le T}$ from $(\varsigma_j + x)_{j \in E}$ (in which case we have $\varsigma_s^j = \varsigma_j + x + \eta(W_s - W_t)$, for $s \in [t, T]$ and $j \in E$).

Remark 3.11 It is absolutely crucial to understand that, whenever the MFG system (3.19) and the common noise are initialized at time t as in the statement, the random variable u_t^i is almost-surely constant. This is the cornerstone of the construction of the master field for MFGs with common noise (and the additional substantial difficulty in comparison with MFGs without common noise). To wit, the reader can take $t = 0$. Then, u_0^i is required to be $\mathcal{F}_0$-measurable, but the latter is nothing but (recall that $\mathbb{F}$ is the completion of the filtration generated by W) the completion of the trivial σ-field, whence the fact that u_0^i is almost-surely a constant.

When $t > 0$, the same argument holds true, but then the common noise has to be understood as $(x + \eta(W_s - W_t))_{t \le s \le T}$, as suggested by the Markov property for the Brownian motion. And, again, it makes sense to assume that the σ-field at time t is almost-surely trivial. In other words, we then forget the past before t, which is again consistent with the practical meaning of the Markov property.

Remark 3.12 The reader should not confuse the role of x here with the role of x in [6]. In [6], the argument x in the master field is used to denote the state of the tagged player in the population. Here, x is used to denote the initial location of the common noise.

As a by-product of the stability estimate proven in Lemma 3.3, we get

Proposition 3.3 *There exists a constant C, only depending on the underlying constants in the assumption stated in Sect. 3.2.3, such that, for any $(t, x, p) \in [0, T] \times \mathbb{R} \times \mathcal{S}_d$ and $(x', p') \in [0, T] \times \mathbb{R} \times \mathcal{S}_d$,*

$$\sup_{i \in E} |U^i(t, x, p) - U^i(t, x', p')| \le C(|x - x'| + |p - p'|).$$

Proof Whenever $x = x'$, the result is a mere consequence of Lemma 3.3 and Remark 3.10. When $x \ne x'$, it is a formal application of Lemma 3.3 by choosing $\tilde{f} = 0$ and

$$\tilde{f}_s^{\prime,i} = f\left(s, \varsigma_i + x' + \eta(W_s - W_t), \sum_{j \in E} p_s^{\prime,j} \delta_{\varsigma_j + x' + \eta(W_s - W_t)}\right)$$
$$- f\left(s, \varsigma_i + x + \eta(W_s - W_t), \sum_{j \in E} p_s^{\prime,j} \delta_{\varsigma_j + x + \eta(W_s - W_t)}\right),$$

which is to say that we handle the fact that x' is not equal to x by choosing a convenient form for the additional term $\tilde{f}'$, and similarly for $\tilde{g}$ and $\tilde{g}'$. However, this is not correct because this violates the condition in the statement of Lemma 3.3 that $\tilde{f}'$ must be bounded by $(1 - \lambda)\|f\|_\infty$, and similarly for $\tilde{g}'$. Fortunately, the latter is just needed to ensure that everything in the proof works as if the Hamiltonian were strictly concave, see Remark 3.9. Here, there is no need to have this extra guarantee since we already know that u and u' (the respective solutions of the backward equation) take indeed values in the domain where H is strictly concave. The proof of Lemma 3.3 is hence easily adapted. □

3.4.2   Representation of the Value Function Through the Master Field

The following property makes clear the role of the master field:

Proposition 3.4 *For an initial condition* $p^{\mathrm{init}} \in \mathcal{S}_d$, *call* $(\boldsymbol{p}, \boldsymbol{u}, \boldsymbol{v})$ *the unique solution to the MFG system* (3.19) *(say to simplify with* $W_0 = 0$*), as given by Theorem 3.1. Then, for all* $i \in E$ *and* $t \in [0, T]$, *with probability 1,*

$$u_t^i = U^i(t, \eta W_t, p_t). \tag{3.24}$$

Proof For a given $\delta > 0$, we consider a countable partition $(A^{(n)})_{n \in \mathbb{N}}$ of $\mathbb{R} \times \mathcal{S}_d$, the diameter of each $A^{(n)}$ being less than δ. For each $n \in \mathbb{N}$, we then choose an element $(x^{(n)}, q^{(n)}) \in A^{(n)}$ and we denote by $((p_s^{(n),i}, u_s^{(n),i}, v_s^{(n),i})_{i \in E})_{t \leq s \leq T}$ the solution to the MFG system (3.19) with $p_t^{(n)} = q^{(n)}$ and $\eta W_t = x^{(n)}$ as initialization.

By a generalization of Lemma 3.3 (the proof of which is absolutely similar), we could get, for each $n \in \mathbb{N}$,

$$\sum_{i \in E} \mathbb{E}\big[|u_t^i - u_t^{n,i}|^2 \,|\, \mathcal{F}_t\big] \leq C\big(|p_t - q^{(n)}|^2 + |\eta W_t - x^{(n)}|^2\big).$$

The conditional expectation in the left-hand side is in fact purely useless. It is here for pedagogical reasons only. Indeed, we know that $u_t^i - u_t^{(n),i}$ is $\mathcal{F}_t$-adapted. In fact, the rationale for adding the conditional expectation is just to emphasize that, in the analogue of Lemma 3.3, we should replace expectations by conditional expectations. Multiplying both sides by $\mathbf{1}_{\{(\eta W_t, p_t) \in A^{(n)}\}}$ and then taking expectation, we get

$$\sum_{i \in E} \mathbb{E}\big[|u_t^i - u_t^{(n),i}|^2 \mathbf{1}_{\{(W_t, p_t) \in A^{(n)}\}}\big] \leq C\delta^2 \mathbb{P}\big((\eta W_t, p_t) \in A^{(n)}\big).$$

Observe that $u_t^{(n),i}$ writes, by definition, as $U^i(t, x^{(n)}, q^{(n)})$. Therefore, by modifying the value of the constant C in Proposition 3.3, we finally obtain

$$\sum_{i \in E} \mathbb{E}\big[|u_t^i - U^i(t, \eta W_t, p_t)|^2 \mathbf{1}_{\{(\eta W_t, p_t) \in A^{(n)}\}}\big] \leq C\delta^2 \mathbb{P}\big((\eta W_t, p_t) \in A^{(n)}\big).$$

Summing over $n \in \mathbb{N}$ and letting δ tend to 0, we complete the proof. $\square$

As a corollary, we obtain:

Corollary 3.1 *The master field* U *is* $1/2$-*Hölder continuous in time, uniformly in the other variables. In particular, it is (jointly) continuous on* $[0, T] \times \mathbb{R} \times \mathcal{S}_d$ *and, in the statement of Proposition 3.4, almost surely, for all* $i \in E$ *and* $t \in [0, T]$, *the representation property* (3.24) *holds true.*

Proof For a given $(t, x, p) \in [0, T] \times \mathbb{R} \times \mathcal{S}_d$, we call (p, u, v) the solution to (3.19) whenever $(\eta W, p)$ starts from (x, p) at time t.

By taking expectation in $(u_s^i - u_t^i)_{t \le s \le T}$, for any $i \in E$, we get, for a given $h > 0$ such that $t + h \le T$,

$$\mathbb{E}[u_{t+h}^i - u_t^i] = -\mathbb{E} \int_t^{t+h} \left(\sum_{j \in E} H(u_s^j - u_s^i) + \widetilde{F}_s^i(p_s) \right) ds.$$

Since the driver of the backward equation in (3.19) is bounded, there exists a constant C, only depending on the underlying constants in the assumption stated in Sect. 3.2.3, such that

$$\forall i \in E, \quad \left| \mathbb{E}[u_{t+h}^i - u_t^i] \right| \le Ch.$$

By Proposition 3.4, the above inequality reads

$$\forall i \in E, \quad \left| \mathbb{E}[U^i(t + h, \eta W_{t+h}, p_{t+h}) - U^i(t, x, p)] \right| \le Ch.$$

Using the Lipschitz property stated in Proposition 3.3 and then using the fact that the time derivative of p is bounded, we finally obtain:

$$\forall i \in E, \quad \left| U^i(t + h, x, p) - U^i(t, x, p)] \right| \le Ch^{1/2},$$

which, together with Proposition 3.3, permits to conclude.   □

3.4.3   Form of the Master Equation

We now have all the ingredients to derive, at least intuitively, the form of the master equation.

3.4.3.1   Informal Derivation of the Equation

Assuming for a while that the master field U is smooth enough, we may expand $((U^i(t, \eta W_t, p_t))_{i \in E})_{0 \le t \le T}$ by means of Itô's formula, for a solution (p, u) to (3.19), and then compare the resulting expansion with the formula for $(du_t^i)_{0 \le t \le T}$.

We then get the following version the master equation:

$$\partial_t U^i(t, x, p) + \tfrac{1}{2}\eta^2 \partial_x^2 U^i(t, x, p) + \sum_{j \in E} H\big((U^j - U^i)(t, x, p)\big) + f\big(t, \varsigma_i + x, \mu^{\varsigma+x}[p]\big)$$

$$+ \sum_{j,k \in E} p_k\big(\gamma + (U^k - U^j)(t, x, p)\big)_+ \big(\partial_{p_j} U^i - \partial_{p_k} U^i\big)(t, x, p) = 0, \qquad (3.25)$$

$$U^i(T, x, p) = g\big(\varsigma_i + x, \mu^{\varsigma+x}[p]\big),$$

for $(t, x, p) \in [0, T] \times \mathbb{R} \times \mathcal{S}_d$, with the obvious notation $\varsigma + x = (\varsigma_1 + x, \cdots, \varsigma_d + x)$.

In fact, it should be clear for the reader that (3.25) can only be a preliminary unrigorous form of the master equation. Indeed, strictly speaking, it does not make any sense to speak about the derivatives of U with respect to $p_1, \cdots, p_d$, since the variable p in U only belongs to the simplex $\mathcal{S}_d$, which has empty interior in $\mathbb{R}^d$. Instead, we must see $\mathcal{S}_d$ as a $(d-1)$-dimensional manifold or, equivalently, identify any d-tuple $(p_1, \cdots, p_d)$ of $\mathcal{S}_d$ with the $(d-1)$-tuple $(p_1, \cdots, p_{d-1})$, the latter being regarded as an element of the subset of $\mathbb{R}^{d-1}$ with a non-empty interior made of vectors $(q_1, \cdots, q_{d-1})$ with non-negative entries and with sum $\sum_{i=1}^{d-1} q_i$ less than 1. Instead of introducing a notation for the collection of such $(q_1, \cdots, q_{d-1})$'s, we feel easier to use the same notation, whether we see the simplex as a collection of d-tuples or as a collection of $(d-1)$-tuples. This prompts us to let, for such a $(d-1)$-tuple $(q_1, \cdots, q_{d-1})$:

$$\hat{U}^i\big(t, x, (q_1, \cdots, q_{d-1})\big) = U^i\Big(t, x, \big(q_1, \cdots, q_{d-1}, 1 - (q_1 + \cdots + q_{d-1})\big)\Big).$$

It then makes sense to speak about the derivatives of $\hat{U}^i$ with respect to $(q_1, \cdots, q_{d-1})$. If U in the right-hand side were defined on $[0, T] \times \mathbb{R} \times O$, for O an open subset of $\mathbb{R}^d$ containing $\mathcal{S}_d$, we would get

$$\partial_{q_j} \hat{U}^i(t, x, q) = \big(\partial_{p_j} U^i - \partial_{p_d} U^i\big)(t, x, p),$$

for $q = (q_l)_{l=1,\cdots,d-1}$ and $p = (p_l)_{l=1,\cdots,d}$ denoting the same element of $\mathcal{S}_d$. This is absolutely enough to give a rigorous meaning to the derivatives with respect to p in (3.25), provided that we understand them as

$$\big(\partial_{p_j} U^i - \partial_{p_k} U^i\big)(t, x, p) = \begin{cases} \big(\partial_{q_j} \hat{U}^i - \partial_{q_k} \hat{U}^i\big)(t, x, q), & j, k \in \{1, \cdots, d-1\}, \ j \neq k, \\ \partial_{q_j} \hat{U}^i(t, x, q), & j \in \{1, \cdots, d-1\}, \quad k = d. \end{cases}$$

Observe that it makes sense to assume $j \neq k$, as otherwise the difference in the left-hand side is obviously taken as 0.

We let the reader reformulate (3.25) in coordinates (t, x, q) instead of (t, x, p). The key point here is that, whenever we invoke Itô's formula as we did to derive (3.25) and as we do quite often below, it is absolutely the same to expand $(U^i(t, x +$

$\eta W_t, p_t))_{0 \le t \le T}$ if U is defined on the same $[0, T] \times \mathbb{R} \times O$ as before or to expand $(\hat{U}^i(t, x + \eta W_t, q_t))_{0 \le t \le T}$, for $(q_t = (p_t^1, \cdots, p_t^{d-1}))_{0 \le t \le T}$. Obviously, the first expansion leads exactly to Eq. (3.25), whilst the second one leads to the version in coordinates (t, x, q), which is in fact the right-one when U is just defined on $[0, T] \times \mathbb{R} \times S_d$. Anyway, we feel easier to stick to the version (3.25) and not to rewrite the master equation in coordinates (t, x, q).

3.4.3.2 Solvability Result

Here is now the second main statement of this chapter:

Theorem 3.2 *Under the standing assumption (including the condition $\gamma \ge 2M$), the master equation has a unique bounded classical solution with bounded derivatives of order one in t, x and p and of order two in x.*

The proof of Theorem 3.2 is mostly based on a linearization argument that is taken from [6], except that it suffices here to linearize once (namely to get the existence of $(\partial_{p_j} U^i)_{i,j \in E}$) instead of twice as done in [6]. The reason is that, while the master equation in [6] is directly formulated in terms of the second order derivatives with respect to the measure argument, we here avoid any second order derivatives in the variable p as we handle separately the influence of the common noise through the additional variable x. As we already explained in introduction of Sect. 3.4.1, the latter x does not show up in [6] (or, at least, x has a different meaning therein).

Actually, it must be implicitly understood from the previous paragraph that the most difficult part in the proof of Theorem 3.2 is to show that the master field, as given by Definition 3.7, is smooth. In this regard, the main difficulty is to get the existence of the first-order derivatives $(\partial_{p_j} U^i)_{i,j \in E}$. As for the existence of the derivatives in t and x, we give below a tailored-made argument (that is specific to our form of MFG) based on the fact that the equation features a Laplace operator in the direction x.

As for uniqueness of the solution to the master equation, this follows from a standard argument in the theory of forward-backward SDEs. Once the master field (which is called decoupling field in the theory of forward-backward SDEs) is known to be smooth, it must be the unique classical solution of the nonlinear PDE associated (through the method of *stochastic* characteristics) with the forward-backward SDE. We here give a sketch of the proof. More details may be found in [18, 38]. Assume indeed that $V = (V^i)_{i \in E}$ is another solution of the master equation and say that we want to identify $V(0, 0, p)$ with $U(0, 0, p)$ for a given $p \in S_d$ (obviously, the argument would be similar if we replaced $(0, 0, p)$ by (t, x, p) for $t \in [0, T]$ and $x \in \mathbb{R}$, the choice of $(0, 0)$ being here for convenience only). We then consider the solution (p, u, v) to the MFG system (3.19), whenever p starts from p at time 0 and ηW starts from 0 at time 0.

For a given $i \in E$, we then apply Itô's formula to $((V^i(t, \eta W_t, p_t))_{i \in E})_{0 \leq t \leq T}$. We simply denote the latter by $((V_t^i)_{i \in E})_{0 \leq t \leq T}$. By using the PDE (3.25) satisfied by V (replace U by V in (3.25)), we get

$$dV_t^i = \sum_{j,k \in E} p_k \left[\left(\gamma + u_t^k - u_t^j \right)_+ - \left(\gamma + V_t^k - V_t^j \right)_+ \right] \left(\partial_{p_j} V_t^i - \partial_{p_k} V_t^i \right) dt$$

$$- \left[\sum_{j \in E} H \left(V_t^j - V_t^i \right) + f \left(t, \varsigma_i + \eta W_t, \mu^{\varsigma_t}[p_t] \right) \right] dt + \partial_x V_t^i dW_t,$$

for $t \in [0, T]$, with the simpler notation $\partial_{p_j} V_t^i = \partial_{p_j} V^i(t, \eta W_t, p_t)$ and $\partial_x V_t^i = \partial_x V^i(t, \eta W_t, p_t)$. We then subtract $(u_t^i)_{0 \leq t \leq T}$ to both sides. We obtain

$$d\left[V_t^i - u_t^i \right] = \sum_{j,k \in E} p_k \left[\left(\gamma + u_t^k - u_t^j \right)_+ - \left(\gamma + V_t^k - V_t^j \right)_+ \right] \left(\partial_{p_j} V_t^i - \partial_{p_k} V_t^i \right) dt$$

$$+ \sum_{j \in E} \left[H \left(u_t^j - u_t^i \right) - H \left(V_t^j - V_t^i \right) \right] dt + \left(\partial_x V_t^i - v_t^i \right) dW_t.$$

Observe now that $V_T^i - u_T^i = 0$. Integrate from t to T, take conditional expectation given $\mathcal{F}_t$ and deduce from the fact that V and its gradient in (x, p) are bounded that, for some constant $C \geq 0$, and for all $i \in E$ and $t \in [0, T]$,

$$\mathbb{E}\left[|V_t^i - u_t^i| \right] \leq C \int_t^T \sum_{j \in E} \mathbb{E}\left[|V_s^j - u_s^j| \right] ds.$$

Summing over $i \in E$ and then by applying Gronwall's lemma, we get the announced equality: $V_0^i = u_0^i$, that is $V^i(0, 0, p) = U^i(0, 0, p)$, for any $i \in E$.

3.5 Proof of the Smoothness of the Master Field

We now address the main part of the proof of Theorem 3.2, namely the proof of the smoothness of the master field U. We proceed in two steps:

1. The first one is to show the differentiability of U with respect to p and then the continuity of the derivative; to do so, we mostly follow the linearization approach developed in [6], but, fortunately, it is easier to implement in our setting.
2. The second is to prove the differentiability of U with respect to t (at order 1) and to x (at order 2); the argument is here tailored made to the form of the master equation in our setting and relies on standard Schauder estimates for the heat equation.

3.5.1 Linearized System

Generally speaking, the strategy to prove the continuous differentiability of U with respect to p is to linearize the system (3.19) with respect to the initial condition. Equivalently, the system (3.19) is here regarded as the system of characteristics of the master equation (3.25); in order to establish the regularity of the solution to (3.25), we hence prove that the flow generated by (3.19) is differentiable with respect to the parameter p (which stands for the initial condition of the forward component).

3.5.1.1 Form of the Linearized System

Before we provide the form of the linearized system, we feel useful to emphasize the following subtlety: In (3.19), the drift coefficient in the forward equation is not differentiable because of the positive part therein; However, recalling that $\gamma \geq 2M$ and that u^i and u^j therein are bounded by M, see Remark 3.3, we may easily remove the positive part and hence reduce to the case when the drift is smooth. In fact, a similar remark applies to the Hamiltonian in the backward equation: We may reduce it to a mere quadratic function. As result, we get the following formal expression for the linearized system:

$$
dq_t^i = \Big[\sum_{j \in E: j \neq i} q_t^j \big(\gamma + u_t^j - u_t^i \big) - q_t^i \sum_{j \in E: j \neq i} \big(\gamma + u_t^i - u_t^j \big) \Big] dt
$$
$$
+ \Big[\sum_{j \in E: j \neq i} p_t^j \big(y_t^j - y_t^i \big) - p_t^i \sum_{j \in E: j \neq i} \big(y_t^i - y_t^j \big) \Big] dt
$$
$$
dy_t^i = \Big[\sum_{j \in E} \big(\gamma + u_t^i - u_t^j \big)\big(y_t^i - y_t^j \big) - \sum_{j \in E} \frac{\delta \widetilde{F}_t^i}{\delta m}(p_t)(\varsigma_t^j) q_t^j \Big] dt + z_t^i dW_t,
$$
$$
y_T^i = \sum_{j \in E} \frac{\delta \widetilde{G}^i}{\delta m}(p_T)(\varsigma_T^j) q_T^j,
$$

$$(3.26)$$

for $t \in [0, T]$ and $i \in E$. Above, (p, u) denotes the solution to (3.19) for some initial condition. Also, we have used the notation (compare with Remark 3.4):

$$
\frac{\delta \widetilde{F}_t^i}{\delta m}(p)(v) = \frac{\delta f}{\delta m}\big(t, \varsigma_t^i, \mu^{\varsigma_t}[p]\big)(v), \qquad \frac{\delta \widetilde{G}_t^i}{\delta m}(p)(v) = \frac{\delta g}{\delta m}\big(\varsigma_T^i, \mu^{\varsigma_T}[p]\big)(v),
$$

for $p \in S_d$ and $v \in \mathbb{R}$.

In order to study the solvability of (3.26) and then make the connection with (3.19), we are to use the same method of continuation as in Sect. 3.3.3, which prompts us to introduce similar notation. For any $\lambda \in [0, 1]$, any bounded $\mathbb{F}$-progressively measurable processes $\tilde{\boldsymbol{b}} = ((\tilde{b}_t^i)_{i \in E})_{0 \le t \le T}$ and $\tilde{\boldsymbol{f}} = ((\tilde{f}_t^i)_{i \in E})_{0 \le t \le T}$, with

$$\mathbb{P} \otimes \text{Leb}_1\left((\omega, t) : \sum_{i \in E} \tilde{b}_t^i \ne 0\right) = 0, \tag{3.27}$$

and any bounded $\mathcal{F}_T$-measurable random variable $\tilde{\boldsymbol{g}} = (\tilde{g}_T^i)_{i \in E}$, we hence denote by $\mathcal{D}(\lambda, \tilde{\boldsymbol{b}}, \tilde{\boldsymbol{f}}, \tilde{\boldsymbol{g}})$, the forward-backward system:

$$\begin{aligned}
dq_t^i &= \Big[\sum_{j \in E: j \ne i} q_t^j (\gamma + u_t^j - u_t^i) - q_t^i \sum_{j \in E: j \ne i} (\gamma + u_t^i - u_t^j) \Big] dt \\
&\quad + \Big[\sum_{j \in E: j \ne i} p_t^j (y_t^j - y_t^i) - p_t^i \sum_{j \in E: j \ne i} (y_t^i - y_t^j) \Big] dt + \tilde{b}_t^i dt \\
dy_t^i &= \Big[\sum_{j \in E} (\gamma + u_t^i - u_t^j)(y_t^i - y_t^j) - \lambda \sum_{j \in E} \frac{\delta \tilde{F}_t^i}{\delta m}(p_t)(\varsigma_t^j) q_t^j - \tilde{f}_t^i \Big] dt + z_t^i dW_t, \\
y_T^i &= \lambda \sum_{j \in E} \frac{\delta \tilde{G}_T^i}{\delta m}(p_T)(\varsigma_T^j) q_T^j + \tilde{g}_T^i,
\end{aligned} \tag{3.28}$$

for $t \in [0, T]$ and $i \in E$. Solutions are required to satisfy:

$$\sum_{i \in E} \mathbb{E}\Big[\sup_{0 \le t \le T} (|q_t^i|^2 + |y_t^i|^2) + \int_0^T |z_t^i|^2 dt \Big] < \infty. \tag{3.29}$$

They are denoted in the form $(\boldsymbol{q}, \boldsymbol{y}, \boldsymbol{z})$ or $(\boldsymbol{q}, \boldsymbol{y})$.

Remark 3.13 In [6], the condition (3.27) manifests in the form of an additional divergence in the dynamics, see for instance (4.37) therein. Since very few is said about it in [6], we feel that it is worth saying more about its role here. Obviously, the key point is that, whenever $\boldsymbol{q}$ solves the forward equation in (3.28), it satisfies, $\mathbb{P}$-almost surely,

$$\forall t \in [0, T], \quad \sum_{i \in E} q_t^i = \sum_{i \in E} q_0^i. \tag{3.30}$$

We make use of the above equality in (3.31), when invoking the Lasry–Lions monotonicity condition.

3.5.1.2 Stability Lemma

Similar to the analysis of the MFG system (3.19) performed in Sect. 3.3.3, the analysis of the linearized systems (3.26) and (3.28) goes through a stability lemma of the same type as Lemma 3.3. This stability lemma is the main ingredient for proving existence and uniqueness of a solution to (3.26) by a continuation method very much like Lemma 3.2 and also for proving that (3.26) provides indeed the derivative of the master field U with respect to the variable p. As for the latter point, we refer to the next section for the details.

For the time being, we focus on the formulation and the proof of the stability lemma. In our framework, we indeed have the following analogue of Lemma 3.3:

Lemma 3.4 *For any* $\lambda \in [0, 1)$, *consider* $\tilde{\boldsymbol{b}} = ((\tilde{b}_t^i)_{i \in E})_{0 \leq t \leq T}$, $\tilde{\boldsymbol{f}} = ((\tilde{f}_t^i)_{i \in E})_{0 \leq t \leq T}$, $\tilde{\boldsymbol{b}}' = ((\tilde{b}_t'^{,i})_{i \in E})_{0 \leq t \leq T}$ *and* $\tilde{\boldsymbol{f}}' = ((\tilde{f}_t'^{,i})_{i \in E})_{0 \leq t \leq T}$ *four bounded* $\mathbb{F}$-*progressively measurable processes, with* $\sum_{i \in E} \tilde{b}_t^i = \sum_{i \in E} \tilde{b}_t'^{,i} = 0$ $\mathbb{P} \otimes \mathrm{Leb}_1$ *almost everywhere, and* $\tilde{\boldsymbol{g}} = (\tilde{g}_T^i)_{i \in E}$ *and* $\tilde{\boldsymbol{g}}' = (\tilde{g}_T'^{,i})_{i \in E}$ *two bounded* $\mathcal{F}_T$-*measurable random variables. Assume that* $(\boldsymbol{q}, \boldsymbol{y})$ *and* $(\boldsymbol{q}', \boldsymbol{y}')$ *solve respectively* $\mathcal{D}(\lambda, \tilde{\boldsymbol{b}}, \tilde{\boldsymbol{f}}, \tilde{\boldsymbol{g}})$ *and* $\mathcal{D}(\lambda, \tilde{\boldsymbol{b}}', \tilde{\boldsymbol{f}}', \tilde{\boldsymbol{g}}')$ *(with possibly two different initial conditions but for the same pair* $(\boldsymbol{p}, \boldsymbol{u})$). *Then, there exists a constant* C *only depending on the assumption in Sect. 3.2.3 such that*

$$\mathbb{E}\left[\sum_{i \in E} \sup_{t \in [0,T]} \left(|q_t^i - q_t'^{,i}|^2 + |y_t^i - y_t'^{,i}|^2\right) + \sum_{i \in E} \int_0^T |z_t^i - z_t'^{,i}|^2 dt\right]$$

$$\leq C\mathbb{E}\left[\sum_{i \in E} |q_0^i - q_0'^{,i}|^2 + \sum_{i \in E} |\tilde{g}_T^i - \tilde{g}_T'^{,i}|^2 + \sum_{i \in E} \int_0^T \left(|\tilde{b}_t^i - \tilde{b}_t'^{,i}|^2 + |\tilde{f}_t^i - \tilde{f}_t'^{,i}|^2\right) dt\right].$$

Remark 3.14 The reader should notice that, in comparison with the statement of Lemma 3.3, there is here no constraint on the bounds of $((\tilde{f}_t^i)_{i \in E})_{0 \leq t \leq T}$ and of $(\tilde{g}_T^i)_{i \in E}$. This is due to the fact that, here, we directly work on the domain where the Hamiltonian H is quadratic. Indeed, we already know that $\boldsymbol{u}^i$ is bounded by M, for each $i \in E$. In particular, differently from Lemma 3.3, there is no need to prove any a priori bound on the solution $(\boldsymbol{q}, \boldsymbol{y}, \boldsymbol{z})$.

Proof We consider two solutions $(\boldsymbol{q}, \boldsymbol{y}, \boldsymbol{z})$ and $(\boldsymbol{q}', \boldsymbol{y}', \boldsymbol{z}')$ as in the statement.

First Step Following the proof of Lemma 3.3, we compute $d[\sum_{i\in E}(q^i_t - q'^{,i}_t)(y^i_t - y'^{,i}_t)]$. We get

$$
d\left(\sum_{i\in E}(q^i_t - q'^{,i}_t)(y^i_t - y'^{,i}_t)\right)
$$

$$
= \sum_{i,j\in E}\left[\left(q^j_t(\gamma + u^j_t - u^i_t) - q'^{,j}_t(\gamma + u^j_t - u^i_t)\right)\left(y^i_t - y'^{,i}_t - (y^j_t - y'^{,j}_t)\right)\right]dt
$$

$$
+ \sum_{i,j\in E}\left[\left(p^j_t(\gamma + y^j_t - y^i_t) - p^j_t(\gamma + y'^{,j}_t - y'^{,i}_t)\right)\left(y^i_t - y'^{,i}_t - (y^j_t - y'^{,j}_t)\right)\right]dt
$$

$$
+ \sum_{i\in E}(q^i_t - q'^{,i}_t)\sum_{j\in E}\left[(\gamma + u^i_t - u^j_t)(y^j_t - y^i_t) - (\gamma + u^i_t - u^j_t)(y'^{,i}_t - y'^{,j}_t)\right]dt
$$

$$
- \lambda\sum_{i\in E}(q^i_t - q'^{,i}_t)\sum_{j\in E}\frac{\delta\widetilde{F}^i_t}{\delta m}(p_t)(\varsigma^j_t)(q^j_t - q'^{,j}_t)dt - \sum_{i\in E}(q^i_t - q'^{,i}_t)(\widetilde{f}^i_t - \widetilde{f}'^{,i}_t)dt
$$

$$
+ \sum_{i\in E}(\widetilde{b}^i_t - \widetilde{b}'^{,i}_t)(y^i_t - y'^{,i}_t)dt + \sum_{i\in E}(q^i_t - q'^{,i}_t)(z^i_t - z'^{,i}_t)dW_t.
$$

It is easy to see that the first and third terms in the right-hand side cancel. Integrating from 0 to T and taking expectation (by (3.29), observe that the stochastic integral is a true martingale), we get

$$
\mathbb{E}\sum_{i\in E}(q^i_T - q'^{,i}_T)(y^i_T - y'^{,i}_T) + \lambda\mathbb{E}\int_0^T\sum_{i\in E}(q^i_t - q'^{,i}_t)\sum_{j\in E}\frac{\delta\widetilde{F}^i_t}{\delta m}(p_t)(\varsigma^j_t)(q^j_t - q'^{,j}_t)dt
$$

$$
+ \mathbb{E}\int_0^T\sum_{i,j\in E}p^j_t\left(y^i_t - y'^{,i}_t - (y^j_t - y'^{,j}_t)\right)^2 dt
$$

$$
\leq \mathbb{E}\sum_{i\in E}\left[|q^i_0 - q'^{,i}_0||y^i_0 - y'^{,i}_0| + \int_0^T\left(|q^i_t - q'^{,i}_t||\widetilde{f}^i_t - \widetilde{f}'^{,i}_t| + |y^i_t - y'^{,i}_t||\widetilde{b}^i_t - \widetilde{b}'^{,i}_t|\right)dt\right].
$$

Now, the monotonicity of f together with identity (3.30) yield (see for instance [6, Subsection 2.3], paying attention that the measure μ therein must be centred, which is here the rationale for subtracting the initial conditions of q and q')

$$
\mathbb{E}\int_0^T\sum_{i\in E}(q^i_t - q^i_0 - (q'^{,i}_t - q'^{,i}_0))\sum_{j\in E}\frac{\delta\widetilde{F}^i_t}{\delta m}(p_t)(\varsigma^j_t)(q^j_t - q^j_0 - (q'^{,j}_t - q'^{,j}_0))dt \geq 0,
$$

which can be reformulated as[3]

$$\mathbb{E} \int_0^T \sum_{i \in E} (q_t^i - q_t^{',i}) \sum_{j \in E} \frac{\delta \widetilde{F}_t^i}{\delta m}(p_t)(\varsigma_t^j)(q_t^j - q_t^{',j}) dt$$

$$\geq -C\mathbb{E}\left(\sum_{i \in E} |q_0^i - q_0^{',i}|^2 + \sum_{i,j \in E} |q_0^i - q_0^{',i}| \int_0^T |q_t^j - q_t^{',j}| dt \right),$$

(3.31)

for a constant C that only depends on the details of the assumption in Sect. 3.2.3. Similarly, we have

$$\mathbb{E} \sum_{i \in E} (q_T^i - q_T^{',i})(y_T^i - y_T^{',i})$$

$$= \lambda \mathbb{E} \sum_{i \in E} (q_T^i - q_T^{',i}) \sum_{j \in E} \frac{\delta \widetilde{G}_T^i}{\delta m}(p_t)(\varsigma_T^j)(q_T^j - q_T^{',j}) + \mathbb{E} \sum_{i \in E} (q_T^i - q_T^{',i})(\tilde{g}_T^i - \tilde{g}_T^{',i})$$

$$\geq \mathbb{E} \sum_{i \in E} (q_T^i - q_T^{',i})(\tilde{g}_T^i - \tilde{g}_T^{',i}) - C\mathbb{E}\left(\sum_{i \in E} |q_0^i - q_0^{',i}|^2 + \sum_{i,j \in E} |q_0^i - q_0^{',i}||q_T^j - q_T^{',j}| \right).$$

So, we end up this first step with

$$\mathbb{E} \int_0^T \sum_{i,j \in E} p_t^j \left(y_t^i - y_t^{',i} - (y_t^j - y_t^{',j}) \right)^2 dt$$

$$\leq \mathbb{E} \sum_{i \in E} \left[|q_0^i - q_0^{',i}|^2 + |q_0^i - q_0^{',i}||y_0^i - y_0^{',i}| + |q_T^i - q_T^{',i}|\left(|\tilde{g}_T^i - \tilde{g}_T^{',i}| + |q_0^i - q_0^{',i}| \right) \right.$$

$$\left. + \int_0^T \left[|q_t^i - q_t^{',i}|\left(|q_0^i - q_0^{',i}| + |\tilde{f}_t^i - \tilde{f}_t^{',i}| \right) + |y_t^i - y_t^{',i}||\tilde{b}_t^i - \tilde{b}_t^{',i}| \right] dt \right].$$

Second Step Returning to the forward equation in (3.28) and allowing the value of the constant C to vary from line to line below, we now observe that

$$\mathbb{E} \sum_{i \in E} \sup_{t \in [0,T]} |q_t^i - q_t^{',i}|^2$$

$$\leq C \sum_{i \in E} \left[|q_0^i - q_0^{',i}|^2 + \mathbb{E} \int_0^T \left[\sum_{j \in E} p_t^j \left(y_t^i - y_t^{',j} - (y_t^i - y_t^{',i}) \right)^2 + |\tilde{b}_t^i - \tilde{b}_t^{',i}|^2 \right] dt \right].$$

---

[3]The reader who is willing to compare with [6] may observe that, in pages 119 and 120 therein, the square of the norm of the difference of the two initial conditions of the forward equation is missing; obviously, this does not change the final result of the proof of [6, Proposition 4.4.5] since this square is injected in the computations that come next.

In fact, we can proceed in a similar way with the backward equation in (3.28). Forming the difference $(y_t^i - y_t^{\prime,i})_{0 \le t \le T}$, applying Itô's formula to the square of it and then taking expectation, we get

$$
\mathbb{E}\big(|y_t^i - y_t^{\prime,i}|^2\big) \le C\bigg(\sum_{j \in E} \mathbb{E}\big(|q_T^j - q_T^{\prime,j}|^2\big) + \mathbb{E}\big(|\tilde{g}_T^i - \tilde{g}_T^{\prime,i}|^2\big)
$$

$$
+ \mathbb{E}\int_t^T \Big[\sum_{j \in E} \big(|q_s^j - q_s^{\prime,j}|^2 + |y_s^j - y_s^{\prime,j}|^2\big) + |\tilde{f}_s^i - \tilde{f}_s^{\prime,i}|^2\Big]ds \bigg).
$$

By Gronwall's lemma, we get

$$
\sup_{0 \le t \le T} \sum_{i \in E} \mathbb{E}\big(|y_t^i - y_t^{\prime,i}|^2\big)
$$

$$
\le C\mathbb{E}\sum_{i \in E} \bigg(\sup_{t \in [0,T]} |q_t^i - q_t^{\prime,i}|^2 + |\tilde{g}_T^i - \tilde{g}_T^{\prime,i}|^2 + \int_0^T |\tilde{f}_t^i - \tilde{f}_t^{\prime,i}|^2 dt \bigg).
$$

Conclusion By collecting the last inequality in the first step and the first and third inequalities in the second step and then by using Young's inequality $2ab \le \varepsilon a^2 + \varepsilon^{-1}b^2$, for $\varepsilon > 0$ small enough, we derive the announced bound.

□

3.5.1.3 Existence and Uniqueness

By arguing as in Lemma 3.2, we can increase step by step the value of λ in (3.28) and then prove existence and uniqueness for any value of $\lambda \in [0, 1]$. We end up with the following statement:

Proposition 3.5 *Consider an initial condition $p \in S_d$ for the forward component p at time 0 in (3.19), namely $p_0 = p$, and call (p, u) the corresponding solution. Then, for any initial condition $q \in \mathbb{R}^d$, for any bounded $\mathbb{F}$-progressively measurable processes $\tilde{b} = ((\tilde{b}_t^i)_{i \in E})_{0 \le t \le T}$ and $\tilde{f} = ((\tilde{f}_t^i)_{i \in E})_{0 \le t \le T}$, with $\tilde{b}$ satisfying (3.27), and any bounded $\mathcal{F}_T$-measurable random variable $\tilde{g} = (\tilde{g}_T^i)_{i \in E}$, the system $\mathcal{D}(1, \tilde{b}, \tilde{f}, \tilde{g})$ with $q_0 = q$ as initial condition at time 0 is uniquely solvable among the class of triples (q, y, z) satisfying the integrability constraint (3.29).*

Remark 3.15 Implicitly, the initial condition of the common noise is understood as 0 in the above proposition, namely $W_0 = 0$. In fact, this choice is for convenience only, as it permits to make the statement consistent with the framework of Proposition 3.2.

However, it must be clear for the reader that, on the model of Definition 3.7, we could allow the system (3.19) to be initialized at any time $t \in [0, T]$ and, accordingly, the common noise to start from any $x \in \mathbb{R}$ (which means that ηW

has to be replaced by $(x + \eta(W_s - W_t))_{t \leq s \leq T}$. Obviously, both Lemma 3.4 and Proposition 3.5 remain true in this more general setting.

3.5.2 Differentiability in p

We now come back to our original objective, which is to prove that the master field is continuously differentiable with respect to p. In this respect, the reader could object that, so far, we have not made clear the connection between the linearized system (3.26) and the original MFG system (3.19). This is precisely the purpose of the following statement to clarify the latter fact:

Proposition 3.6 *For given $x \in \mathbb{R}$ and p, $p' \in S_d$ and for any $\varepsilon \in [0, 1]$, denote by $(\boldsymbol{p}^{(\varepsilon)}, \boldsymbol{u}^{(\varepsilon)})$ the solution to (3.19) whenever the forward component therein starts from $(1 - \varepsilon)p + \varepsilon p'$ at time 0 and the common noise ηW starts from x (also at time 0). Merely writing $(\boldsymbol{p}, \boldsymbol{u})$ for $(\boldsymbol{p}^{(0)}, \boldsymbol{u}^{(0)})$, call then $(\boldsymbol{q}, \boldsymbol{y})$ the solution to the linearized system (3.26) when $\boldsymbol{q}$ therein starts from $p' - p$ at time 0 and again the common noise ηW starts from x (also at time 0). Then,*

$$\lim_{\varepsilon \searrow 0} \mathbb{E}\left[\sup_{t \in [0,T]} \left(\left| \frac{p_t^{(\varepsilon)} - p_t}{\varepsilon} - q_t \right|^2 + \left| \frac{u_t^{(\varepsilon)} - u_t}{\varepsilon} - y_t \right|^2 \right) \right] = 0.$$

Remark 3.16 We could also address the asymptotic behaviour of the martingale representation term $v^{(\varepsilon)}$ (with obvious notation) as ε tends to 0. We would have

$$\lim_{\varepsilon \searrow 0} \mathbb{E} \int_0^T \left| \frac{v_t^{(\varepsilon)} - v_t}{\varepsilon} - z_t \right|^2 dt = 0,$$

but there would not be any specific interest for us to do so.

Proof *First Step* We introduce the following useful notation:

$$q_t^{(\varepsilon)} = \frac{p_t^{(\varepsilon)} - p_t}{\varepsilon}, \quad y_t^{(\varepsilon)} = \frac{u_t^{(\varepsilon)} - u_t}{\varepsilon}, \quad t \in [0, T],$$

for $\varepsilon \in [0, 1]$. The key point in the proof is to write $(\boldsymbol{q}^{(\varepsilon)}, \boldsymbol{y}^{(\varepsilon)}) = (q_t^\varepsilon, y_t^\varepsilon)_{0 \leq t \leq T}$ as a solution to $\mathcal{D}(1, \tilde{\boldsymbol{b}}^{(\varepsilon)}, \tilde{\boldsymbol{f}}^{(\varepsilon)}, \tilde{\boldsymbol{g}}^{(\varepsilon)})$ for well chosen $\tilde{\boldsymbol{b}}^{(\varepsilon)}$, $\tilde{\boldsymbol{f}}^{(\varepsilon)}$ and $\tilde{\boldsymbol{g}}^{(\varepsilon)}$. To this end, it is worth recalling that, in (3.19), we can remove the positive parts in the forward equation and restrict H in the backward equation to the domain where it is quadratic. It is plain to see that a natural choice for $\tilde{\boldsymbol{b}}^{(\varepsilon)}$, $\tilde{\boldsymbol{f}}^{(\varepsilon)}$ and $\tilde{\boldsymbol{g}}^{(\varepsilon)}$ is

$$\tilde{b}_t^{(\varepsilon),i} = \varepsilon\bigg[\sum_{j\in E:j\neq i} q_t^{(\varepsilon),j}\big(y_t^{(\varepsilon),j} - y_t^{(\varepsilon),i}\big) - q_t^{(\varepsilon),i}\sum_{j\in E:j\neq i}\big(y_t^{(\varepsilon),i} - y_t^{(\varepsilon),j}\big)\bigg],$$

$$\tilde{f}_t^{(\varepsilon),i} = -\tfrac{1}{2}\varepsilon\sum_{j\in E}|y_t^{(\varepsilon),i} - y_t^{(\varepsilon),j}|^2$$

$$+\sum_{j\in E} q_t^{(\varepsilon),j}\int_0^1\bigg(\frac{\delta f}{\delta m}\big(t,\varsigma_t^i,\mu^{\varsigma_t}[\sigma p_t^{(\varepsilon)}+(1-\sigma)p_t]\big) - \frac{\delta f}{\delta m}\big(t,\varsigma_t^i,\mu^{\varsigma_t}[p_t]\big)\bigg)\big(\varsigma_t^j\big)d\sigma,$$

$$\tilde{g}_T^{(\varepsilon),i}$$

$$=\sum_{j\in E} q_T^{(\varepsilon),j}\int_0^1\bigg(\frac{\delta g}{\delta m}\big(\varsigma_T^i,\mu^{\varsigma_T}[\sigma p_T^{(\varepsilon)}+(1-\sigma)p_T]\big) - \frac{\delta g}{\delta m}\big(\varsigma_T^i,\mu^{\varsigma_T}[p_T]\big)\bigg)\big(\varsigma_T^j\big)d\sigma,$$

where ς_t is here equal to $(\varsigma_i + x + \eta W_t)_{i\in E}$. Since $\delta f/\delta m$ and $\delta g/\delta m$ are Lipschitz continuous with respect to d_{BL}, we deduce that there exists a constant C, only depending on the details of the assumption in Sect. 3.2.3, such that

$$\sum_{i\in E}\big(|\tilde{b}_t^{(\varepsilon),i}| + |\tilde{f}_t^{(\varepsilon),i}|\big) \leq C\varepsilon\sum_{i\in E}\big(|y_t^{(\varepsilon),i}|^2 + |q_t^{(\varepsilon),i}|^2\big), \quad t\in[0,T],$$

$$\sum_{i\in E}|\tilde{g}_T^{(\varepsilon),i}| \leq C\varepsilon\sum_{i\in E}|q_T^{(\varepsilon),i}|^2.$$

Observing that $q_0^{(\varepsilon)} = q_0$ and recalling that $(\boldsymbol{q},\boldsymbol{y})$ is here a solution of $\mathcal{D}(1,0,0,0)$ (for the prescribed choice of initial conditions), we deduce from Lemma 3.4 that

$$\sum_{i\in E}\mathbb{E}\Big[\sup_{t\in[0,T]}\big(|q_t^{(\varepsilon),i} - q_t^i|^2 + |y_t^{(\varepsilon),i} - y_t^i|^2\big)\Big]$$

$$\leq C\varepsilon^2\sum_{i\in E}\mathbb{E}\Big[|q_T^{(\varepsilon),i}|^4 + \int_0^T\big(|q_t^{(\varepsilon),i}|^4 + |y_t^{(\varepsilon),i}|^4\big)dt\Big].$$

Second Step The difficulty here is to sort out the fourth moment in the above inequality. Obviously, we could think of making use of Lemma 3.3, but the problem is precisely that the estimate therein is just for the second moment.

In order to bypass this difficulty, we use the definition of the master field and its regularity properties. Indeed, we know that $\boldsymbol{p}$ solves the following ODE (with random coefficients):

$$\dot{p}_t^i = \sum_{j\in E} p_t^j\big(\gamma + (U^j - U^i)(t,x+\eta W_t,p_t)\big)_+$$

$$-\sum_{j\in E} p_t^i\big(\gamma + (U^i - U^j)(t,x+\eta W_t,p_t)\big)_+,$$

for $t \in [0, T]$ and $i \in E$, with $p_0 = p$ as initial condition and similarly for $\boldsymbol{p}^{(\varepsilon)}$ but with $p_0^{(\varepsilon)} = (1 - \varepsilon)p + \varepsilon p'$ as initial condition. Since we know from Proposition 3.3 that U is Lipschitz continuous in the last argument, we get

$$\sup_{t \in [0,T]} \sum_{i \in E} |p_t^{(\varepsilon),i} - p_t^i| \le C\varepsilon \sum_{i \in E} |p_i' - p_i|, \tag{3.32}$$

with probability 1, for a new value of the constant C, which is allowed to vary from line to line below. In turn, by composing by $U(t, x + \eta W_t, \cdot)$ on both sides and by using Proposition 3.3, we obtain, $\mathbb{P}$ almost-surely,

$$\sup_{t \in [0,T]} \sum_{i \in E} |u_t^{(\varepsilon),i} - u_t^i| \le C\varepsilon \sum_{i \in E} |p_i' - p_i|. \tag{3.33}$$

Hence, by dividing by ε in both (3.32) and (3.33), we finally have

$$\mathbb{P}\left(\sup_{t \in [0,T]} \sum_{i \in E} \left(|q_t^{(\varepsilon),i}| + |y_t^{(\varepsilon),i}| \right) \le C \right) = 1. \tag{3.34}$$

Inserting the above bound in the conclusion of the first step, we complete the proof.
□

The following corollary is a key step in the proof of Theorem 3.2.

Corollary 3.2 *The master U field is continuously differentiable in p (when p is regarded as an element of the $(d - 1)$-dimensional simplex) and the $(d - 1)$-dimensional gradient is Lipschitz continuous with respect to (x, p) and $1/2$-Hölder continuous in time.*

Proof *First Step* We first prove differentiability of U at $t = 0$, the more general case $t \in [0, T]$ being treated in the same way. To do so, it is worth pointing out, see Remark 3.11, that y_0^i in (3.26) is almost surely constant. Therefore, we deduce from Proposition 3.6 that, for any $i \in E$,

$$\lim_{\varepsilon \to 0} \frac{U^i(o, x, (1 - \varepsilon)p + \varepsilon p') - U^i(o, x, p)}{\varepsilon} = y_0^i, \tag{3.35}$$

which proves the existence of a directional derivative. The fact that the direction is given by $p' - p$, the coordinates of which have a sum equal to 0 (namely $\sum_{i \in E}(p_i' - p_i) = 0$), is consistent with the fact the above derivative has to be regarded in dimension $d - 1$. Also, it is worth emphasizing that y_0^i in the right-hand side depends on x, p and the difference $p' - p$, but, since (3.26) is a linear, it should be in fact a linear function of $p' - p$. To make it clear, we may call $(\boldsymbol{q}[j, x, p], \boldsymbol{y}[j, x, p])$, for $j \in E \setminus \{d\}$, the solution to (3.26) when $q_0^k = \delta_{k,j} - \delta_{k,d}$ (with $\delta_{k,j}$ being the standard Kronecker symbol) and when, as before, the forward component in (3.19) starts from p and the common noise starts from x at time 0. Then, observing that

$p' - p = \sum_{j\in E:j\neq d}(p'_j - p_j)q_0[j, x, p]$, (3.35) may be written in the form

$$\lim_{\varepsilon\searrow 0} \frac{U^i(o, x, (1-\varepsilon)p + \varepsilon p') - U^i(o, x, p)}{\varepsilon} = \sum_{j\in E:j\neq d} (p'_j - p_j)y_0^i[j, x, p].$$

(3.36)

Second Step We now prove that the mapping $(x, p) \mapsto y_0[j, x, p] = (y_0^i[j, x, p])_{i\in E}$ is Lipschitz continuous in $(x, p) \in \mathbb{R} \times S_d$ for any $j \in E \setminus \{d\}$. In order to proceed, we consider two points (x, p) and (x', p') in $\mathbb{R} \times S_d$. We then denote by $(\boldsymbol{p}, \boldsymbol{u})$ and $(\boldsymbol{p}', \boldsymbol{u}')$ the two solutions of (3.19), the first one being initialized from p at time 0 and the common noise therein being initialized from x, and the second one being initialized from p' at time 0 and the common noise therein being initialized from x'. With $(\boldsymbol{p}, \boldsymbol{u})$ (and with the same initial condition x for the common noise), we associate $(\boldsymbol{q}, \boldsymbol{y})$ the solution to (3.26) with $q_0^k = \delta_{k,j} - \delta_{k,d}$, for $k \in E$. Similarly, we call $(\boldsymbol{q}', \boldsymbol{y}')$ the solution to (3.26) that is associated to $(\boldsymbol{p}', \boldsymbol{u}')$ (and with the initial condition x' for the common noise and with $q_0^{',k} = \delta_{k,j} - \delta_{k,d}$, for $k \in E$). We then make use of Lemma 3.4 to compare the two of them, it being understood that we write both systems with respect to $(\boldsymbol{p}, \boldsymbol{u})$ and then with some remainders $(\tilde{\boldsymbol{b}}, \tilde{\boldsymbol{f}}, \tilde{\boldsymbol{g}})$ and $(\tilde{\boldsymbol{b}}', \tilde{\boldsymbol{f}}', \tilde{\boldsymbol{g}}')$. Obviously, it is plain to see that $(\tilde{\boldsymbol{b}}, \tilde{\boldsymbol{f}}, \tilde{\boldsymbol{g}})$ is in fact equal to $(0, 0, 0)$. As for $(\tilde{\boldsymbol{b}}', \tilde{\boldsymbol{f}}', \tilde{\boldsymbol{g}}')$, we here follow the same argument as in the proof of Proposition 3.6 and then choose it as

$$\tilde{b}_t^{',i} = \sum_{k\in E:k\neq i} q_t^{',k}\left(\delta u_t^k - \delta u_t^i\right) - q_t^{',i}\sum_{k\in E:k\neq i}\left(\delta u_t^i - \delta u_t^k\right)$$
$$+ \sum_{k\in E:k\neq i} \delta p_t^{',k}\left(y_t^{',k} - y_t^{',i}\right) - \delta p_t^{',i}\sum_{k\in E:k\neq i}\left(y_t^{',i} - y_t^{',k}\right),$$

$$\tilde{f}_t^{',i} = -\sum_{k\in E:k\neq i}\left(\delta u_t^i - \delta u_t^k\right)\left(y_t^i - y_t^k\right)$$
$$+ \sum_{k\in E} q_t^{',k}\frac{\delta f}{\delta m}\left(t, \varsigma_i + x' + \eta W_t, \sum_{\ell\in E} p_t^{',\ell}\delta_{\varsigma_\ell + x' + \eta W_t}\right)(\varsigma_k + x' + \eta W_t)$$
$$- \sum_{k\in E} q_t^{',k}\frac{\delta f}{\delta m}\left(t, \varsigma_i + x + \eta W_t, \sum_{\ell\in E} p_t^\ell\delta_{\varsigma_\ell + x + \eta W_t}\right)(\varsigma_k + x + \eta W_t),$$

$$\tilde{g}_T^{',i} = \sum_{k\in E} q_T^{',k}\frac{\delta g}{\delta m}\left(\varsigma_i + x' + \eta W_T, \sum_{\ell\in E} p_T^{',\ell}\delta_{\varsigma_\ell + x' + \eta W_T}\right)(\varsigma_k + x' + \eta W_T)$$
$$- \sum_{k\in E} q_T^{',k}\frac{\delta g}{\delta m}\left(\varsigma_i + x + \eta W_T, \sum_{\ell\in E} p_T^\ell\delta_{\varsigma_\ell + x + \eta W_T}\right)(\varsigma_k + x + \eta W_T),$$

where

$$\delta u_t^i = u_t^{\prime,i} - u_t^i, \quad \delta p_t^i = p_t^{\prime,i} - p_t^i.$$

We deduce that, for any $i \in E$ and $t \in [0, T]$,

$$|\tilde{b}_t^i| + |\tilde{f}_t^i| \leq \sum_{\ell \in E} (|q_t^{\prime,\ell}| + |y_t^{\prime,\ell}|) \sum_{k \in E} (|x_k' - x_k| + |\delta p_t^k| + |\delta u_t^k|),$$

$$|\tilde{g}_T^i| \leq \sum_{\ell \in E} |q_T^{\prime,\ell}| \sum_{k \in E} (|x_k' - x_k| + |\delta p_T^k|).$$

Following the proofs of (3.32) and (3.33), we have

$$\sup_{t \in [0,T]} \sum_{i \in E} (|\delta p_t^{\prime,i}| + |\delta u_t^{\prime,i}|) \leq C(|x - x'| + |p - p'|),$$

with probability 1, for a constant C only depending on the details of the assumption in Sect. 3.2.3. Also, returning to (3.34) and letting ε tend to 0 therein (with $p' - p$ in the statement of Proposition 3.6 being given by q_0, which is licit since the sum of the latter is 0), it clear that q' and y' are bounded by C, for a possibly new value of C. By invoking Lemma 3.4, this shows that, for $j \in E \setminus \{d\}$, the mapping $(x, p) \mapsto y_0[j, x, p]$ is Lipschitz continuous in $(x, p) \in \mathbb{R} \times \mathcal{S}_d$.

Returning to the conclusion of the first step, we deduce that U is differentiable with respect to p in $\mathcal{S}_d$ in the sense explained in Sect. 3.4.3 and that $\partial_p U(0, \cdot, \cdot)$ (with the same meaning as therein) is Lipschitz continuous in (x, p). To make it clear, if, for some $i \in E$, $j \in E \setminus \{d\}$, $p \in \mathcal{S}_d$ such that $p_j, p_d \in (0, 1)$ and $\varrho \in \mathbb{R}$ such that $|\varrho| \leq \min(p_j, p_d)$, we choose $p_j' - p_j = \varrho$, $p_d' - p_d = -\varrho$ and $p_k' - p_k = 0$ if $k \in E \setminus \{j, d\}$, we get in (3.36), using the same notation as in Sect. 3.4.3,

$$\lim_{\varepsilon \to 0} \frac{1}{\varepsilon} \frac{\hat{U}^i(t, x, (p_1, \cdots, p_j + \varepsilon, \cdots, p_{d-1})) - \hat{U}^i(t, x, (p_1, \cdots, p_{d-1}))}{\varepsilon} = y_0^i[j, x, p].$$

This permits to identify $y_0^i[j, x, p]$ with $\partial_{q_j} \hat{U}^i(0, x, (p_1, \cdots, p_{d-1}))$ whenever $p_j, p_d \in (0, 1)$. By continuity of $y_0^i[j, x, p]$ with respect to p, we deduce that $\hat{U}^i(0, x, \cdot)$ is differentiable on the interior of the $(d-1)$-dimensional simplex; since the derivative is Lipschitz continuous up to the boundary of the simplex, differentiability holds on the entire $(d-1)$-dimensional simplex (obviously, differentiability at the boundary holds true along admissible directions only).

We let the reader verify that we can proceed similarly at any time $t \in [0, T]$ and then check that the Lipschitz constant of $\partial_p U(t, \cdot, \cdot)$ is uniform in time.

Third Step It then remains to show that $\partial_p U$ is 1/2-Hölder continuous in time. The proof is very much in the spirit of Corollary 3.1. Indeed, as we explained in the second step, we have deterministic bounds for $\boldsymbol{q}$ and $\boldsymbol{p}$ when $q_0^k = \delta_{k,j} - \delta_{k,d}$ for any $k \in E$ and for some $j \in E \setminus \{d\}$. Then, it is easy to see that, for any $i \in E$,

$$\left| \mathbb{E}\left[y_h^i[j, x, p] \right] - y_0^i[j, x, p] \right|^2 \le Ch,$$

for a constant C only depending on the various parameters in the assumption in Sect. 3.2.3. Now, by arguing as in the construction of the master field, $y_h^i[j, x, p]$ must be equal to $\partial_{q_j} \hat{U}^i(h, x + \eta W_h, (p_h^1, \cdots, p_h^{d-1}))$, where $\boldsymbol{p}$ is the forward component of (3.19) when starting from p at time 0 (and when the common noise starts from x). By the second step, we then have, for any $i \in E$,

$$\left| \mathbb{E}\left[y_h^i[j, x, p] \right] - \partial_{q_j} \hat{U}^i\left(h, x, (p_1, \cdots, p_{d-1}) \right) \right|^2 \le Ch,$$

which proves that

$$\left| \partial_{q_j} \hat{U}^i\left(h, x, (p_1, \cdots, p_{d-1}) \right) - \partial_{q_j} \hat{U}^i\left(0, x, (p_1, \cdots, p_{d-1}) \right) \right| \le Ch^{1/2}.$$

The argument holds in fact at any starting time $t \in [0, T]$ instead of 0, which suffices to conclude.

$\square$

3.5.3 Heat Kernel and Differentiability in x

It now remains to address the regularity in x. Whilst we could perform a similar analysis as before by means of the characteristics, we feel better to use here the finite-dimensional character of the PDE. In comparison with the approach used in [6], this permits to simplify the argument. In this respect, the key point is the following proposition:

Proposition 3.7 Denote by $\Gamma(t, x) = (2\pi \eta^2 t)^{-1/2} \exp(-x^2/(2\eta^2 t))$ the η-rescaled Gaussian kernel. Then, for any $(t, x, p) \in [0, T] \times \mathbb{R} \times \mathcal{S}_d$,

$$U^i(t, x, p)$$

$$= \int_{\mathbb{R}} \Gamma(T - t, x - y) g(\varsigma_i + y, p) dy + \int_t^T \int_{\mathbb{R}} \Psi^i(s, y, p) \Gamma(s - t, x - y) ds dy,$$

$$(3.37)$$

where

$$\Psi^i(t, x, p) = \sum_{j \in E} H\big((U^j - U^i)(t, x, p)\big) + f\big(t, \varsigma_i + x, \mu^{\varsigma + x}[p]\big)$$

$$+ \sum_{j,k \in E} p_k\big(\gamma + (U^k - U^j)(t, x, p)\big)_+ \big(\partial_{p_j} U^i - \partial_{p_k} U^i\big)(t, x, p).$$

Proof *First Step* By combining the representation formula (3.24) with the backward equation in (3.19) and by initializing the latter from (t, x, p) as in the statement, we get, for any time $S \in [t, T]$,

$$U^i(t, x, p) = \mathbb{E}\bigg[U^i\big(S, x + \eta(W_S - W_t), p_S\big)$$

$$+ \sum_{j \in E} \int_t^S H\big((U^j - U^i)(s, x + \eta(W_s - W_t), p_s)\big) ds$$

$$+ \int_t^S f\big(s, \varsigma_i + x + \eta(W_s - W_t), \mu^{\varsigma + x + \eta(W_s - W_t)}[p_s]\big) ds \bigg],$$

$$(3.38)$$

where $p = (p_s)_{t \le s \le T}$ here solves the forward equation in (3.19) when the latter is initialized from (t, x, p). In words, p solves the SDE

$$\dot{p}_s^i = \sum_{j \in E} p_s^j \big(\gamma + (U^j - U^i)(s, x + \eta(W_s - W_t), p_s)\big)_+$$

$$- \sum_{j \in E} p_s^i \big(\gamma + (U^i - U^j)(s, x + \eta(W_s - W_t), p_s)\big)_+.$$

We now expand the first term in the right-hand side in (3.38), but with respect to the last component only, by taking benefit of the regularity result established in the previous section. We get

$$U^i(t, x, p) = \mathbb{E}\bigg[U^i\big(S, x + \eta(W_S - W_t), p\big) + \sum_{j \in E} \int_t^S \Psi_0^i\big(S, x + \eta(W_S - W_t), p_s\big) ds$$

$$+ \sum_{j \in E} \int_t^S H\big((U^j - U^i)(s, x + \eta(W_s - W_t), p_s)\big) ds$$

$$+ \int_t^S f\big(s, \varsigma_i + x + \eta(W_s - W_t), \mu^{\varsigma + x + \eta(W_s - W_t)}[p_s]\big) ds \bigg],$$

where, for simplicity, we have let

$$\Psi_0^i(t, x, p) = \sum_{j,k \in E} p_k \big(\gamma + (U^k - U^j)(t, x, p)\big)_+ \big(\partial_{p_j} U^i - \partial_{p_k} U^i\big)(t, x, p).$$

We then observe that, there exists a constant $C \geq 0$, only depending on the various parameters in the assumption in Sect. 3.2.3, such that all the three terms below

$$\mathbb{E}\Big[\big|\Psi_0^i\big(S, x + \eta(W_S - W_t), p_s\big) - \Psi_0^i\big(s, x + \eta(W_s - W_t), p)\big|\Big],$$

$$\mathbb{E}\Big[\Big|H\big((U^j - U^i)(s, x + \eta(W_s - W_t), p_s)\big) - H\big((U^j - U^i)(s, x + \eta(W_s - W_t), p)\big)\Big|\Big],$$

$$\mathbb{E}\Big[\Big|f\big(s, \varsigma_i + x + \eta(W_s - W_t), \mu^{s+x+\eta(W_s-W_t)}[p_s]\big)$$

$$- f\big(s, \varsigma_i + x + \eta(W_s-W_t), \mu^{s+x+\eta(W_s-W_t)}[p]\big)\Big|\Big],$$

are bounded by $C(S - t)^{1/2}$, from which we obtain that

$$U^i(t, x, p) = \mathbb{E}\Big[U^i\big(S, x + \eta(W_S - W_t), p\big) + \int_t^S \Psi^i\big(s, x + \eta(W_s - W_t), p)ds\Big]$$

$$+ O\big((S - t)^{3/2}\big), \tag{3.39}$$

where $|O((S - t)^{3/2})| \leq C(S - t)^{3/2}$.

Second Step We now choose a subdivision $\pi = \{t = t_0 < t_1 < \cdots < t_N = T\}$ of the interval $[t, T]$ with $|\pi|$ as step size. We then consider a given (t, x, p) as in the first step, but we are to apply (3.39) for several possible values of the first two arguments in U^i.

To make it clear, for any $k \in \{0, \cdots, N - 1\}$, we apply (3.39) to $U^i(t_k, x + \eta(W_{t_k} - W_t), p)$ on the interval $[t_k, t_{k+1}]$. Obviously, this requires some care since the starting point is then random. We may get a similar conclusion to (3.39) by replacing the expectation in the right-hand side by a conditional expectation given $\mathcal{F}_t$. Another way is to take an additional expectation in left-hand side, namely

$$\mathbb{E}\Big[U^i\big(t_k, x + \eta(W_{t_k} - W_t), p\big)\Big]$$

$$= \mathbb{E}\Big[U^i\big(t_{k+1}, x + \eta(W_{t_{k+1}} - W_t), p\big) + \int_{t_k}^{t_{k+1}} \Psi^i\big(s, x + \eta(W_s - W_t), p\big)ds\Big]$$

$$+ O\big(|\pi|^{3/2}\big).$$

By summing over k and then by letting $|\pi|$ tend to 0, we complete the proof.

$\square$

Corollary 3.3 *The function U is twice differentiable in x and once in t and the functions $\partial_t U$ and $\partial_x^2 U$ are continuous on $[0, T] \times \mathbb{R} \times \mathcal{S}_d$.*

Proof We recall that, for any $i \in E$, the functions $[0, T] \times \mathbb{R} \ni (t, x) \mapsto \Psi^i(t, x, p), \mathbb{R} \ni x \mapsto g(\varsigma_i + x, p), \mathbb{R} \ni x \mapsto \partial_x g(\varsigma_i + x, p), \mathbb{R} \ni x \mapsto \partial_x^2 g(\varsigma_i + x, p)$ are bounded and Hölder continuous (for some Hölder exponent which we do not specify here), uniformly in $p \in \mathcal{S}_d$. It is then a standard fact from Schauder's theory for the heat equation that the left-hand side in (3.37) is once differentiable in t and twice in space and that the derivatives $[0, T] \times \mathbb{R} \ni (t, x) \mapsto \partial_x U^i(t, x, p)$, $[0, T] \times \mathbb{R} \ni (t, x) \mapsto \partial_x^2 U^i(t, x, p)$ and $[0, T] \times \mathbb{R} \ni (t, x) \mapsto \partial_t U^i(t, x, p)$ are bounded and Hölder continuous (for some Hölder exponent), uniformly in p.

Take now a sequence $(p_n)_{n \in \mathbb{N}}$ with values in $\mathcal{S}_d$ converging to some $p \in \mathcal{S}_d$. Then, the functions $([0, T] \times \mathbb{R} \ni (t, x) \mapsto \partial_x U^i(t, x, p_n))_{n \in \mathbb{N}}$, $([0, T] \times \mathbb{R} \ni (t, x) \mapsto \partial_x^2 U^i(t, x, p_n))_{n \in \mathbb{N}}$ and $([0, T] \times \mathbb{R} \ni (t, x) \mapsto \partial_t U^i(t, x, p_n))_{n \in \mathbb{N}}$ are uniformly bounded and uniformly continuous. Up to a subsequence, they have a limit, but passing to the limit in (3.37) (replacing p by p_n therein, taking the first and second-order derivatives in x, removing the singularity of the second-order derivative of Γ by using the regularity of Ψ^i and then letting n tend to ∞), we deduce there should be only one possible limit for each of the two first sequences and that those limits should be the functions $[0, T] \times \mathbb{R} \ni (t, x) \mapsto \partial_x U^i(t, x, p)$ and $[0, T] \times \mathbb{R} \ni (t, x) \mapsto \partial_x^2 U^i(t, x, p)$. In turn, noticing from the representation (3.37) that Eq. (3.25) must hold true, we get that $([0, T] \times \mathbb{R} \ni (t, x) \mapsto \partial_t U^i(t, x, p_n))_{n \in \mathbb{N}}$ converges to $[0, T] \times \mathbb{R} \ni (t, x) \mapsto \partial_t U^i(t, x, p)$. We deduce that $\partial_x U$, $\partial_x^2 U$ and $\partial_t U$ are continuous on $[0, T] \times \mathbb{R} \times \mathcal{S}_d$. $\square$

Acknowledgments François Delarue thanks the organizers of the 2019 CIME Summer School on Mean Field Games for the opportunity he was offered to give a course there. He is also grateful to ANR-16-CE40-0015-01 (ANR MFG) and to the Institut Universitaire de France for the financial support he has received from the two of them for his research on mean field games.

References

1. L. Ambrosio, N. Gigli, G. Savaré, *Gradient Flows in Metric Spaces and in the Space of Probability Measures*. Lectures in Mathematics. ETH Zürich (Birkhäuser Verlag, Basel, 2005)
2. E. Bayraktar, A. Cohen, Analysis of a finite state many player game using its master equation. SIAM J. Control Optim. **56**(5), 3538–3568 (2018)
3. E. Bayraktar, A. Cecchin, A. Cohen, F. Delarue, Finite state mean field games with Wright-Fisher common noise (2019). http://arxiv.org/abs:1912.06701
4. C. Bertucci, J.-M. Lasry, P.-L. Lions, Some remarks on mean field games. Commun. Partial Differ. Equ. **44**(3), 205–227 (2019)
5. P. Cardaliaguet, Notes on mean field games (2013). https://www.ceremade.dauphine.fr/~cardalia/MFG20130420.pdf/
6. P. Cardaliaguet, F. Delarue, J.-M. Lasry, P.-L. Lions, *The Master Equation and the Convergence Problem in Mean Field Games*. Annals of Mathematics Studies, vol. 201 (Princeton University Press, Princeton, NJ, 2019)

7. P. Cardaliaguet, M. Cirant, A. Porretta, Splitting methods and short time existence for the master equations in mean field games (2020). http://arxiv.org/abs:2001.10406
8. R. Carmona, F. Delarue, *Probabilistic Theory of Mean Field Games with Applications. I.* Probability Theory and Stochastic Modelling, vol. 83 (Springer, Cham, 2018)
9. R. Carmona, F. Delarue, *Probabilistic Theory of Mean Field Games with Applications. II.* Probability Theory and Stochastic Modelling, vol. 84 (Springer, Cham, 2018)
10. R. Carmona, P. Wang, An alternative approach to mean field game with major and minor players, and applications to herders impacts. Appl. Math. Optim. **76**(1), 5–27 (2017)
11. R. Carmona, X. Zhu, A probabilistic approach to mean field games with major and minor players. Ann. Appl. Probab. **26**(3), 1535–1580 (2016)
12. R. Carmona, J. Fouque, L. Sun, Mean field games and systemic risk: a toy model. Commun. Math. Sci. **13**, 911–933 (2015)
13. R. Carmona, F. Delarue, D. Lacker, Probabilistic analysis of mean field games with a common noise. Ann. Probab. **44**, 3740–3803 (2016)
14. A. Cecchin, G. Pelino, Convergence, fluctuations and large deviations for finite state mean field games via the master equation. Stoch. Process. Appl. **129**(11), 4510–4555 (2019)
15. A. Cecchin, P. Dai Pra, M. Fischer, G. Pelino, On the convergence problem in mean field games: a two state model without uniqueness. SIAM J. Control Optim. **57**(4), 2443–2466 (2019)
16. J.-F. Chassagneux, D. Crisan, F. Delarue, A probabilistic approach to classical solutions of the master equation for large population equilibria. Mem. AMS (to appear)
17. D. Dawson, J. Vaillancourt, Stochastic McKean-Vlasov equations. NoDEA Nonlinear Differ. Equ. Appl. **2**(2), 199–229 (1995)
18. F. Delarue, On the existence and uniqueness of solutions to FBSDEs in a non-degenerate case. Stoch. Process. Appl. **99**(2), 209–286 (2002)
19. F. Delarue, Restoring uniqueness to mean-field games by randomizing the equilibria. Stoch. Partial Differ. Equ. Anal. Comput. **7**(4), 598–678 (2019)
20. F. Delarue, R. Foguen Tchuendom, Selection of equilibria in a linear quadratic mean-field game. Stoch. Process. Appl. **130**(2), 1000–1040 (2020)
21. F. Delarue, D. Lacker, K. Ramanan, From the master equation to mean field game limit theory: a central limit theorem. Electron. J. Probab. **24**(51), 54 (2019)
22. F. Delarue, D. Lacker, K. Ramanan, From the master equation to mean field game limit theory: Large deviations and concentration of measure. Ann. Probab. **48**(1), 211–263 (2020)
23. R.M. Dudley, Distances of probability measures and random variables. Ann. Math. Stat. **39**(5), 1563–1572 (1968)
24. G. Fu, P. Graewe, U. Horst, A. Popier, A mean field game of optimal portfolio liquidation (2018). http://arxiv.org/abs/1804.04911
25. W. Gangbo, A. Swiech, Existence of a solution to an equation arising from the theory of mean field games. J. Differ. Equ. **259**(11), 6573–6643 (2015)
26. D.A. Gomes, J. Saúde, Mean field games models—a brief survey. Dyn. Games Appl. **4**(2), 110–154 (2014)
27. D.A. Gomes, J. Mohr, R.R. Souza, Discrete time, finite state space mean field games. J. Math. Pures Appl. (9) **93**(3), 308–328 (2010)
28. D.A. Gomes, J. Mohr, R.R. Souza, Continuous time finite state mean field games. Appl. Math. Optim. **68**(1), 99–143 (2013)
29. O. Guéant, Existence and uniqueness result for mean field games with congestion effect on graphs. Appl. Math. Optim. **72**(2), 291–303 (2015)
30. O. Guéant, J.-M. Lasry, P.-L. Lions, Mean field games and applications. In *Paris-Princeton Lectures on Mathematical Finance 2010*. Lecture Notes in Mathematics, vol. 2003 (Springer, Berlin, 2011), pp. 205–266
31. M. Huang, Large-population LQG games involving a major player: the Nash equivalence principle. SIAM J. Control Optim. **48**, 3318–3353 (2010)
32. I. Karatzas, S.E. Shreve, *Brownian Motion and Stochastic Calculus*. Graduate Texts in Mathematics, 2nd edn., vol. 113 (Springer, New York, 1991)

33. T.G. Kurtz, J. Xiong, Particle representations for a class of nonlinear SPDEs. Stoch. Process. Appl. **83**(1), 103–126 (1999)
34. T.G. Kurtz, J. Xiong, A stochastic evolution equation arising from the fluctuations of a class of interacting particle systems. Commun. Math. Sci. **2**(3), 325–358 (2004)
35. P. Lions, Théorie des jeux à champs moyen et applications. Lectures at the Collège de France (2006–2012). https://www.college-de-france.fr/site/en-pierre-louis-lions/_course.htm
36. P. Lions, Estimées nouvelles pour les équations quasilinéaires. Seminar in Applied Mathematics at the Collège de France (2014). https://www.college-de-france.fr/site/en-pierre-louis-lions/seminar-2014-11-14-11h15.htm
37. J. Lott, Some geometric calculations on Wasserstein space. Commun. Math. Phys. **277**(2), 423–437 (2008)
38. J. Ma, P. Protter, J. Yong, Solving forward-backward stochastic differential equations explicitly — a four step scheme. Probab. Theory Relat. Fields **98**(3), 339–359 (1994)
39. S. Nguyen, M. Huang, Linear-quadratic-Gaussian mixed games with continuum-parametrized minor players. SIAM J. Control Optim. **50**, 2907–2937 (2012)
40. S. Nguyen, M. Huang, Mean field LQG games with mass behavior responsive to a major player, in *51th IEEE Conference on Decision and Control* (2012), pp. 5792–5797
41. M. Nourian, P. Caines, ϵ-Nash mean field game theory for nonlinear stochastic dynamical systems with major and minor agents. SIAM J. Control Optim. **50**, 2907–2937 (2013)
42. F. Otto, The geometry of dissipative evolution equations: the porous medium equation. Commun. Partial Differ. Equ. **26**(1-2), 101–174 (2001)
43. E. Pardoux, S.G. Peng, Adapted solution of a backward stochastic differential equation. Syst. Control Lett. **14**(1), 55–61 (1990)
44. E. Pardoux, A. Răşcanu, *Stochastic Differential Equations, Backward SDEs, Partial Differential Equations*. Stochastic Modelling and Applied Probability, vol. 69 (Springer, Cham, 2014)
45. S. Peng, Z. Wu, Fully coupled forward-backward stochastic differential equations and applications to optimal control. SIAM J. Control Optim. **37**(3), 825–843 (1999)
46. R.F. Tchuendom, Uniqueness for linear-quadratic mean field games with common noise. Dyn. Games Appl. **8**(1), 199–210 (2018)
47. J. Vaillancourt, On the existence of random McKean-Vlasov limits for triangular arrays of exchangeable diffusions. Stoch. Anal. Appl. **6**(4), 431–446 (1988)
48. J. Zhang, *Backward Stochastic Differential Equations*. Probability Theory and Stochastic Modelling, vol. 86 (Springer, New York, 2017)

Chapter 4
Mean Field Games and Applications: Numerical Aspects

Yves Achdou and Mathieu Laurière

Abstract The theory of mean field games aims at studying deterministic or stochastic differential games (Nash equilibria) as the number of agents tends to infinity. Since very few mean field games have explicit or semi-explicit solutions, numerical simulations play a crucial role in obtaining quantitative information from this class of models. They may lead to systems of evolutive partial differential equations coupling a backward Bellman equation and a forward Fokker–Planck equation. In the present survey, we focus on such systems. The forward-backward structure is an important feature of this system, which makes it necessary to design unusual strategies for mathematical analysis and numerical approximation. In this survey, several aspects of a finite difference method used to approximate the previously mentioned system of PDEs are discussed, including convergence, variational aspects and algorithms for solving the resulting systems of nonlinear equations. Finally, we discuss in details two applications of mean field games to the study of crowd motion and to macroeconomics, a comparison with mean field type control, and present numerical simulations.

4.1 Introduction

The theory of mean field games (*MFGs* for short), has been introduced in the pioneering works of J.-M. Lasry and P.-L. Lions [1–3], and aims at studying deterministic or stochastic differential games (Nash equilibria) as the number of agents tends to infinity. It supposes that the rational agents are indistinguishable

Y. Achdou (✉)
Université de Paris, Laboratoire Jacques-Louis Lions (LJLL), Paris, France
e-mail: achdou@ljll-univ-paris-diderot.fr

M. Laurière
Princeton University, Operations Research and Financial Engineering (ORFE) department, Sherrerd Hall, Princeton, NJ, USA
e-mail: lauriere@princeton.edu

© The Editor(s) (if applicable) and The Author(s), under exclusive license to Springer Nature Switzerland AG 2020

P. Cardaliaguet, A. Porretta (eds.), *Mean Field Games*,
Lecture Notes in Mathematics 2281, https://doi.org/10.1007/978-3-030-59837-2_4

and individually have a negligible influence on the game, and that each individual strategy is influenced by some averages of quantities depending on the states (or the controls) of the other agents. The applications of MFGs are numerous, from economics to the study of crowd motion. On the other hand, very few MFG problems have explicit or semi-explicit solutions. Therefore, numerical simulations of MFGs play a crucial role in obtaining quantitative information from this class of models.

The paper is organized as follows: in Sect. 4.2, we discuss finite difference schemes for the system of forward-backward PDEs. In Sect. 4.3, we focus on the case when the latter system can be interpreted as the system of optimality of a convex control problem driven by a linear PDE (in this case, the terminology *variational MFG* is often used), and put the stress on primal-dual optimization methods that may be used in this case. Section 4.4 is devoted to multigrid preconditioners that prove to be very important in particular as ingredients of the latter primal-dual methods for stochastic variational MFGs (when the volatility is positive). In Sect. 4.5, we address numerical algorithms that may be used also for non-variational MFGs. Sections 4.6, 4.7, and 4.8 are devoted to examples of applications of mean field games and their numerical simulations. Successively, we consider a model for a pedestrian flow with two populations, a comparison between mean field games and mean field control still for a model of pedestrian flow, and applications of MFGs to the field of macroeconomics.

In what follows, we suppose for simplicity that the state space is the d-dimensional torus $\mathbb{T}^d$, and we fix a finite time horizon $T > 0$. The periodic setting makes it possible to avoid the discussion on non periodic boundary conditions which always bring additional difficulties. The results stated below may be generalized to other boundary conditions, but this would lead us too far. Yet, Sects. 4.6, 4.7, and 4.8 below, which are devoted to some applications of MFGs and to numerical simulations, deal with realistic boundary conditions. In particular, boundary conditions linked to state constraints play a key role in Sect. 4.8.

We will use the notation $Q_T = [0, T] \times \mathbb{T}^d$, and $\langle \cdot, \cdot \rangle$ for the inner product of two vectors (of compatible sizes).

Let $f : \mathbb{T}^d \times \mathbb{R} \times \mathbb{R}^d \to \mathbb{R}, (x, m, \gamma) \mapsto f(x, m, \gamma)$ and $\phi : \mathbb{T}^d \times \mathbb{R} \to \mathbb{R}, (x, m) \mapsto \phi(x, m)$ be respectively a running cost and a terminal cost, on which assumptions will be made later on. Let $v > 0$ be a constant parameter linked to the volatility. Let $b : \mathbb{T}^d \times \mathbb{R} \times \mathbb{R}^d \to \mathbb{R}^d, (x, m, \gamma) \mapsto b(x, m, \gamma)$ be a drift function.

We consider the following MFG: find a flow of probability densities $\hat{m} : Q_T \to \mathbb{R}$ and a feedback control $\hat{v} : Q_T \to \mathbb{R}^d$ satisfying the following two conditions:

1. $\hat{v}$ minimizes

$$J_{\hat{m}} : v \mapsto J_{\hat{m}}(v) = \mathbb{E}\left[\int_0^T f(X_t^v, \hat{m}(t, X_t^v), v(t, X_t^v))dt + \phi(X_T^v, \hat{m}(T, X_T^v)) \right]$$

under the constraint that the process $X^v = (X_t^v)_{t \geq 0}$ solves the stochastic differential equation (SDE)

$$dX_t^v = b(X_t^v, \hat{m}(t, X_t^v), v(t, X_t^v))dt + \sqrt{2\nu}dW_t, \qquad t \geq 0, \tag{4.1}$$

and X_0^v has distribution with density m_0;

2. For all $t \in [0, T]$, $\hat{m}(t, \cdot)$ is the law of $X_t^{\hat{v}}$.

It is useful to note that for a given feedback control v, the density m_t^v of the law of X_t^v following (4.1) solves the Kolmogorov–Fokker–Planck (KFP) equation:

$$\begin{cases} \dfrac{\partial m^v}{\partial t}(t, x) - \nu \Delta m^v(t, x) + \mathrm{div}\left(m^v(t, \cdot)b(\cdot, \hat{m}(t, \cdot), v(t, \cdot))\right)(x) = 0, & \text{in } (0, T] \times \mathbb{T}^d, \\ m^v(0, x) = m_0(x), & \text{in } \mathbb{T}^d. \end{cases} \tag{4.2}$$

Let $H : \mathbb{T}^d \times \mathbb{R} \times \mathbb{R}^d \ni (x, m, p) \mapsto H(x, m, p) \in \mathbb{R}$ be the Hamiltonian of the control problem faced by an infinitesimal agent in the first point above, which is defined by

$$H : \mathbb{T}^d \times \mathbb{R} \times \mathbb{R}^d \ni (x, m, p) \mapsto H(x, m, p) = \max_{\gamma \in \mathbb{R}^d} -L(x, m, \gamma, p) \in \mathbb{R},$$

where L is the Lagrangian, defined by

$$L : \mathbb{T}^d \times \mathbb{R} \times \mathbb{R}^d \times \mathbb{R}^d \ni (x, m, \gamma, p) \mapsto L(x, m, \gamma, p) = f(x, m, \gamma) + \langle b(x, m, \gamma), p \rangle \in \mathbb{R}.$$

In the sequel, we will assume that the running cost f and the drift b are such that H is well-defined, C^1 with respect to (x, p), and strictly convex with respect to p.

From standard optimal control theory, one can characterize the best strategy through the value function u of the above optimal control problem for a typical agent, which satisfies a Hamilton–Jacobi–Bellman (HJB) equation. Together with the equilibrium condition on the distribution, we obtain that the equilibrium best response $\hat{v}$ is characterized by

$$\hat{v}(t, x) = \arg\max_{a \in \mathbb{R}^d} \left\{ -f(x, m(t, x), a) - \langle b(x, m(t, x), a), \nabla u(t, x) \rangle \right\},$$

where (u, m) solves the following forward-backward PDE system:

$$\begin{cases} -\dfrac{\partial u}{\partial t}(t, x) - \nu \Delta u(t, x) + H(x, m(t, x), \nabla u(t, x)) = 0, & \text{in } [0, T) \times \mathbb{T}^d, \quad (4.3a) \\ \dfrac{\partial m}{\partial t}(t, x) - \nu \Delta m(t, x) - \mathrm{div}\left(m(t, \cdot)H_p(\cdot, m(t, \cdot), \nabla u(t, \cdot))\right)(x) = 0, & \text{in } (0, T] \times \mathbb{T}^d, \quad (4.3b) \\ u(T, x) = \phi(x, m(T, x)), \qquad m(0, x) = m_0(x), & \text{in } \mathbb{T}^d. \quad (4.3c) \end{cases}$$

Example 1 (f Depends Separately on γ and m) Consider the case where the drift is the control, i.e. $b(x, m, \gamma) = \gamma$, and the running cost is of the form $f(x, m, \gamma) = L_0(x, \gamma) + f_0(x, m)$ where $L_0(x, \cdot) : \mathbb{R}^d \ni \gamma \mapsto L_0(x, \gamma) \in \mathbb{R}$ is strictly convex and such that $\lim_{|\gamma| \to \infty} \min_{x \in \mathbb{T}^d} \frac{L_0(x, \gamma)}{|\gamma|} = +\infty$. We set $H_0(x, p) = \max_{\gamma \in \mathbb{R}^d} \langle -p, \gamma \rangle - L_0(x, \gamma)$, which is convex with respect to p. Then

$$H(x, m, p) = \max_{\gamma \in \mathbb{R}^d} \{ -L_0(x, \gamma) - \langle \gamma, p \rangle \} - f_0(x, m) = H_0(x, p) - f_0(x, m).$$

In particular, if $H_0(x, \cdot) = H_0^*(x, \cdot) = \frac{1}{2} | \cdot |^2$, then the maximizer in the above expression is $\hat{\gamma}(p) = -p$, the Hamiltonian reads $H(x, m, p) = \frac{1}{2}|p|^2 - f_0(x, m)$ and the equilibrium best response is $\hat{v}(t, x) = -\nabla u(t, x)$ where (u, m) solves the PDE system

$$\begin{cases} -\dfrac{\partial u}{\partial t}(t, x) - \nu \Delta u(t, x) + \dfrac{1}{2}|\nabla u(t, x)|^2 = f_0(x, m(t, x)), & \text{in } [0, T) \times \mathbb{T}^d, \\[2mm] \dfrac{\partial m}{\partial t}(t, x) - \nu \Delta m(t, x) - \operatorname{div}\left(m(t, \cdot) \nabla u(t, \cdot) \right)(x) = 0, & \text{in } (0, T] \times \mathbb{T}^d, \\[2mm] u(T, x) = \phi(x, m(T, x)), \qquad m(0, x) = m_0(x), & \text{in } \mathbb{T}^d. \end{cases}$$

Remark 1 The setting presented above is somewhat restrictive and does not cover the case when the Hamiltonian depends non locally on m. Nevertheless, the latter situation makes a lot of sense from the modeling viewpoint. The case when the Hamiltonian depends in a separate manner on ∇u and m and when the coupling cost continuously maps probability measures on $\mathbb{T}^d$ to smooth functions plays an important role in the theory of mean field games, because it permits to obtain the most elementary existence results of strong solutions of the system of PDEs, see [3]. Concerning the numerical aspects, all what follows may be easily adapted to the latter case, see the numerical simulations at the end of Sect. 4.2. Similarly, the case when the volatility $\sqrt{2\nu}$ is a function of the state variable x can also be dealt with by finite difference schemes.

Remark 2 Deterministic mean field games (i.e. for $\nu = 0$) are also quite meaningful. One may also consider volatilities as functions of the state variable that may vanish. When the Hamiltonian depends separately on ∇u and m and the coupling cost is a smoothing map, see Remark 1, the Hamilton–Jacobi equation (respectively the Fokker–Planck equation) should be understood in viscosity sense (respectively in the sense of distributions), see the notes of P. Cardaliaguet [4]. When the coupling costs depends locally on m, then under some assumptions on the growth at infinity of the coupling cost as a function of m, it is possible to propose a notion of weak solutions to the system of PDEs, see [3, 5, 6]. The numerical schemes discussed below can also be applied to these situations, even if the convergence results in the available literature are obtained under the hypothesis that ν is bounded from below by a positive constant. The results and methods presented in Sect. 4.3 below for variational MFGs, i.e. when the system of PDEs can be seen as the optimality conditions of an optimal control problem driven by a PDE, hold if ν vanishes. In Sect. 4.8 below, we present some applications and numerical simulations for which the viscosity is zero.

Remark 3 The study of the so-called *master equation* plays a key role in the theory of MFGs, see [7, 8]: the mean field game is described by a single first or second order time dependent partial differential equation in the set of probability measures, thus in an infinite dimensional space in general. When the state space is finite (with say N admissible states), the distribution of states is a linear combination of N Dirac masses, and the master equation is posed in $\mathbb{R}^N$; numerical simulations are then possible, at least if N is not too large. We will not discuss this aspect in the present notes.

4.2   Finite Difference Schemes

In this section, we present a finite-difference scheme first introduced in [9]. We consider the special case described in Example 1, with $H(x, m, p) = H_0(x, p) - f_0(x, m)$, although similar methods have been proposed, applied and at least partially analyzed in situations when the Hamiltonian does not depend separately on m and p, for example models addressing congestion, see for example [10].

   To alleviate the notation, we present the scheme in the one-dimensional setting, i.e. $d = 1$ and the domain is $\mathbb{T}$.

Remark 4 Although we focus on the time dependent problem, a similar scheme has also been studied for the ergodic MFG system, see [9].

Discretization

Let N_T and N_h be two positive integers. We consider (N_T+1) and (N_h+1) points in time and space respectively. Let $\Delta t = T/N_T$ and $h = 1/N_h$, and $t_n = n \times \Delta t$, $x_i = i \times h$ for $(n, i) \in \{0, \ldots, N_T\} \times \{0, \ldots, N_h\}$.

   We approximate u and m respectively by vectors U and $M \in \mathbb{R}^{(N_T+1) \times (N_h+1)}$, that is, $u(t_n, x_i) \approx U_i^n$ and $m(t_n, x_i) \approx M_i^n$ for each $(n, i) \in \{0, \ldots, N_T\} \times \{0, \ldots, N_h\}$. We use a superscript and a subscript respectively for the time and space indices. Since we consider periodic boundary conditions, we slightly abuse notation and for $W \in \mathbb{R}^{N_h+1}$, we identify W_{N_h+1} with W_1, and W_{-1} with W_{N_h-1}.

   We introduce the finite difference operators

$$(D_t W)^n = \frac{1}{\Delta t}(W^{n+1} - W^n), \qquad n \in \{0, \ldots N_T - 1\}, \qquad W \in \mathbb{R}^{N_T+1},$$

$$(DW)_i = \frac{1}{h}(W_{i+1} - W_i), \qquad i \in \{0, \ldots N_h\}, \qquad W \in \mathbb{R}^{N_h+1},$$

$$(\Delta_h W)_i = -\frac{1}{h^2}\left(2W_i - W_{i+1} - W_{i-1}\right), \qquad i \in \{0, \ldots N_h\}, \qquad W \in \mathbb{R}^{N_h+1},$$

$$[\nabla_h W]_i = \left((DW)_i, (DW)_{i-1}\right)^\top, \qquad i \in \{0, \ldots N_h\}, \qquad W \in \mathbb{R}^{N_h+1}.$$

Discrete Hamiltonian

Let $\tilde{H} : \mathbb{T} \times \mathbb{R} \times \mathbb{R} \to \mathbb{R}$, $(x, p_1, p_2) \mapsto \tilde{H}(x, p_1, p_2)$ be a discrete Hamiltonian, assumed to satisfy the following properties:

1. $(\tilde{\mathbf{H}}_1)$ Monotonicity: for every $x \in \mathbb{T}$, $\tilde{H}$ is nonincreasing in p_1 and nondecreasing in p_2.
2. $(\tilde{\mathbf{H}}_2)$ Consistency: for every $x \in \mathbb{T}$, $p \in \mathbb{R}$, $\tilde{H}(x, p, p) = H_0(x, p)$.
3. $(\tilde{\mathbf{H}}_3)$ Differentiability: for every $x \in \mathbb{T}$, $\tilde{H}$ is of class C^1 in p_1, p_2
4. $(\tilde{\mathbf{H}}_4)$ Convexity: for every $x \in \mathbb{T}$, $(p_1, p_2) \mapsto \tilde{H}(x, p_1, p_2)$ is convex.

Example 2 For instance, if $H_0(x, p) = \frac{1}{2}|p|^2$, then one can take $\tilde{H}(x, p_1, p_2) = \frac{1}{2}|P_K(p_1, p_2)|^2$ where P_K denotes the projection on $K = \mathbb{R}_- \times \mathbb{R}_+$.

Remark 5 Similarly, for d-dimensional problems, the discrete Hamiltonians that we consider are real valued functions defined on $\mathbb{T}^d \times (\mathbb{R}^2)^d$.

Discrete HJB Equation

We consider the following discrete version of the HJB equation (4.3a):

$$
\begin{cases}
-(D_t U_i)^n - \nu(\Delta_h U^n)_i + \tilde{H}(x_i, [\nabla_h U^n]_i) = f_0(x_i, M_i^{n+1}), & i \in \{0, \ldots, N_h\},\, n \in \{0, \ldots, N_T - 1\}, & (4.4a) \\[2mm]
U_0^n = U_{N_h}^n, & n \in \{0, \ldots, N_T - 1\}, & (4.4b) \\[2mm]
U_i^{N_T} = \phi(M_i^{N_T}), & i \in \{0, \ldots, N_h\}. & (4.4c)
\end{cases}
$$

Note that it is an implicit scheme since the equation is backward in time.

Discrete KFP Equation

To define an appropriate discretization of the KFP equation (4.3b), we consider the weak form. For a smooth test function $w \in C^\infty([0, T] \times \mathbb{T})$, it involves, among other terms, the expression

$$
-\int_{\mathbb{T}} \partial_x \big(H_p(x, \partial_x u(t, x))m(t, x)\big) w(t, x)\, dx = \int_{\mathbb{T}} H_p(x, \partial_x u(t, x))m(t, x)\, \partial_x w(t, x)\, dx,
$$
(4.5)

where we used an integration by parts and the periodic boundary conditions. In view of what precedes, it is quite natural to propose the following discrete version of the right hand side of (4.5):

$$
h \sum_{i=0}^{N_h-1} M_i^{n+1} \left(\tilde{H}_{p_1}(x_i, [\nabla_h U^n]_i)\frac{W_{i+1}^n - W_i^n}{h} + \tilde{H}_{p_2}(x_i, [\nabla_h U^n]_i)\frac{W_i^n - W_{i-1}^n}{h} \right),
$$

and performing a discrete integration by parts, we obtain the discrete counterpart of the left hand side of (4.5) as follows: $-h \sum_{i=0}^{N_h-1} \mathcal{T}_i(U^n, M^{n+1}) W_i^n$, where $\mathcal{T}_i$ is the following discrete transport operator:

$$\mathcal{T}_i(U, M) = \frac{1}{h}\Big(M_i \tilde{H}_{p_1}(x_i, [\nabla_h U]_i) - M_{i-1}\tilde{H}_{p_1}(x_{i-1}, [\nabla_h U]_{i-1})$$

$$+ M_{i+1}\tilde{H}_{p_2}(x_{i+1}, [\nabla_h U]_{i+1}) - M_i \tilde{H}_{p_2}(x_i, [\nabla_h U]_i)\Big).$$

Then, for the discrete version of (4.3b), we consider

$$\begin{cases} (D_t M_i)^n - \nu(\Delta_h M^{n+1})_i - \mathcal{T}_i(U^n, M^{n+1}) = 0, & i \in \{0, \dots, N_h\}, n \in \{0, \dots, N_T - 1\}, & (4.6a) \\[2mm] M_0^n = M_{N_h}^n, & n \in \{1, \dots, N_T\}, & (4.6b) \\[2mm] M_i^0 = \bar{m}_0(x_i), & i \in \{0, \dots, N_h\}, & (4.6c) \end{cases}$$

where, for example,

$$\bar{m}_0(x_i) = \int_{|x - x_i| \le h/2} m_0(x)dx. \tag{4.7}$$

Here again, the scheme is implicit since the equation is forward in time.

Remark 6 (Structure of the Discrete System) The finite difference system (4.4)–(4.6) preserves the structure of the PDE system (4.3) in the following sense: The operator $M \mapsto -\nu(\Delta_h M)_i - \mathcal{T}_i(U, M)$ is the adjoint of the linearization of the operator $U \mapsto -\nu(\Delta_h U)_i + \tilde{H}(x_i, [\nabla_h U]_i)$ since

$$\sum_i \mathcal{T}_i(U, M)W_i = -\sum_i M_i \Big(\tilde{H}_p(x_i, [\nabla_h U]_i), [\nabla_h W]_i\Big).$$

Convergence Results

Existence and uniqueness for the discrete system has been proved in [9, Theorems 6 and 7]. The monotonicity properties ensure that the grid function M is nonnegative. By construction of $\mathcal{T}$, the scheme preserves the total mass $h \sum_{i=0}^{N_h - 1} M_i^n$. Note that there is no restriction on the time step since the scheme is implicit. Thus, this method may be used for long horizons and the scheme can be very naturally adapted to ergodic MFGs, see [9].

Furthermore, convergence results are available. A first type of convergence theorems see [9, 11, 12] (in particular [9, Theorem 8] for finite horizon problems) make the assumption that the MFG system of PDEs has a unique classical solution and strong versions of Lasry-Lions monotonicity assumptions, see [1–3]. Under such assumptions, the solution to the discrete system converges towards the classical solution as the grid parameters tend to zero.

In what follows, we discuss a second type of results that were obtained in [13], namely, the convergence of the solution of the discrete problem to weak solutions of the system of forward-backward PDEs.

Theorem 1 ([13, Theorem 3.1]) *We make the following assumptions:*

- $\nu > 0$;
- $\phi(x, m) = u_T(x)$ where u_T is a continuous function on $\mathbb{T}^d$;
- m_0 is a nonnegative function in $L^\infty(\mathbb{T}^d)$ such that $\int_{\mathbb{T}^d} m_0(x)dx = 1$;
- f_0 is a continuous function on $\mathbb{T}^d \times \mathbb{R}^+$, which is bounded from below;
- *The Hamiltonian* $(x, p) \mapsto H_0(x, p)$ *is assumed to be continuous, convex and* C^1 *regular w.r.t.* p;
- *The discrete Hamiltonian* $\tilde{H}$ *satisfies* $(\tilde{\mathbf{H}}_1)$–$(\tilde{\mathbf{H}}_4)$ *and the following further assumption:*

$(\tilde{\mathbf{H}}_5)$ growth conditions: *there exist positive constants* c_1, c_2, c_3, c_4 *such that*

$$\langle \tilde{H}_q(x, q), q \rangle - \tilde{H}(x, q) \geq c_1 |\tilde{H}_q(x, q)|^2 - c_2, \qquad (4.8)$$

$$|\tilde{H}_q(x, q)| \leq c_3 |q| + c_4. \qquad (4.9)$$

Let (U^n), (M^n) *be a solution of the discrete system (4.6a)–(4.6c) (more precisely of its d-dimensional counterpart). Let* $u_{h,\Delta t}$, $m_{h,\Delta t}$ *be the piecewise constant functions which take the values* U_I^{n+1} *and* M_I^n, *respectively, in* $(t_n, t_{n+1}) \times \omega_I$, *where* ω_I *is the d-dimensional cube centered at* x_I *of side h and I is any multi-index in* $\{0, \dots, N_h - 1\}^d$.

Under the assumptions above, there exists a subsequence of h and Δt *(not relabeled) and functions* $\tilde{u}$, $\tilde{m}$, *which belong to* $L^\alpha(0, T; W^{1,\alpha}(\mathbb{T}^d))$ *for any* $\alpha \in [1, \frac{d+2}{d+1})$, *such that* $u_{h,\Delta t} \to \tilde{u}$ *and* $m_{h,\Delta t} \to \tilde{m}$ *in* $L^\beta(Q_T)$ *for all* $\beta \in [1, \frac{d+2}{d})$, *and* $(\tilde{u}, \tilde{m})$ *is a weak solution to the system*

$$\begin{cases} -\dfrac{\partial \tilde{u}}{\partial t}(t, x) - \nu \Delta \tilde{u}(t, x) + H_0(x, \nabla \tilde{u}(t, x)) = f_0(x, \tilde{m}(t, x)), & \text{in } [0, T) \times \mathbb{T}^d, & (4.10a) \\[2mm] \dfrac{\partial \tilde{m}}{\partial t}(t, x) - \nu \Delta \tilde{m}(t, x) - \operatorname{div}\left(\tilde{m}(t, \cdot) D_p H_0 \nabla \tilde{u}(t, \cdot)\right)(x) = 0, & \text{in } (0, T] \times \mathbb{T}^d, & (4.10b) \\[2mm] \tilde{u}(T, x) = u_T(x), \qquad \tilde{m}(0, x) = m_0(x), & \text{in } \mathbb{T}^d, & (4.10c) \end{cases}$$

in the following sense:

(i) $H_0(\cdot, \nabla \tilde{u}) \in L^1(Q_T)$, $\tilde{m} f_0(\cdot, \tilde{m}) \in L^1(Q_T)$, $\tilde{m}[D_p H_0(\cdot, \nabla \tilde{u}) \cdot \nabla \tilde{u} - H_0(\cdot, \nabla \tilde{u})] \in L^1(Q_T)$
(ii) $(\tilde{u}, \tilde{m})$ *satisfies (4.10a)–(4.10b) in the sense of distributions*
(iii) $\tilde{u}, \tilde{m} \in C^0([0, T]; L^1(\mathbb{T}^d))$ *and* $\tilde{u}|_{t=T} = u_T$, $\tilde{m}|_{t=0} = m_0$.

Remark 7

- Theorem 1 does not suppose the existence of a (weak) solution to (4.10a)–(4.10c), nor Lasry-Lions monotonicity assumptions, see [1–3]. It can thus be seen as an alternative proof of existence of weak solutions of (4.10a)–(4.10c).
- The assumption made on the coupling cost f_0 is very general.

- If uniqueness holds for the weak solutions of (4.10a)–(4.10c), which is the case if Lasry-Lions monotonicity assumptions hold, then the whole sequence of discrete solutions converges.
- It would not be difficult to generalize Theorem 1 to the case when the terminal condition at T is of the form (4.3c).
- Similarly, non-local coupling cost of the type $F[m](x)$ could be addressed, where for instance, F maps bounded measures on $\mathbb{T}^d$ to functions in $C^0(\mathbb{T}^d)$.

An Example Illustrating the Robustness of the Scheme in the Deterministic Limit

To illustrate the robustness of the finite difference scheme described above, let us consider the following ergodic problem

$$
\begin{cases}
\rho - \nu \Delta u + H(x, \nabla u) = F[m](x), & \text{in } \mathbb{T}^2, \\
-\nu \Delta m - \operatorname{div}\left(m \dfrac{\partial H}{\partial p}(x, \nabla u)\right) = 0, & \text{in } \mathbb{T}^2,
\end{cases}
$$

where the ergodic constant $\rho \in \mathbb{R}$ is also an unknown of the problem, and

$$
\nu = 0.001, \qquad H(x, p) = \sin(2\pi x_2) + \sin(2\pi x_1) + \cos(4\pi x_1) + |p|^{3/2},
$$
$$
F[m](x) = \left((1 - \Delta)^{-1}(1 - \Delta)^{-1} m\right)(x).
$$

Note the very small value of the viscosity parameter ν. Since the coupling term is a smoothing nonlocal operator, nothing prevents m from becoming singular in the deterministic limit when $\nu \to 0$. In Fig. 4.1, we display the solution. We see that the value function tends to be singular (only Lipschitz continuous) and that the density of the distribution of states tends to be the sum of two Dirac masses located at the minima of the value function. The monotonicity of the scheme has made it possible to capture the singularities of the solution.

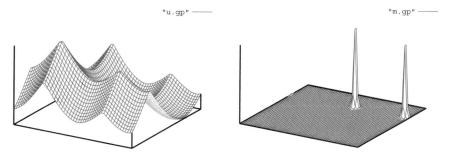

Fig. 4.1 Left: the value function. Right: the distribution of states

4.3 Variational Aspects of Some MFGs and Numerical Applications

In this section, we restrict our attention to the setting described in the Example 1. We assume that L_0 is smooth, strictly convex in the variable γ and such that there exists positive constants c and C, and $r > 1$ such that

$$-c + \frac{1}{C}|\gamma|^r \leq L_0(x, \gamma) \leq c + C|\gamma|^r, \quad \forall x \in \mathbb{T}^d, \gamma \in \mathbb{R}^d, \tag{4.11}$$

and that H_0 is smooth and has a superlinear growth in the gradient variable. We also assume that the function f_0 is smooth and non-decreasing with respect to m. We make the same assumption on ϕ.

4.3.1 An Optimal Control Problem Driven by a PDE

Let us consider the functions

$$F : \mathbb{T}^d \times \mathbb{R} \to \mathbb{R}, \quad (x, m) \mapsto F(x, m) = \begin{cases} \displaystyle\int_0^m f_0(x, s)ds, & \text{if } m \geq 0, \\ +\infty, & \text{if } m < 0, \end{cases} \tag{4.12}$$

and

$$\Phi : \mathbb{T}^d \times \mathbb{R} \to \mathbb{R}, \quad (x, m) \mapsto \Phi(x, m) = \begin{cases} \displaystyle\int_0^m \phi(x, s)ds, & \text{if } m \geq 0, \\ +\infty, & \text{if } m < 0. \end{cases} \tag{4.13}$$

Let us also introduce

$$L : \mathbb{T}^d \times \mathbb{R} \times \mathbb{R}^d \to \mathbb{R} \cup \{+\infty\}, \quad (x, m, w) \mapsto L(x, m, w) = \begin{cases} m L_0\left(x, \dfrac{w}{m}\right), & \text{if } m > 0, \\ 0, & \text{if } m = 0, \text{ and } w = 0, \\ +\infty, & \text{otherwise.} \end{cases} \tag{4.14}$$

Note that since $f_0(x, \cdot)$ is non-decreasing, $F(x, \cdot)$ is convex and l.s.c. with respect to m. Likewise, since $\phi(x, \cdot)$ is non-decreasing, then $\Phi(x, \cdot)$ is convex and l.s.c. with respect to m. Moreover, it can be checked that the assumptions made on L_0 imply that $L(x, \cdot, \cdot)$ is convex and l.s.c. with respect to (m, w).

We now consider the following minimization problem expressed in a formal way:
Minimize, when it is possible, $\mathcal{J}$ defined by

$$(m, w) \mapsto \mathcal{J}(m, w) = \int_0^T \int_{\mathbb{T}^d} [L(x, m(t, x), w(t, x)) + F(x, m(t, x))] \, dx dt + \int_{\mathbb{T}^d} \Phi(x, m(T, x)) dx$$

(4.15)

on pairs (m, w) such that $m \geq 0$ and

$$\begin{cases} \dfrac{\partial m}{\partial t} - \nu \Delta m + \mathrm{div}_x w = 0, & \text{in } (0, T] \times \mathbb{T}^d, \\ m|_{t=0} = m_0, \end{cases}$$

(4.16)

holds in the sense of distributions.

Note that, thanks to our assumptions and to the structure in terms of the variables (m, w), this problem is the minimization of a convex functional under a linear constraint. The key insight is that the optimality conditions of this minimization problem yield a weak version of the MFG system of PDEs (4.3).

Remark 8 If $(m, w) \in L^1(Q_T) \times L^1(Q_T; \mathbb{R}^d)$ is such that $\int_0^T \int_{\mathbb{T}^d} L(x, m(t, x), w(t, x)) dx dt < +\infty$, then it is possible to write that $w = m\gamma$ with $\int_0^T \int_{\mathbb{T}^d} m(t, x)|\gamma(t, x)|^r dx dt < +\infty$ thanks to the assumptions on L_0. From this piece of information and (4.16), we deduce that $t \mapsto m(t)$ for $t < 0$ is Hölder continuous a.e. for the weak$*$ topology of $\mathcal{P}(\mathbb{T}^d)$, see [5, Lemma 3.1]. This gives a meaning to $\int_{\mathbb{T}^d} \Phi(x, m(T, x)) dx$ in the case when $\Phi(x, m(T, x)) = u_T(x)m(T, x)$ for a smooth function $u_T(x)$. Therefore, at least in this case, the minimization problem is defined rigorously in $L^1(Q_T) \times L^1(Q_T; \mathbb{R}^d)$.

Remark 9 Although we consider here the case of a non-degenerate diffusion, similar problems have been considered in the first-order case or with degenerate diffusion in [5, 14].

Remark 10 Note that, in general, the system of forward-backward PDEs cannot be seen as the optimality conditions of a minimization problem. This is never the case if the Hamiltonian does not depend separately on m and ∇u.

4.3.2 Discrete Version of the PDE Driven Optimal Control Problem

We turn our attention to a discrete version of the minimization problem introduced above. To alleviate notation, we restrict our attention to the one dimensional case, i.e., $d = 1$. Let us introduce the following spaces, respectively for the discrete counterpart of m, w, and u:

$$\mathcal{M} = \mathbb{R}^{(N_T+1) \times N_h}, \qquad \mathcal{W} = (\mathbb{R}^2)^{N_T \times N_h}, \qquad \mathcal{U} = \mathbb{R}^{N_T \times N_h}.$$

Note that at each space-time grid point, we will have two variables for the value of w, which will be useful to define an upwind scheme.

We consider a discrete Hamiltonian $\tilde{H}$ satisfying the assumptions $(\tilde{\mathbf{H}}_1)$–$(\tilde{\mathbf{H}}_4)$, and introduce its convex conjugate w.r.t. the p variable:

$$\tilde{H}^* : \mathbb{T}^d \times (\mathbb{R}^2)^d \to \mathbb{R} \cup \{+\infty\}, (x, \gamma) \mapsto \tilde{H}^*(x, \gamma) = \max_{p \in \mathbb{R}^2} \left\{ \langle \gamma, p \rangle - \tilde{H}(x, p) \right\}. \tag{4.17}$$

Example 3 If $\tilde{H}(x, p_1, p_2) = \frac{1}{2} |P_K(p_1, p_2)|^2$ with $K = \mathbb{R}_- \times \mathbb{R}_+$, then

$$\tilde{H}^*(x, \gamma_1, \gamma_2) = \begin{cases} \frac{1}{2} |\gamma|^2, & \text{if } \gamma \in K, \\ +\infty, & \text{otherwise.} \end{cases}$$

A discrete counterpart Θ to the functional $\mathcal{J}$ introduced in (4.15) can be defined as

$$\Theta : \mathcal{M} \times \mathcal{W} \to \mathbb{R}, \qquad (M, W) \mapsto \sum_{n=1}^{N_T} \sum_{i=0}^{N_h-1} \left[\tilde{L}(x_i, M_i^n, W_i^{n-1}) + F(x_i, M_i^n) \right] + \frac{1}{\Delta t} \sum_{i=0}^{N_h-1} \Phi(x_i, M_i^{N_T}), \tag{4.18}$$

where $\tilde{L}$ is a discrete version of L introduced in (4.14) defined as

$$\tilde{L} : \mathbb{T} \times \mathbb{R} \times \mathbb{R}^2 \to \mathbb{R} \cup \{+\infty\}, \qquad (x, m, w) \mapsto \tilde{L}(x, m, w) = \begin{cases} m\tilde{H}^*\left(x, -\frac{w}{m}\right), & \text{if } m > 0 \text{ and } w \in K, \\ 0, & \text{if } m = 0, \text{ and } w = 0, \\ +\infty, & \text{otherwise.} \end{cases} \tag{4.19}$$

Furthermore, a discrete version of the linear constraint (4.16) can be written as

$$\Sigma(M, W) = (0_{\mathcal{U}}, \bar{M}^0) \tag{4.20}$$

where $0_{\mathcal{U}} \in \mathcal{U}$ is the vector with 0 on all coordinates, $\bar{M}^0 = (\bar{m}_0(x_0), \ldots, \bar{m}_0(x_{N_h-1})) \in \mathbb{R}^{N_h}$, see (4.7) for the definition of $\bar{m}_0$, and

$$\Sigma : \mathcal{M} \times \mathcal{W} \to \mathcal{U} \times \mathbb{R}^{N_h}, \qquad (M, W) \mapsto \Sigma(M, W) = (\Lambda(M, W), M^0), \tag{4.21}$$

with $\Lambda(M, W) = AM + BW$, A and B being discrete versions of respectively the heat operator and the divergence operator, defined as follows:

$$A : \mathcal{M} \to \mathcal{U}, \qquad (AM)_i^n = \frac{M_i^{n+1} - M_i^n}{\Delta t} - \nu(\Delta_h M^{n+1})_i, \qquad 0 \le n < N_T, 0 \le i < N_h, \tag{4.22}$$

and

$$B : \mathcal{W} \to \mathcal{U}, \qquad (BW)_i^n = \frac{W_{i+1,2}^n - W_{i,2}^n}{h} + \frac{W_{i,1}^n - W_{i-1,1}^n}{h} \qquad 0 \leq n \leq N_T, 0 \leq i < N_h.$$
(4.23)

We define the transposed operators $A^* : \mathcal{U} \to \mathcal{M}$ and $B^* : \mathcal{U} \to \mathcal{W}$, i.e. such that

$$\langle M, A^*U \rangle = \langle AM, U \rangle, \qquad \langle W, B^*U \rangle = \langle BW, U \rangle.$$

This yields

$$(A^*U)_i^n = \begin{cases} -\dfrac{U_i^0}{\Delta t}, & n = 0, \ 0 \leq i < N_h, \\[2ex] \dfrac{U_i^{n-1} - U_i^n}{\Delta t} - \nu(\Delta_h U^{n-1})_i, & 0 < n < N_T - 1, \ 0 \leq i < N_h, \\[2ex] \dfrac{U_i^{N_T-1}}{\Delta t} - \nu(\Delta_h U^{N_T-1})_i, & n = N_T - 1, \ 0 \leq i < N_h, \end{cases}$$
(4.24)

and

$$(B^*U)_i^n = -[\nabla_h U^n]_i = -\left(\frac{U_{i+1}^n - U_i^n}{h}, \frac{U_i^n - U_{i-1}^n}{h} \right), \qquad 0 \leq n < N_T, 0 \leq i < N_h.$$
(4.25)

This implies that

$$\mathrm{Im}(B) = \mathrm{Ker}(B^*)^{\perp} = \left\{ U \in \mathcal{U} : \forall 0 \leq n < N_T, \sum_i U_i^n = 0 \right\}.$$
(4.26)

The discrete counterpart of the variational problem (4.15)–(4.16) is therefore

$$\inf_{\substack{(M,W) \in \mathcal{M} \times \mathcal{W}, \\ \Sigma(M,W) = (0_{\mathcal{U}}, \bar{M}^0)}} \Theta(M, W).$$
(4.27)

The following results provide a constraint qualification property and the existence of a minimizer. See e.g. [15, Theorem 3.1] for more details on a special case (see also [16, Section 6] and [17, Theorem 2.1] for similar results respectively in the context of the planning problem and in the context of an ergodic MFG).

Proposition 1 (Constraint Qualification) *There exists a pair* $(\tilde{M}, \tilde{W}) \in M \times W$ *such that*

$$\Theta(\tilde{M}, \tilde{W}) < \infty, \qquad and \qquad \Sigma(\tilde{M}, \tilde{W}) = (0_{\mathcal{U}}, \bar{M}^0).$$

Theorem 2 *There exists a minimizer* $(M, W) \in M \times W$ *of* (4.27). *Moreover, if* $\nu > 0$, $M_i^n > 0$ *for all* $i \in \{0, \ldots, N_h\}$ *and all* $n > 0$.

4.3.3 Recovery of the Finite Difference Scheme by Convex Duality

We will now show that the finite difference scheme introduced in the previous section corresponds to the optimality conditions of the discrete minimization problem (4.27) introduced above. To this end, we first rewrite the discrete problem by embedding the constraint in the cost functional as follows:

$$\min_{(M,W) \in M \times W} \{\Theta(M, W) + \chi_0(\Sigma(M, W))\}, \tag{4.28}$$

where $\chi_0 : \mathcal{U} \times \mathbb{R}^{N_h} \to \mathbb{R} \cup \{+\infty\}$ is the characteristic function (in the convex analytical sense) of the singleton $\{(0_{\mathcal{U}}, \bar{M}^0)\}$, i.e., $\chi_0(U, \varphi)$ vanishes if $(U, \varphi) = (0_{\mathcal{U}}, \bar{M}^0)$ and otherwise it takes the value $+\infty$.

By Fenchel-Rockafellar duality theorem, see [18], we obtain that

$$\min_{(M,W) \in M \times W} \{\Theta(M, W) + \chi_0(\Sigma(M, W))\} = - \min_{(U,\varphi) \in \mathcal{U} \times \mathbb{R}^{N_h}} \{\Theta^*(\Sigma^*(U, \varphi)) + \chi_0^*(-U, -\varphi)\},$$

$$\tag{4.29}$$

where Σ^* is the adjoint operator of Σ, and χ_0^* and Θ^* are the convex conjugates of χ_0 and Θ respectively, to wit,

$$\left\langle \Sigma^*(U, \varphi), (M, W) \right\rangle = \langle A^*U, M \rangle + \langle B^*U, W \rangle + \sum_{i=0}^{N_h-1} \varphi_i M_i^0 \tag{4.30}$$

and

$$\chi_0^*(-U, -\varphi) = - \sum_{i=0}^{N_h-1} \bar{M}_i^0 \varphi_i. \tag{4.31}$$

Therefore

$$\Theta^*(\Sigma^*(U, \varphi))$$

$$= \max_{(M,W)\in\mathcal{M}\times\mathcal{W}} \left(\langle A^*U, M \rangle + \langle B^*U, W \rangle + \sum_{i=0}^{N_h-1} \varphi_i M_i^0 - \sum_{n=1}^{N_T}\sum_{i=0}^{N_h-1} \left[\tilde{L}(x_i, M_i^n, W_i^{n-1}) + F(x_i, M_i^n) \right] - \frac{1}{\Delta t}\sum_{i=0}^{N_h-1} \Phi(x_i, M_i^{N_T}) \right)$$

We use the fact that

$$\max_{W\in\mathcal{W}} \left(\langle B^*U, W \rangle - \sum_{n=1}^{N_T}\sum_{i=0}^{N_h-1} \tilde{L}(x_i, M_i^n, W_i^{n-1}) \right) = \begin{cases} \sum_{n=1}^{N_T}\sum_{i=0}^{N_h-1} M_i^n \tilde{H}\left(x_i, -(B^*U)_i^{n-1}\right), & \text{if } M^n \geq 0 \text{ for all } n \geq 1, \\ -\infty, & \text{otherwise.} \end{cases}$$

We deduce the following:

$$\Theta^*(\Sigma^*(U, \varphi))$$

$$= \max_{M\in\mathcal{M}} \left(\langle A^*U, M \rangle - \sum_{i=0}^{N_h-1} \varphi_i M_i^0 + \sum_{n=1}^{N_T}\sum_{i=0}^{N_h-1} \left[M_i^n \tilde{H}\left(x_i, -(B^*U)_i^{n-1}\right) - F(x_i, M_i^n) \right] - \frac{1}{\Delta t}\sum_{i=0}^{N_h-1} \Phi(x_i, M_i^{N_T}) \right)$$

$$= \sum_{n=1}^{N_T-1}\sum_{i=0}^{N_h-1} F^*\left(x_i, (A^*U)_i^n + \tilde{H}\left(x_i, [\nabla_h U^{n-1}]_i\right)\right) + \sum_{i=0}^{N_h-1}\left(F + \frac{1}{\Delta t}\Phi\right)^*\left(x_i, (A^*U)_i^{N_T} + \tilde{H}\left(x_i, [\nabla_h U^{N_T-1}]_i\right)\right)$$

$$+ \max_{M\in\mathcal{M}} M_i^0\left((A^*U)_i^0 + \varphi_i\right),$$

(4.32)

where for every $x \in \mathbb{T}$, $F^*(x, \cdot)$ and $\left(F + \frac{1}{\Delta t}\Phi\right)^*(x, \cdot)$ are the convex conjugate of $F(x, \cdot)$ and $F(x, \cdot) + \frac{1}{\Delta t}\Phi(x, \cdot)$ respectively. Note that from the definition of F, it has not been not necessary to impose the constraint $M^n \geq 0$ for all $n \geq 1$ in (4.32). Note also that the last term in (4.32) can be written $\max_{M\in\mathcal{M}} M_i^0\left(-\frac{1}{\Delta t}U_i^0 + \varphi_i\right)$. Using (4.32), in the dual minimization problem given in the right hand side of (4.29), we see that the minimization with respect to φ yields that $M_i^0 = \tilde{M}_i^0$. Therefore, the the dual problem can be rewritten as

$$- \min_{U\in\mathcal{U}} \left\{ \sum_i \left(F + \frac{1}{\Delta t}\Phi\right)^*\left(x_i, \frac{U_i^{N_T-1}}{\Delta t} - \nu(\Delta_h U^{N_T-1})_i + \tilde{H}(x_i, [\nabla_h U^{N_T-1}]_i)\right) \right.$$

$$\left. + \sum_{n=0}^{N_T-2}\sum_i F^*\left(x_i, \frac{U_i^n - U_i^{n+1}}{\Delta t} - \nu(\Delta_h U^n)_i + \tilde{H}(x_i, [\nabla_h U^n]_i)\right) - \frac{1}{\Delta t}\sum_i \tilde{M}_i^0 U_i^0 \right\},$$

(4.33)

which is the discrete version of

$$- \min_u \left\{ \int_0^T \int_{\mathbb{T}} F^*\left(x, -\partial_t u(t, x) - \nu\Delta u(t, x) + H_0(x, \nabla u(t, x))\right) dx\, dt + \int_{\mathbb{T}} \Phi^*(x, u(T, x)) dx - \int_{\mathbb{T}} m_0(x)u(0, x) dx \right\}.$$

The arguments above lead to the following:

Theorem 3 *The solutions* $(M, W) \in \mathcal{M} \times \mathcal{W}$ *and* $U \in \mathcal{U}$ *of the primal/dual problems are such that:*

- *if* $\nu > 0$, $M_i^n > 0$ *for all* $0 \leq i \leq N_h - 1$ *and* $0 < n \leq N_T$;
- $W_i^n = M_i^{n+1} P_K([\nabla_h U^n]_i)$ *for all* $0 \leq i \leq N_h - 1$ *and* $0 \leq n < N_T$;
- (U, M) *satisfies the discrete MFG system* (4.4)–(4.6) *obtained by the finite difference scheme.*

Furthermore, the solution is unique if f_0 *and* ϕ *are strictly increasing.*

Proof The proof follows naturally for the arguments given above. See e.g. [11] in the case of planning problems, [15, Theorem 3.1], or [17, Theorem 2.1] in the stationary case. □

In the rest of this section, we describe two numerical methods to solve the discrete optimization problem.

4.3.4 Alternating Direction Method of Multipliers

Here we describe the Alternating Direction Method of Multipliers (ADMM) based on an Augmented Lagrangian approach for the dual problem. The interested reader is referred to e.g. the monograph [19] by Fortin and Glowinski and the references therein for more details. This idea was made popular by Benamou and Brenier for optimal transport, see [20], and first used in the context of MFGs by Benamou and Carlier in [21]; see also [22] for an application to second order MFG using multilevel preconditioners and [23] for an application to mean field type control.

The dual problem (4.33) can be rewritten as

$$\min_{U \in \mathcal{U}} \left\{ \Psi(\Lambda^* U) + \Gamma(U) \right\}$$

where

$$\Lambda^* : \mathcal{U} \to \mathcal{M} \times \mathcal{W}, \qquad \Lambda^* U = (A^* U, B^* U),$$

$$\Psi(M, W) = \sum_i \left(F + \frac{1}{\Delta t} \Phi \right)^* \left(x_i, M_i^{N_T} + \tilde{H}(x_i, W_i^{N_T-1}) \right) + \sum_{n=0}^{N_T-2} \sum_i F^* \left(x_i, M_i^n + \tilde{H}(x_i, W_i^{n-1}) \right)$$

and

$$\Gamma(U) = -\frac{1}{\Delta t} \sum_i \bar{M}_i^0 U_i^0.$$

Note that Ψ is convex and l.s.c., and Γ is linear. Moreover, Ψ depend in a separate manner on the pairs $(M_i^n, W_i^{n-1})_{n,i}$, and a similar observation can be made for Λ.

We can artificially introduce an extra variable $Q \in M \times W$ playing the role of $\Lambda^* U$. Introducing a Lagrange multiplier $\sigma = (M, W)$ for the constraint $Q = \Lambda^* U$, we obtain the following equivalent problem:

$$\min_{U \in \mathcal{U}, Q \in M \times W} \quad \sup_{\sigma \in M \times W} \quad \mathcal{L}(U, Q, \sigma), \tag{4.34}$$

where the Lagrangian $\mathcal{L} : \mathcal{U} \times M \times W \times M \times W \to \mathbb{R}$ is defined as

$$\mathcal{L}(U, Q, \sigma) = \Psi(Q) + \Gamma(U) - \langle \sigma, Q - \Lambda^* U \rangle.$$

Given a parameter $r > 0$, it is convenient to introduce an equivalent problem involving the augmented Lagrangian $\mathcal{L}_r$ obtained by adding to $\mathcal{L}$ a quadratic penalty term in order to obtain strong convexity:

$$\mathcal{L}_r(U, Q, \sigma) = \mathcal{L}(U, Q, \sigma) + \frac{r}{2} \| Q - \Lambda^* U \|_2^2.$$

Since the saddle-points of $\mathcal{L}$ and $\mathcal{L}_r$ coincide, equivalently to (4.34), we will seek to solve

$$\min_{U \in \mathcal{U}, Q \in M \times W} \quad \sup_{\sigma \in M \times W} \quad \mathcal{L}_r(U, Q, \sigma).$$

In this context, the Alternating Direction Method of Multipliers (see e.g. the method called ALG2 in [19]) is described in Algorithm 1 below. We use $\partial \Gamma$ and $\partial \Psi$ to denote the subgradients of Γ and Ψ respectively.

Algorithm 1 Alternating Direction Method of Multipliers

---

   function ADMM(U, Q, σ)
      Initialize $(U^{(0)}, Q^{(0)}, \sigma^{(0)}) \leftarrow (U, Q, \sigma)$
      for $k = 0, \ldots, K - 1$ **do**
          Find $U^{(k+1)} \in \arg\min_{U \in \mathcal{U}} \mathcal{L}_r(U, Q^{(k)}, \sigma^{(k)})$; the first order optimality condition yields:

$$-r\Lambda \left(\Lambda^* U^{(k+1)} - Q^{(k)} \right) - \Lambda \sigma^{(k)} \in \partial \Gamma \left(U^{(k+1)} \right).$$

          Find $Q^{(k+1)} \in \arg\min_{Q \in M \times W} \mathcal{L}_r \left(U^{(k+1)}, Q, \sigma^{(k+1)} \right)$; the first order optimality condition yields:

$$\sigma^{(k)} + r \left(\Lambda^* U^{(k+1)} - Q^{(k+1)} \right) \in \partial \Psi \left(Q^{(k+1)} \right).$$

          Perform a gradient step: $\sigma^{(k+1)} = \sigma^{(k)} + r \left(\Lambda^* U^{(k+1)} - Q^{(k+1)} \right)$.
      return $(U^{(K)}, Q^{(K)}, \sigma^{(K)})$

Remark 11 (Preservation of the Sign) An important consequence of this method is that, thanks to the second and third steps above, the non-negativity of the discrete approximations to the density is preserved. Indeed, $\sigma^{(k+1)} \in \partial \Psi \left(Q^{(k+1)} \right)$ and $\Psi \left(Q^{(k+1)} \right) < +\infty$, hence $\Psi^* \left(\sigma^{(k+1)} \right) = \langle \sigma^{(k+1)}, Q^{(k+1)} \rangle - \Psi \left(Q^{(k+1)} \right)$, which yields $\Psi^* \left(\sigma^{(k+1)} \right) < +\infty$. In particular, denoting $(M^{(k+1)}, W^{(k+1)}) = \sigma^{(k+1)} \in \mathcal{M} \times \mathcal{W}$, this implies that we have $(M^{(k+1)})_i^n \geq 0$ for every $0 \leq i < N_h, 0 \leq n \leq N_T$, and $(W^{(k+1)})_i^n \in K$ vanishes if $(M^{(k+1)})_i^{n+1} = 0$ for every $0 \leq i < N_h, 0 \leq n < N_T$.

Let us provide more details on the numerical solution of the first and second steps in the above method. For the first step, since Γ only acts on U^0, the first order optimality condition amounts to solving the discrete version of a boundary value problem on $(0, T) \times \mathbb{T}$ or more generally $(0, T) \times \mathbb{T}^d$ when the state is in dimension d, with a degenerate fourth order linear elliptic operator (of order four in the state variable if ν is positive and two in time), and with a Dirichlet condition at $t = T$ and a Neumann like condition at $t = 0$. If $d > 1$, it is generally not possible to use direct solvers; instead, one has to use iterative methods such as the conjugate gradient algorithm. An important difficulty is that, since the equation is fourth order with respect to the state variable if $\nu > 0$, the condition number grows like h^{-4}, and a careful preconditionning is mandatory (we will come back to this point in the next section). In the deterministic case, i.e. if $\nu = 0$, the first step consists of solving the discrete version of a second order $d + 1$ dimensional elliptic equation and preconditioning is also useful for reducing the computational complexity.

As for the second step, the first order optimality condition amounts to solving, at each space-time grid node (i, n), a non-linear minimization problem in $\mathbb{R}^{1+2d}$ of the form

$$\min_{a \in \mathbb{R}, b \in \mathbb{R}^2} F^*(x_i, a + \tilde{H}(x_i, b)) - \langle \sigma, (a, b) \rangle + \frac{r}{2} \|(a, b) - (\bar{a}, \bar{b})\|_2^2,$$

where (a, b) plays the role of (M_i^n, W_i^{n-1}), whereas $(\bar{a}, \bar{b})$ plays the role of $(\Lambda^* U)_i^n$.

Note that the quadratic term, which comes from the fact that we consider the augmented Lagrangian $\mathcal{L}_r$ instead of the original Lagrangian $\mathcal{L}$, provides coercivity and strict convexity. The two main difficulties in this step are the following. First, in general, F^* does not admit an explicit formula and is itself obtained by solving a maximization problem. Second, for general F and $\tilde{H}$, computing the minimizer explicitly may be impossible. In the latter case, Newton iterations may be used, see e.g. [23].

The following result ensures convergence of the numerical method; see [24, Theorem 8] for more details.

Theorem 4 (Eckstein and Bertsekas [24]) *Assume that $r > 0$, that $\Lambda\Lambda^*$ is symmetric and positive definite and that there exists a solution to the primal-dual* *extremality. Then the sequence $(U^{(k)}, Q^{(k)}, \sigma^{(k)})$ converges as $k \to \infty$ and*

$$\lim_{k \to +\infty} (U^{(k)}, Q^{(k)}, \sigma^{(k)}) = (\bar{U}, \Lambda^* \bar{U}, \bar{\sigma}),$$

where $\bar{U}$ solves the dual problem and $\bar{\sigma}$ solves the primal problem.

4.3.5   Chambolle and Pock's Algorithm

We now consider a primal-dual algorithm introduced by Chambolle and Pock in [25]. The application to stationary MFGs was first investigated by Briceño-Arias, Kalise and Silva in [17]; see also [15] for an extension to the dynamic setting.

Introducing the notation $\Pi = \chi_0 \circ \Sigma : \mathcal{M} \times \mathcal{W} \ni (M, W) \mapsto \chi_0(\Sigma(M, W)) \in \mathbb{R} \cup \{+\infty\}$, the primal problem (4.28) can be written as

$$\min_{\sigma \in \mathcal{M} \times \mathcal{W}} \{\Theta(\sigma) + \Pi(\sigma)\},$$

and the dual problem can be written as

$$\min_{Q \in \mathcal{M} \times \mathcal{W}} \{\Theta^*(-Q) + \Pi^*(Q)\}.$$

For $r, s > 0$, the first order optimality conditions at $\hat{\sigma}$ and $\hat{Q}$ can be equivalently written as

$$
\begin{cases}
-\hat{Q} \in \partial\Theta(\hat{\sigma}) \\
\hat{\sigma} \in \partial\Pi^*(\hat{Q})
\end{cases}
\Leftrightarrow
\begin{cases}
\hat{\sigma} - r\hat{Q} \in \hat{\sigma} + r\partial\Theta(\hat{\sigma}) \\
\hat{Q} + s\hat{\sigma} \in \hat{Q} + s\partial\Pi^*(\hat{Q})
\end{cases}
\Leftrightarrow
\begin{cases}
\hat{\sigma} \in \arg\min_\sigma \left\{\Theta(\sigma) + \frac{1}{2r}\|\sigma - (\hat{\sigma} - r\hat{Q})\|^2\right\} \\
\hat{Q} \in \arg\min_Q \left\{\Pi^*(Q) + \frac{1}{2s}\|Q - (\hat{Q} + s\hat{\sigma})\|^2\right\}.
\end{cases}
$$

Given some parameters $r > 0$, $s > 0$, $\tau \in [0, 1]$, the Chambolle-Pock method is described in Algorithm 2. The algorithm has been proved to converge if $rs < 1$.

Note that the first step is similar to the first step in the ADMM method described in Sect. 4.3.4 and amounts to solving a linear fourth order PDE. The second step is easier than in ADMM because Θ admits an explicit formula.

4.4   Multigrid Preconditioners

4.4.1   General Considerations on Multigrid Methods

Before explaining how multigrid methods can be applied in the context of numerical solutions for MFG, let us recall the main ideas. We refer to [26] for an introduction

---

Algorithm 2 Chambolle-Pock Method

---

function CHAMBOLLEPOCK($\sigma, \tilde{\sigma}, Q$)
    Initialize $(\sigma^{(0)}, \tilde{\sigma}^{(0)}, Q^{(0)}) \leftarrow (\sigma, \tilde{\sigma}, Q)$
    for $k = 0, \ldots, K - 1$ **do**
        Find

$$Q^{(k+1)} \in \arg\min_Q \left\{ \Pi^*(Q) + \frac{1}{2s} \| Q - (Q^{(k)} + s\tilde{\sigma}^{(k)}) \|^2 \right\} \qquad (4.35)$$

        Find

$$\sigma^{(k+1)} \in \arg\min_\sigma \left\{ \Theta(\sigma) + \frac{1}{2r} \| \sigma - (\sigma^{(k)} - rQ^{(k+1)}) \|^2 \right\} \qquad (4.36)$$

        Set

$$\tilde{\sigma}^{(k+1)} = \sigma^{(k+1)} + \tau \left(\sigma^{(k+1)} - \sigma^{(k)} \right).$$

    return $(\sigma^{(K)}, \tilde{\sigma}^{(K)}, Q^{(K)})$

---

to multigrid methods and more details. In order to solve a linear system which corresponds to the discretisation of an equation on a given grid, we can use coarser grids in order to get approximate solutions. Intuitively, a multigrid scheme should be efficient because solving the system on a coarser grid is computationally easier and the solution on this coarser grid should provide a good approximation of the solution on the original grid. Indeed, using a coarser grid should help capturing quickly the low frequency modes (i.e., the modes corresponding to the smallest eigenvalues of the differential operator from which the linear system comes), which takes more iterations on a finer grid.

When the linear system stems from a well behaved second order elliptic operator for example, one can find simple iterative methods (e.g. Jacobi or Gauss–Seidel algorithms) such that a few iterations of these methods are enough to damp the higher frequency components of the residual, i.e. to make the error smooth. Typically, these iterative methods have bad convergence properties, but they have good smoothing properties and are hence called *smoothers*. The produced residual is smooth on the grid under consideration (i.e., it has small fast Fourier components on the given grid), so it is well represented on the next coarser grid. This suggests to transfer the residual to the next coarser grid (in doing so, half of the low frequency components on the finer grid become high frequency components on the coarser one, so they will be damped by the smoother on the coarser grid). These principles are the basis for a recursive algorithm. Note that in such an algorithm, using a direct method for solving systems of linear equations is required only on the coarsest grid, which contains much fewer nodes than the initial grid. On the grids of intermediate sizes, one only needs to perform matrix multiplications.

To be more precise, let us consider a sequence of nested grids $(\mathcal{G}_\ell)_{\ell=0,\ldots,L}$, i.e. such that $\mathcal{G}_\ell \subseteq \mathcal{G}_{\ell+1}$, $\ell = 0, \ldots, L-1$. Denote the corresponding number of points by $\tilde{N}_\ell = \tilde{N}2^{d\ell}$, $\ell = 0, \ldots, L$, where $\tilde{N}$ is a positive integer representing the number of points in the coarsest grid. Assume that the linear system to be solved is

$$M_L x_L = b_L \tag{4.37}$$

where the unknown is $x_L \in \mathbb{R}^{\tilde{N}_L}$ and with $b_L \in \mathbb{R}^{\tilde{N}_L}$, $M_L \in \mathbb{R}^{\tilde{N}_L \times \tilde{N}_L}$. In order to perform intergrid communications, we introduce

- Prolongation operators, which represent a grid function on the next finer grid:
  $$P_\ell^{\ell+1} : \mathcal{G}_\ell \to \mathcal{G}_{\ell+1}.$$
- Restriction operators, which interpolate a grid function on the next coarser grid:
  $$R_\ell^{\ell-1} : \mathcal{G}_\ell \to \mathcal{G}_{\ell-1}.$$

Using these operators, we can define on each grid $\mathcal{G}_\ell$ a matrix corresponding to an approximate version of the linear system to be solved:

$$M_\ell = R_{\ell+1}^\ell M_{\ell+1} P_\ell^{\ell+1}.$$

Then, in order to solve $M_\ell x_\ell = b_\ell$, the method is decomposed into three main steps. First, a pre-smoothing step is performed: starting from an initial guess $\tilde{x}_\ell^{(0)}$, a few smoothing iterations, say η_1, i.e. Jacobi or Gauss–Seidel iterations for example. This produces an estimate $\tilde{x}_\ell = \tilde{x}_\ell^{(\eta_1)}$. Second, an (approximate) solution $x_{\ell-1}$ on the next coarser grid is computed for the equation $M_{\ell-1} x_{\ell-1} = R_\ell^{\ell-1}(b_\ell - M_\ell \tilde{x}_\ell)$. This is performed either by calling recursively the same function, or by means of a direct solver (using Gaussian elimination) if it is on the coarsest grid. Third, a post-smoothing step is performed: $\tilde{x}_\ell + P_{\ell-1}^\ell x_{\ell-1}$ is used as an initial guess, from which η_2 iterations of the smoother are applied, for the problem with right-hand side b_ℓ. To understand the rationale behind this method, it is important to note that

$$R_\ell^{\ell-1} M_\ell \left(\tilde{x}_\ell + P_{\ell-1}^\ell x_{\ell-1} \right) = R_\ell^{\ell-1} M_\ell \tilde{x}_\ell + M_{\ell-1} x_{\ell-1}$$

$$\approx R_\ell^{\ell-1} M_\ell \tilde{x}_\ell + R_\ell^{\ell-1}(b_\ell - M_\ell \tilde{x}_\ell)$$

$$= R_\ell^{\ell-1} b_\ell.$$

In words, the initial guess (namely, $\tilde{x}_\ell + P_{\ell-1}^\ell x_{\ell-1}$) for the third step above is a good candidate for a solution to the equation $M_\ell x_\ell = b_\ell$, at least on the coarser grid $\mathcal{G}_{\ell-1}$.

Algorithm 3 provides a pseudo-code for the method described above. Here, $S_\ell(x, b, \eta)$ can be implemented by performing η steps of Gauss–Seidel algorithm starting with x and with b as right-hand side. The method as presented uses once the multigrid scheme on the coarser grid, which is called a V-cycle. Other approaches are possible, such as W-cycle (in which the multigrid scheme is called twice) or

F-cycle (which is intermediate between the V-cycle and the W-cycle). See e.g. [26] for more details.

Algorithm 3 Multigrid Method for $M_L x_L = b_L$ with V-cycle

---

function MULTIGRIDSOLVER$(\ell, \tilde{x}_\ell^{(0)}, b_\ell)$
  if $\ell = 0$ **then**
    $x_0 \leftarrow M_0^{-1} b_0$ *// exact solver at level 0*
  else
    $\tilde{x}_\ell \leftarrow S_\ell\left(\tilde{x}_\ell^{(0)}, b_\ell, \eta_1\right)$ *// pre-smoothing with η_1 steps of smoothing*
    $\tilde{x}_{\ell-1}^{(0)} \leftarrow 0$
    $x_{\ell-1} \leftarrow$ MultigridSolver$\left(\ell - 1, \tilde{x}_{\ell-1}^{(0)}, R_\ell^{\ell-1}(b_\ell - M_\ell \tilde{x}_\ell)\right)$ *// coarser grid*
correction
    $x_\ell \leftarrow S_\ell\left(x_\ell + P_{\ell-1}^\ell x_{\ell-1}, b_\ell, \eta_2\right)$ *// post-smoothing with η_2 steps of smoothing*
    return x_ℓ

4.4.2 Applications in the Context of Mean Field Games

Multigrid methods can be used for a linearized version of the MFG PDE system, see [27], or as a key ingredient of the ADDM or the primal-dual algorithms, see [17, 22]. In the latter case, it corresponds to taking $M_L = \Lambda \Lambda^*$ in (4.37). A straightforward application of the multigrid scheme described above leads to coarsening the space-time grid which does not distinguish between the space and time dimensions. This is called *full coarsening*. However, in the context of second-order MFG, this approach leads to poor performance. See e.g. [27]. We reproduce here one table contained in [27]: the multigrid method is used as a preconditioner in a preconditioned BiCGStab iterative method, see [28], in order to solve a linearized version of the MFG system of PDEs. In Table 4.1, we display the number of iterations of the preconditioned BiCGStab method: we see that the number of iterations grows significantly as the number of nodes is increased.

The reason for this poor behavior can be explained by the fact that the usual smoothers actually make the error smooth in the hyperplanes $t = n\Delta t$, i.e. with respect to the variable x, but not with respect to the variable t. Indeed, the unknowns

Table 4.1 Full coarsening multigrid preconditioner with 4 levels and several values of the viscosity ν: average number of preconditioned BiCGStab iterations to decrease the residual by a factor 0.01

$\nu\backslash$ grid	$32 \times 32 \times 32$	$64 \times 64 \times 64$	$128 \times 128 \times 64$
0.6	40	92	Fail
0.36	24	61	Fail
0.2	21	45	Fail

are more strongly coupled in the hyperplanes $\{(t, x) : t = n\Delta t\}, n = 0, \ldots, N_T$ (fourth order operator w.r.t. x) than on the lines $\{(t, x) : x = ih\}, i = 0, \ldots, N_h$ (second order operator w.r.t. t). This leads to the idea of using semi-coarsening: the hierarchy of nested grids should be obtained by coarsening the grids in the x directions only, but not in the t direction. We refer the reader to [26] for semi-coarsening multigrid methods in the context of anisotropic operators.

In the context of the primal-dual method discussed in Sect. 4.3.5, the first step (4.35) amounts to solving a discrete version of the PDE with operator $-\partial_{tt}^2 + v^2 \Delta^2 - \Delta$ where Δ^2 denotes the bi-Laplacian operator. In other words, one needs to solve a system of the form (4.37) where M_L corresponds to the discretization of this operator on the (finest) grid under consideration. One can thus use one cycle of the multigrid algorithm, which is a linear operator as a function of the residual on the finest grid, as a preconditioner for solving (4.37) with the `BiCGStab` method.

We now give details on the restriction and prolongation operators when $d = 1$. Using the notations introduced above, we consider that N_h is of the form $N_h = n_0 2^L$ for some integer n_0. Remember that $\Delta t = T/N_T$ and $h = 1/N_h$ since we are using the one-dimensional torus as the spatial domain. The number of points on the coarsest grid G_0 is $\tilde{N} = (N_T + 1) \times n_0$ while on the ℓ-th grid G_ℓ, it is $\tilde{N}_\ell = (N_T + 1) \times n_0 \times 2^\ell$.

For the restriction operator $R_\ell^{\ell-1} : G_\ell \to G_{\ell-1}$, following [27], we can use the second-order full-weighting operator defined by

$$(R_\ell^{\ell-1} x)_i^n := \frac{1}{4} \left(2X_{2i}^n + X_{2i+1}^n + X_{2i-1}^n \right),$$

for $n = 0, \ldots, N_T, i = 1, \ldots, 2^{\ell-1} n_0$.

As for the prolongation operator $P_\ell^{\ell+1} : G_\ell \to G_{\ell+1}$, one can take the standard linear interpolation which is second order accurate. An important aspect in the analysis of multigrid methods is that the sum of the accuracy orders of the two intergrid transfer operators should be not smaller than the order of the partial differential operator. Here, both are 4. In this case, multigrid theory states that convergence holds even with a single smoothing step, i.e. it suffices to take η_1, η_2 such that $\eta_1 + \eta_2 = 1$.

4.4.3   Numerical Illustration

In this paragraph, we borrow a numerical example from [15]. We assume that $d = 2$ and that given $q > 1$, with conjugate exponent denoted by $q' = q/(q - 1)$, the Hamiltonian $H_0 : \mathbb{T}^2 \times \mathbb{R}^2 \to \mathbb{R}$ has the form

$$H_0(x, p) = \frac{1}{q'} |p|^{q'}, \quad \forall x \in \mathbb{T}^2, \; p \in \mathbb{R}^2.$$

In this case the function L defined in (4.14) takes the form

$$L(x, m, w) = \begin{cases} \frac{|w|^q}{qm^{q-1}}, & \text{if } m > 0, \\ 0, & \text{if } (m, w) = (0, 0), \\ +\infty, & \text{otherwise.} \end{cases}$$

Furthermore, recalling the notation (4.12)–(4.13), we consider $\phi \equiv 0$ and

$$f(x, m) = m^2 - \overline{H}(x), \qquad \overline{H}(x) = \sin(2\pi x_2) + \sin(2\pi x_1) + \cos(2\pi x_1),$$

for all $x = (x_1, x_2) \in \mathbb{T}^2$ and $m \in \mathbb{R}_+$. This means that in the underlying differential game, a typical agent aims to get closer to the maxima of $\bar{H}$ and, at the same time, she is adverse to crowded regions (because of the presence of the m^2 term in f).

Figure 4.2 shows the evolution of the mass at four different time steps. Starting from a constant initial density, the mass converges to a steady state, and then, when t gets close to the final time T, the mass is influenced by the final cost and converges to a final state. This behavior is referred to as *turnpike phenomenon* in the literature

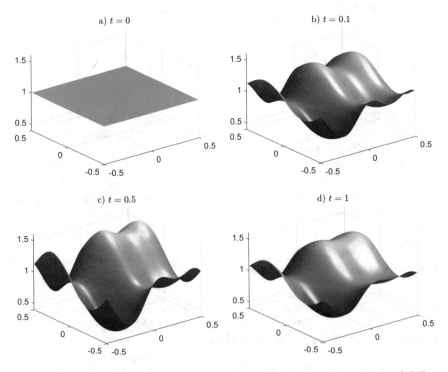

Fig. 4.2 Evolution of the density m obtained with the multi-grid preconditioner for $v = 0.5$, $T = 1$, $N_T = 200$ and $N_h = 128$. At $t = 0.12$ the solution is close to the solution of the associated stationary MFG. (**a**) $t = 0$. (**b**) $t = 0.1$. (**c**) $t = 0.5$. (**d**) $t = 1$

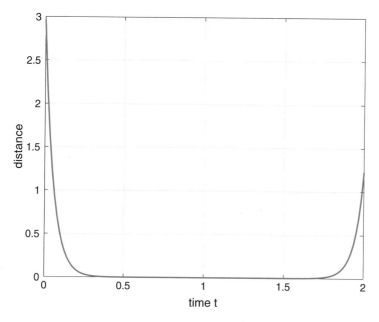

Fig. 4.3 Distance to the stationary solution at each time $t \in [0, T]$, for $\nu = 0.5$, $T = 2$, $N_T = 200$ and $N_h = 128$. The distance is computed using the ℓ^2 norm as explained in the text. The turnpike phenomenon is observed as for a long time frame the time-dependent mass approaches the solution of the stationary MFG

on optimal control [29]. Theoretical results on the long time behavior of MFGs can be found in [30, 31]. It is illustrated by Fig. 4.3, which displays as a function of time t the distance of the mass at time t to the stationary state computed as in [17]. In other words, denoting by $M^\infty \in \mathbb{R}^{N_h \times N_h}$ the solution to the discrete stationary problem and by $M \in \mathcal{M}$ the solution to the discrete evolutive problem, Fig. 4.3 displays the graph of $n \mapsto \|M^\infty - M^n\|_{\ell_2} = \left(h^2 \sum_{i,j} (M_{i,j}^\infty - M_{i,j}^n)^2 \right)^{1/2}$, $n \in \{0, \ldots, N_T\}$.

For the multigrid preconditioner, Table 4.2 shows the computation times for different discretizations (i.e. different values of N_h and N_T in the coarsest grid). It has been observed in [15, 17] that finer meshes with 128^3 degrees of freedom are solvable within CPU times which outperfom several other methods such as Conjugate Gradient or BiCGStab unpreconditioned or preconditioned with modified incomplete Cholesky factorization. Furthermore, the method is robust with respect to different viscosities.

From Table 4.2 we observe that most of the computational time is used for solving (4.36), which does not use a multigrid strategy but which is a pointwise operator (see [17, Proposition 3.1]) and thus could be fully parallelizable.

Table 4.3 shows that the method is robust with respect to the viscosity since the average number of iterations of BiCGStab does not increase much as the viscosity decreases. For instance, as shown in Table 4.3b for a grid of size $128 \times 128 \times 128$,

Table 4.2 Time (in seconds) for the convergence of the Chambolle-Pock algorithm, cumulative time of the first proximal operator with the multigrid preconditioner, and number of iterations in the Chambolle-Pock algorithm, for different viscosity values ν and two types of grids

(a) Grid with $64 \times 64 \times 64$ points				(b) Grid with $128 \times 128 \times 128$ points			
ν	Total time	Time first prox	Iterations	ν	Total time	Time first prox	Iterations
0.6	116.3 [s]	11.50 [s]	20	0.6	921.1 [s]	107.2 [s]	20
0.36	120.4 [s]	11.40 [s]	21	0.36	952.3 [s]	118.0 [s]	21
0.2	119.0 [s]	11.26 [s]	22	0.2	1028.8 [s]	127.6 [s]	22
0.12	129.1 [s]	14.11 [s]	22	0.12	1036.4 [s]	135.5 [s]	23
0.046	225.0 [s]	23.28 [s]	39	0.046	1982.2 [s]	260.0 [s]	42

Here we used $\eta_1 = \eta_2 = 2$ in the multigrid methods, $T = 1$. Instead of using a number of iterations K fixed a priori, the iterations have been stopped when the quantity $\|M^{(k+1)} - M^{(k)}\|_{\ell_2} = \left(\Delta t h^2 \sum_{n=0}^{N_T} \sum_{i,j} (M_{i,j}^{(k+1),n} - M_{i,j}^{(k),n})^2 \right)^{1/2}$ became smaller than 10^{-6}, where $M^{(k)}$ denotes the approximation of M at iteration k, which is given by the first component of $\sigma^{(k)}$ in the notation used in Algorithm 2

the average number of iterations increases from 3.38 to 4.67 when ν is decreased from 0.6 to 0.046. On the other hand, Table 4.2 shows that the average number of Chambolle-Pock iterations depends on the viscosity parameter, but this is not related to the use of the multigrid preconditioner.

4.5 Algorithms for Solving the System of Non-Linear Equations

4.5.1 Combining Continuation Methods with Newton Iterations

Following previous works of the first author, see e.g. [16], one may use a continuation method (for example with respect to the viscosity parameter ν) in which every system of nonlinear equations (given the parameter of the continuation method) is solved by means of Newton iterations. With Newton algorithm, it is important to have a good initial guess of the solution; for that, it is possible to take advantage of the continuation method by choosing the initial guess as the solution obtained with the previous value of the parameter. Alternatively, the initial guess can be obtained from the simulation of the same problem on a coarser grid, using interpolation. It is also important to implement the Newton algorithm on a "well conditioned" system. Consider for example the system (4.4)–(4.6): in this case, it proves more convenient to introduce auxiliary unknowns, namely $\left((f_i^n,)_{i,n}, (\Phi)_i \right)$, and see (4.4)–(4.6) as a fixed point problem for the map $\Xi : \left((f_i^n,)_{i,n}, (\Phi)_i \right) \mapsto \left((g_i^n,)_{i,n}, (\Psi)_i \right)$ defined as follows: one solves first the discrete Bellman equation

Table 4.3 Average number of iterations of the preconditioned BiCGStab with $\eta_1 = \eta_2 = 2, T = 1$

(a) Iterations to decrease the residual by a factor 10^{-3}

ν	$32 \times 32 \times 32$	$64 \times 64 \times 64$	$128 \times 128 \times 128$
0.6	1.65	1.86	2.33
0.36	1.62	1.90	2.43
0.2	1.68	1.93	2.59
0.12	1.84	2.25	2.65
0.046	1.68	2.05	2.63

(b) Iterations to solve the system with an error of 10^{-8}

ν	$32 \times 32 \times 32$	$64 \times 64 \times 64$	$128 \times 128 \times 128$
0.6	3.33	3.40	3.38
0.36	3.10	3.21	3.83
0.2	3.07	3.31	4.20
0.12	3.25	3.73	4.64
0.046	2.88	3.59	4.67

with data $\left((f_i^n,)_{i,n}, (\Phi)_i \right)$:

$$-(D_t U_i)^n - \nu(\Delta_h U^n)_i + \tilde{H}(x_i, [\nabla_h U^n]_i) = f_i^n \quad i \in \{0, \ldots, N_h\}, n \in \{0, \ldots, N_T - 1\},$$
$$U_0^n = U_{N_h}^n, \ n \in \{0, \ldots, N_T - 1\},$$
$$U_i^{N_T} = \Phi_i \quad i \in \{0, \ldots, N_h\}$$

then the discrete Fokker–Planck equation

$$(D_t M_i)^n - \nu(\Delta_h M^{n+1})_i - \mathcal{T}_i(U^n, M^{n+1}) = 0, \quad i \in \{0, \ldots, N_h\}, n \in \{0, \ldots, N_T - 1\},$$
$$M_0^n = M_{N_h}^n, \quad n \in \{1, \ldots, N_T\},$$
$$M_i^0 = \bar{m}_0(x_i), \ i \in \{0, \ldots, N_h\},$$

and finally sets

$$g_i^n = f_0(x_i, M_i^{n+1}), \quad i \in \{0, \ldots, N_h\}, \ n \in \{0, \ldots, N_T - 1\},$$
$$\Psi_i = \phi(M_i^{N_T}), \quad i \in \{0, \ldots, N_h\}.$$

The Newton iterations are applied to the fixed point problem $(I_d - \Xi)\left((f_i^n,)_{i,n}, (\Phi)_i\right) = 0$. Solving the discrete Fokker–Planck guarantees that at each Newton iteration, the grid functions M^n are non negative and have the same mass, which is not be the case if the Newton iterations are applied directly to (4.4)–(4.6). Note also that Newton iterations consist of solving systems of linear equations involving the Jacobian of Ξ: for that, we use a nested iterative method (BiCGStab for example), which only requires a function that returns the matrix-vector product by the Jacobian, and not the construction of the Jacobian, which is a huge and dense matrix.

The strategy consisting in combining the continuation method with Newton iterations has the advantage to be very general: it requires neither a variational structure nor monotonicity: it has been successfully applied in many simulations, for example for MFGs including congestion models, MFGs with two populations, see paragraph Sect. 4.6 or MFGs in which the coupling is through the control. It is generally much faster than the methods presented in paragraph Sect. 4.3, when the latter can be used.

On the other hand, to the best of our knowledge, there is no general proof of convergence. Having these methods work efficiently is part of the know-how of the numerical analyst.

Since the strategy has been described in several articles of the first author, and since it is discussed thoroughly in the paragraph Sect. 4.6 devoted to pedestrian flows, we shall not give any further detail here.

4.5.2 A Recursive Algorithm Based on Elementary Solvers on Small Time Intervals

In this paragraph, we consider a recursive method introduced by Chassagneux, Crisan and Delarue in [32] and further studied in [33]. It is based on the following idea. When the time horizon is small enough, mild assumptions allow one to apply Banach fixed point theorem and give a constructive proof of existence and uniqueness for system (4.3). The unique solution can be obtained by Picard iterations, i.e., by updating iteratively the flow of distributions and the value function. Then, when the time horizon T is large, one can partition the time interval into intervals of duration τ, with τ small enough. Let us consider the two adjacent intervals $[0, T - \tau]$ and $[T - \tau, T]$. The solutions in $[0, T - \tau] \times \mathbb{T}^d$ and in $[T - \tau, T] \times \mathbb{T}^d$ are coupled through their initial or terminal conditions: for the former interval $[0, T - \tau]$, the initial condition for the distribution of states is given, but the terminal condition on the value function will come from the solution in $[T - \tau, T]$. The principle of the global solver is to use the elementary solver on $[T - \tau, T]$ (because τ is small enough) and recursive calls of the global solver on $[0, T - \tau]$, (which will in turn call the elementary solver on $[T - 2\tau, T - \tau]$ and recursively the global solver on $[0, T - 2\tau]$, and so on so forth).

We present a version of this algorithm based on PDEs (the original version in [32] is based on forward-backward stochastic differential equations (FBSDEs for short) but the principle is the same). Recall that $T > 0$ is the time horizon, m_0 is the initial density, ϕ is the terminal condition, and that we want to solve system (4.3).

Let K be a positive integer such that $\tau = T/K$ is small enough. Let

$$\texttt{ESolver} : (\tau, \tilde{m}, \tilde{\phi}) \mapsto (m, u)$$

be an elementary solver which, given τ, an initial probability density $\tilde{m}$ defined on $\mathbb{T}^d$ and a terminal condition $\tilde{\phi} : \mathbb{T}^d \to \mathbb{R}$, returns the solution $(u(t), m(t))_{t \in [0, \tau]}$ to the MFG system of forward-backward PDEs corresponding to these data, i.e.,

$$\begin{cases} -\dfrac{\partial u}{\partial t}(t, x) - \nu \Delta u(t, x) = H(x, m(t, x), \nabla u(t, x)), & \text{in } [0, \tau) \times \mathbb{T}^d, \\[2mm] \dfrac{\partial m}{\partial t}(t, x) - \nu \Delta m(t, x) - \operatorname{div}\big(m(t, \cdot)H_p(\cdot, m(t, \cdot), \nabla u(t, \cdot))\big)(x) = 0, & \text{in } (0, \tau] \times \mathbb{T}^d, \\[2mm] u(\tau, x) = \tilde{\phi}(x), \qquad m(0, x) = \tilde{m}(x), & \text{in } \mathbb{T}^d. \end{cases}$$

The solver ESolver may for instance consist of Picard or Newton iterations.

We then define the following recursive function, which takes as inputs the level of recursion k, the maximum recursion depth K, and an initial distribution $\tilde{m} : \mathbb{T}^d \to \mathbb{R}$, and returns an approximate solution of the system of PDEs in $[kT/K, T] \times \mathbb{T}^d$ with initial condition $m(t, x) = \tilde{m}(x)$ and terminal condition $u(T, x) = \phi(m(T, x), x)$. Calling RSolver$(0, K, m_0)$ then returns an approximate solution $(u(t), m(t))_{t \in [0, T]}$ to system (4.3) in $[0, T] \times \mathbb{T}^d$, as desired.

To compute the approximate solution on $[kT/K, T] \times \mathbb{T}^d$, the following is repeated J times, say from $j = 0$ to $J - 1$, after some initialization step (see Algorithm 4 for a pseudo code):

1. Compute the approximate solution on $[(k + 1)T/K, T] \times \mathbb{T}^d$ by a recursive call of the algorithm, given the current approximation of $m((k + 1)T/K, \cdot)$ (it will come from the next point if $j > 0$).
2. Call the elementary solver on $[kT/K, (k+1)T/K] \times \mathbb{T}^d$ given $u((k+1)T/K, \cdot)$ coming from the previous point.

Algorithm 4 Recursive Solver for the system of forward-backward PDEs

---

  function RSOLVER$(k, K, \tilde{m})$

    if $k = K$ **then**

      $(u(t, \cdot), m(t, \cdot))_{t=T} = (\phi(\tilde{m}(\cdot), \cdot), \tilde{m}(\cdot))$ // *last level of recursion*

    else

      $(u(t, \cdot), m(t, \cdot))_{t\in[kT/K,(k+1)T/K)]} \leftarrow (0, \tilde{m}(\cdot))$ // *initialization*

      for $j = 0, \ldots, J$ **do**

        $(u(t, \cdot), m(t, \cdot))_{t\in[(k+1)T/K,T]} \leftarrow \text{RSolver}(k + 1, K, m((k + 1)T/K, \cdot))$

        // *interval* $[(k + 1)T/K, T]$

        $(u(t, \cdot), m(t, \cdot))_{t\in[kT/K,(k+1)T/K]} \leftarrow \text{ESolver}(T/K, m(kT/K, \cdot),$

        $u((k + 1)T/K, \cdot))$ // *interval* $[kT/K, (k + 1)T/K]$

    return $(u(t, \cdot), m(t, \cdot))_{t\in[kT/K,T]}$

In [32], Chassagneux et al. introduced a version based on FBSDEs and proved, under suitable regularity assumptions on the decoupling field, the convergence of the algorithm, with a complexity that is exponential in K. The method has been further tested in [33] with implementations relying on trees and grids to discretize the evolution of the state process.

4.6   An Application to Pedestrian Flows

4.6.1   An Example with Two Populations, Congestion Effects and Various Boundary Conditions

The numerical simulations discussed in this paragraphs somehow stand at the state of the art because they combine the following difficulties:

- The MFG models includes congestion effects. Hence the Hamiltonian does not depend separately on Du and m. In such cases, the MFG can never be interpreted as an optimal control problem driven by a PDE; in other words, there is no variational interpretation, which makes it impossible to apply the methods discussed in Sect. 4.3.

- There are two populations of agents which interact with each other, which adds a further layer of complexity. The now well known arguments due to Lasry and Lions and leading to uniqueness do not apply
- The model will include different kinds of boundary conditions corresponding to walls, entrances and exits, which need careful discretizations.
- We are going to look for stationary equilibria, despite the fact that there is no underlying ergodicity: there should be a balance between exit and entry fluxes. A special numerical algorithm is necessary in order to capture such situations

We consider a two-population mean field game in a complex geometry. It models situations in which two crowds of pedestrians have to share a confined area. In this case, the state space is a domain of $\mathbb{R}^2$. The agents belonging to a given population are all identical, and differ from the agents belonging to the other population because they have for instance different objectives, and also because they feel uncomfortable in the regions where their own population is in minority (xenophobia). In the present example, there are several exit doors and the agents aim at reaching some of these doors, depending on which population they belong to. To reach their targets, the agents may have to cross regions in which their own population is in minority. More precisely, the running cost of each individual is made of different terms:

- a first term only depends of the state variable: it models the fact that a given agent more or less wishes to reach one or several exit doors. There is an exit cost or reward at each doors, which depends on which population the agents belong to. This translates the fact that the agents belonging to different populations have different objectives
- the second term is a cost of motion. In order to model congestion effects, it depends on the velocity and on the distributions of states for both populations
- the third term models xenophobia and aversion to highly crowded regions.

4.6.1.1   The System of Partial Differential Equations

Labelling the two populations with the indexes 0 and 1, the model leads to a system of four forward-backward partial differential equations as follows:

$$\frac{\partial u_0}{\partial t} + \nu \Delta u_0 - H_0(\nabla u_0; m_0, m_1) = -\Phi_0(x, m_0, m_1), \qquad (4.38)$$

$$\frac{\partial m_0}{\partial t} - \nu \Delta m_0 - \text{div}\left(m_0 \frac{\partial H_0}{\partial p}(\nabla u_0; m_0, m_1)\right) = 0, \qquad (4.39)$$

$$\frac{\partial u_1}{\partial t} + \nu \Delta u_1 - H_1(\nabla u_1; m_1, m_0) = -\Phi_1(x, m_1, m_0), \qquad (4.40)$$

$$\frac{\partial m_1}{\partial t} - \nu \Delta m_1 - \text{div}\left(m_1 \frac{\partial H_1}{\partial p}(\nabla u_1; m_1, m_0)\right) = 0. \qquad (4.41)$$

In the numerical simulations discussed below, we have chosen

$$H_i(x, p; m_i, m_j) = \frac{|p|^2}{1 + m_i + 5m_j}, \qquad (4.42)$$

and

$$\Phi_i(x, m_i, m_j) = 0.5 + 0.5 \left(\frac{m_i}{m_i + m_j + \epsilon} - 0.5 \right)_- + (m_i + m_j - 4)_+, \qquad (4.43)$$

where ϵ is a small parameter and $j = 1 - i$. Note that we may replace (4.43) by a smooth function obtained by a regularization involving another small parameter ϵ_2.

- The choice of H_i aims at modelling the fact that motion gets costly in the highly populated regions of the state space. The different factors in front of m_i and m_j in (4.42) aim at modelling the fact that the cost of motion of an agent of a given type, say i, is more sensitive to the density of agents of the different type, say j; indeed, since the agents of different types have different objectives, their optimal controls are unlikely to be aligned, which makes motion even more difficult.
- The coupling cost in (4.43) is made of three terms: the first term, namely 0.5, is the instantaneous cost for staying in the domain; the term $0.5 \left(\frac{m_i}{m_i + m_j + \epsilon} - 0.5 \right)_-$ translates the fact that an agent in population i feels uncomfortable if its population is locally in minority. The last term, namely $(m_i + m_j - 4)_+$, models aversion to highly crowded regions.

4.6.1.2 The Domain and the Boundary Conditions

The domain Ω is displayed in Fig. 4.4. The solid lines stand for walls, i.e. parts of the boundaries that cannot be crossed by the agents. The colored arrows indicate entries or exits depending if they are pointing inward or outward. The two different colors correspond to the two different populations, green for population 0 and orange for population 1. The length of the northern wall is 5, and the north-south diameter is 2.5. The width of the southern exit is 1. The widths of the other exits and entrances are 0.33. The width of the outward arrows stands for the reward for exiting.

- The boundary conditions at walls are as follows:

$$\frac{\partial u_i}{\partial n}(x) = 0, \quad \text{and} \quad \frac{\partial m_i}{\partial n}(x) = 0. \qquad (4.44)$$

The first condition in (4.44) comes from the fact that the stochastic process describing the state of a given agent in population i is reflected on walls. The second condition in (4.44) is in fact $-\nu \frac{\partial m_i}{\partial n} - m_i \, n \cdot \frac{\partial H_i}{\partial p}(\nabla u_i; m_i, m_j) = 0$, where we have taken into account the Neumann condition on u_i.

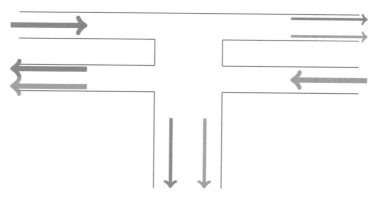

Fig. 4.4 The domain Ω. The colored arrows indicate entries or exits depending if they are pointing inward or outward. The two different colors correspond to the two different populations. The width of the outward arrows stands for the reward for exiting

- At an exit door for population i, the boundary conditions are as follow

$$u_i = \text{exit cost}, \quad \text{and} \quad m_i = 0. \tag{4.45}$$

A negative exit cost means that the agents are rewarded for exiting the domain through this door. The homogeneous Dirichlet condition on m_i in (4.45) can be explained by saying that the agents stop taking part to the game as soon as they reach the exit.
- At an entrance for population i, the boundary conditions are as follows:

$$u_i = \text{exit cost}, \quad \text{and} \quad \nu \frac{\partial m_i}{\partial n} + m_i\, n \cdot \frac{\partial H_i}{\partial p}(\nabla u_i; m_i, m_j) = \text{entry flux.} \tag{4.46}$$

Setting a high exit cost prevents the agents from exiting though the entrance doors.

In our simulations, the exit costs of population 0 are as follows:

1. North-West entrance : 0
2. South-West exit : -8.5
3. North-East exit : -4
4. South-East exit : 0
5. South exit : -7

and the exit costs of population 1 are

1. North-West exit : 0
2. South-West exit : -7
3. North-East exit : -4

4. South-East entrance : 0
5. South exit: -4.

The entry fluxes are as follows:

1. Population 0: at the North-West entrance, the entry flux is 1
2. Population 1: at the South-East entrance, the entry flux is 1.

For equilibria in finite horizon T, the system should be supplemented with an initial Dirichlet condition for m_0, m_1 since the laws of the initial distributions are known, and a terminal Dirichlet-like condition for u_0, u_1 accounting for the terminal costs.

4.6.2   Stationary Equilibria

We look for a stationary equilibrium. For that, we solve numerically (4.38)–(4.41) with a finite horizon T, and with the boundary conditions described in paragraph Sect. 4.6.1.2, see paragraph Sect. 4.6.5 below and use an iterative method in order to progressively diminish the effects of the initial and terminal conditions: starting from $(u_i^0, m_i^0)_{i=0,1}$, the numerical solution of the finite horizon problem described above, we construct a sequence of approximate solutions $(u_i^\ell, m_i^\ell)_{\ell \geq 1}$ by the following induction: $(u_i^{\ell+1}, m_i^{\ell+1})$ is the solution of the finite horizon problem with the same system of PDEs in $(0, T) \times \Omega$, the same boundary conditions on $(0, T) \times \partial\Omega$, and the new initial and terminal conditions as follows:

$$u_i^{\ell+1}(T, x) = u_i^\ell \left(\frac{T}{2}, x \right), \qquad x \in \Omega, \ i = 0, 1, \tag{4.47}$$

$$m_i^{\ell+1}(0, x) = m_i^\ell \left(\frac{T}{2}, x \right), \qquad x \in \Omega, \ i = 0, 1. \tag{4.48}$$

As ℓ tends to $+\infty$, we observe that (u_i^ℓ, m_i^ℓ) converge to time-independent functions. At the limit, we obtain a steady solution of

$$\nu \Delta u_0 - H_0(\nabla u_0; m_0, m_1) = -\Phi_0(x, m_0, m_1), \tag{4.49}$$

$$-\nu \Delta m_0 - \text{div} \left(m_0 \frac{\partial H_0}{\partial p}(\nabla u_0; m_0, m_1) \right) = 0, \tag{4.50}$$

$$\nu \Delta u_1 - H_1(\nabla u_1; m_1, m_0) = -\Phi_1(x, m_1, m_0), \tag{4.51}$$

$$-\nu \Delta m_1 - \text{div} \left(m_1 \frac{\partial H_1}{\partial p}(\nabla u_1; m_1, m_0) \right) = 0, \tag{4.52}$$

with the boundary conditions on $\partial\Omega$ described in paragraph Sect. 4.6.1.2.

4.6.3   A Stationary Equilibrium with $v = 0.30$

We first take a relatively large viscosity coefficient namely $v = 0.30$. In Fig. 4.5, we display the distributions of states for the two populations, see Fig. 4.5a, the value functions for both populations, see Fig. 4.5b, and the optimal feedback controls of population 0 (respectively 1) in Fig. 4.5c (respectively Fig. 4.5d). We see that population 0 enters the domain via the north-west entrance, and most probably exits by the south-west exit door. Population 1 enters the domain via the south-east entrance, and exits by two doors, the south-west and the southern ones. The effect of viscosity is large enough to prevent complete segregation of the two populations.

4.6.4   A Stationary Equilibrium with $v = 0.16$

We decrease the viscosity coefficient to the value $v = 0.16$. We are going to see that the solution is quite different from the one obtained for $v = 0.3$, because the

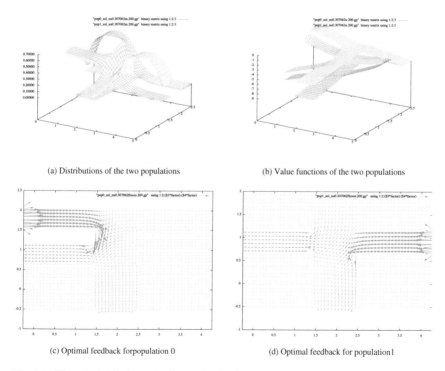

(a) Distributions of the two populations

(b) Value functions of the two populations

(c) Optimal feedback forpopulation 0

(d) Optimal feedback for population1

Fig. 4.5   Numerical Solution to Stationary Equilibrium with $v \sim 0.3$. (**a**) Distributions of the two populations. (**b**) Value functions of the two populations. (**c**) Optimal feedback for population 0. (**d**) Optimal feedback for population 1

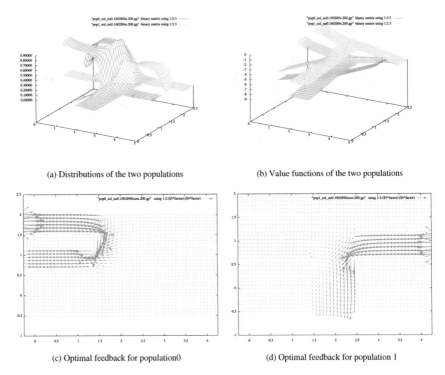

(a) Distributions of the two populations   (b) Value functions of the two populations

(c) Optimal feedback for population0   (d) Optimal feedback for population 1

Fig. 4.6 Numerical Solution to Stationary Equilibrium with $\nu \sim 0.16$. (**a**) Distributions of the two populations. (**b**) Value functions of the two populations. (**c**) Optimal feedback for population 0. (**d**) Optimal feedback for population 1

populations now occupy separate regions. In Fig. 4.6, we display the distributions of states for the two populations, see Fig. 4.6a, the value functions for both populations, see Fig. 4.6b, and the optimal feedback controls of population 0 (respectively 1) in Fig. 4.6c (respectively Fig. 4.6d). We see that population 0 enters the domain via the north-west entrance, and most probably exits by the south-west exit door. Population 1 enters the domain via the south-east entrance, and most probably exits the domain by the southern door. The populations occupy almost separate regions. We have made simulations with smaller viscosities, up to $\nu = 10^{-3}$, and we have observed that the results are qualitatively similar to the ones displayed on Fig. 4.6, i.e. the distribution of the populations overlap less and less and the optimal strategies are similar. As ν is decreased, the gradients of the distributions of states increase in the transition regions between the two populations.

4.6.5   Algorithm for Solving the System of Nonlinear Equations

We briefly describe the iterative method used in order to solve the system of non-linear equations arising from the discretization of the finite horizon problem. Since the latter system couples forward and backward (nonlinear) equations, it cannot be solved by merely marching in time. Assuming that the discrete Hamiltonians are C^2 and the coupling functions are C^1 (this is true after a regularization procedure involving a small regularization parameter) allows us to use a Newton–Raphson method for the whole system of nonlinear equations (which can be huge if $d \geq 2$).

Disregarding the boundary conditions for simplicity, the discrete version of the MFG system reads

$$\frac{U_{i,j}^{k,n+1} - U_{i,j}^{k,n}}{\Delta t} - \nu (\Delta_h U^{k,n+1})_{i,j} + g\left([\nabla_h U^{k,n+1}]_{i,j}, Z_{i,j}^{k,n+1}\right) = Y_{i,j}^{k,n+1} \qquad (4.53)$$

$$\frac{M_{i,j}^{k,n+1} - M_{i,j}^{k,n}}{\Delta t} - \nu (\Delta_h M^{k,n+1})_{i,j} - \mathcal{T}_{i,j}\left(U^{k,n+1}, M^{k,n+1}, Z^{k,n+1}\right) = 0, \qquad (4.54)$$

$$Y_{i,j}^{k,n+1} = -\Phi_k\left(x_{i,j}, M_{i,j}^{k,n+1}, M_{i,j}^{1-k,n+1}\right), \qquad (4.55)$$

$$Z_{i,j}^{k,n+1} = \left(1 + M_{i,j}^{k,n+1} + 5 M_{i,j}^{1-k,n+1}\right)^{-1}, \qquad (4.56)$$

for $k = 0, 1$. The system above is satisfied for internal points of the grid, i.e. $2 \leq i, j \leq N_h - 1$, and is supplemented with suitable boundary conditions. The ingredients in (4.53)–(4.56) are as follows:

$$[\nabla_h U]_{i,j} = ((D_1^+ U)_{i,j}, (D_1^+ U)_{i-1,j}, (D_2^+ U)_{i,j}, (D_2^+ U)_{i,j-1}) \in \mathbb{R}^4,$$

and the Godunov discrete Hamiltonian is

$$\tilde{H}(q_1, q_2, q_3, q_4, z) = z\left([(q_1)^-]^2 + [(q_3)^-]^2 + [(q_2)^+]^2 + [(q_4)^+]^2\right).$$

The transport operator in the discrete Fokker–Planck equation is given by

$$\mathcal{T}_{i,j}(U, M, Z) = \frac{1}{h}\begin{pmatrix} M_{i,j}\partial_{q_1}\tilde{H}[\nabla_h U]_{i,j}, Z_{i,j}) - M_{i-1,j}\partial_{q_1}\tilde{H}[\nabla_h U]_{i-1,j}, Z_{i-1,j}) \\ + M_{i+1,j}\partial_{q_2}\tilde{H}[\nabla_h U]_{i+1,j}, Z_{i+1,j}) - M_{i,j}\partial_{q_2}\tilde{H}[\nabla_h U]_{i,j}, Z_{i,j}) \\ + M_{i,j}\partial_{q_3}\tilde{H}[\nabla_h U]_{i,j}, Z_{i,j}) - M_{i,j-1}\partial_{q_3}\tilde{H}x, [\nabla_h U]_{i,j-1}, Z_{i,j-1}) \\ + M_{i,j+1}\partial_{q_4}\tilde{H}[\nabla_h U]_{i,j+1}, Z_{i,j+1}) - M_{i,j}\partial_{q_4}\tilde{H}[\nabla_h U]_{i,j}, Z_{i,j}) \end{pmatrix},$$

We define the map Θ :

$$\Theta : \quad \left(Y_{i,j}^{0,n}, Y_{i,j}^{1,n}, Z_{i,j}^{0,n}, Z_{i,j}^{1,n}\right)_{i,j,n} \mapsto \left(M_{i,j}^{0,n}, M_{i,j}^{1,n}\right)_{i,j,n},$$

by solving first the discrete HJB equation (4.53) (supplemented with boundary conditions) then the discrete Fokker–Planck equation (4.54) (supplemented with boundary conditions). We then summarize (4.55) and (4.56) by writing

$$\left(Y_{i,j}^{0,n}, Y_{i,j}^{1,n}, Z_{i,j}^{0,n}, Z_{i,j}^{1,n}\right)_{i,j,n} = \Psi\left(\left(M_{i,j}^{0,n}, M_{i,j}^{1,n}\right)_{i,j,n}\right). \qquad (4.57)$$

Here n takes its values in $\{1 \ldots, N\}$ and i, j take their values in $\{1 \ldots, N_h\}$. We then see the full system (4.53)–(4.56) (supplemented with boundary conditions) as a fixed point problem for the map $\Xi = \Psi \circ \Theta$.

Note that in (4.53) the two discrete Bellman equations are decoupled and do not involve $M^{k,n+1}$. Therefore, one can first obtain $U^{k,n}$ $0 \leq n \leq N$, $k = 0, 1$ by marching backward in time (i.e. performing a backward loop with respect to the index n). For every time index n, the two systems of nonlinear equations for $U^{k,n}$, $k = 0, 1$ are themselves solved by means of a nested Newton–Raphson method. Once an approximate solution of (4.53) has been found, one can solve the (linear) Fokker–Planck equations (4.54) for $M^{k,n}$ $0 \leq n \leq N$, $k = 0, 1$, by marching forward in time (i.e. performing a forward loop with respect to the index n). The solutions of (4.53)–(4.54) are such that $M^{k,n}$ are nonnegative.

The fixed point equation for Ξ is solved numerically by using a Newton–Raphson method. This requires the differentiation of both the Bellman and Kolmogorov equations in (4.53)–(4.54), which may be done either analytically (as in the present simulations) or via automatic differentiation of computer codes (to the best of our knowledge, no computer code for MFGs based on automatic differentiation is available yet, but developing such codes seems doable and interesting).

A good choice of an initial guess is important, as always for Newton methods. To address this matter, we first observe that the above mentioned iterative method generally quickly converges to a solution when the value of ν is large. This leads us to use a continuation method in the variable ν: we start solving (4.53)–(4.56) with a rather high value of the parameter ν (of the order of 1), then gradually decrease ν down to the desired value, the solution found for a value of ν being used as an initial guess for the iterative solution with the next and smaller value of ν.

4.7   Mean Field Type Control

As mentioned in the introduction, the theory of mean field games allows one to study Nash equilibria in games with a number of players tending to infinity. In such models, the players are selfish and try to minimize their own individual cost. Another kind of asymptotic regime is obtained by assuming that all the agents use the same distributed feedback strategy and by passing to the limit as $N \to \infty$ before optimizing the common feedback. For a fixed common feedback strategy, the asymptotic behavior is given by the McKean-Vlasov theory, [34, 35]: the dynamics of a representative agent is found by solving a stochastic differential equation with coefficients depending on a mean field, namely the statistical distribution of the states, which may also affect the objective function. Since the feedback strategy is common to all agents, perturbations of the latter affect the mean field (whereas in a mean field game, the other players' strategies are fixed when a given player optimizes). Then, having each agent optimize her objective function amounts to solving a control problem driven by the McKean-Vlasov dynamics. The latter is named control of McKean-Vlasov dynamics by R. Carmona and F. Delarue [36–38] and mean field type control by A. Bensoussan et al. [39, 40]. Besides the aforementioned interpretation as a social optima in collaborative games with a number of agents growing to infinity, mean field type control problems have also found applications in finance and risk management for problems in which the distribution of states is naturally involved in the dynamics or the cost function. Mean field type control problems lead to a system of forward-backward PDEs which has some features similar to the MFG system, but which can always be seen as the optimality conditions of a minimization problem. These problems can be tackled using the methods presented in this survey, see e.g. [23, 41]. For the sake of comparison with mean field games, we provide in this section an example of crowd motion (with a single population) taking into account congestion effects. The material of this section is taken from [42].

4.7.1 Definition of the Problem

Before focusing a specific example, let us present the generic form of a mean field type control problem. To be consistent with the notation used in the MFG setting, we consider the same form of dynamics and the same running and terminal costs functions f and ϕ. However, we focus on a different notion of solution: Instead of looking for a fixed point, we look for a minimizer when the control directly influences the evolution of the population distribution. More precisely, the goal is to find a feedback control $v^* : Q_T \to \mathbb{R}^d$ minimizing the following functional:

$$J : v \mapsto J(v) = \mathbb{E}\left[\int_0^T f(X_t^v, m^v(t, X_t^v), v(t, X_t^v))dt + \phi(X_T^v, m^v(T, X_T^v))\right]$$

under the constraint that the process $X^v = (X_t^v)_{t \geq 0}$ solves the stochastic differential equation (SDE)

$$dX_t^v = b(X_t^v, m^v(t, X_t^v), v(t, X_t^v))dt + \sqrt{2v}dW_t, \qquad t \geq 0, \qquad (4.58)$$

and X_0^v has distribution with density m_0. Here m_t^v is the probability density of the law of X_t^v, so the dynamics of the stochastic process is of McKean-Vlasov type.

For a given feedback control v, m_t^v solves the same Kolmogorov–Fokker–Planck (KFP) equation (4.2) as in the MFG, but the key difference between the two problems lies in the optimality condition. For the mean field type control problem, one cannot rely on standard optimal control theory because the distribution of the controlled process is involved in a potentially non-linear way.

In [40], A. Bensoussan, J. Frehse and P. Yam have proved that a necessary condition for the existence of a smooth feedback function v^* achieving $J(v^*) = \min J(v)$ is that

$$v^*(t, x) = \arg\max_{a \in \mathbb{R}^d} \left\{ -f(x, m(t, x), a) - \langle b(x, m(t, x), a), \nabla u(t, x) \rangle \right\},$$

where (m, u) solve the following system of partial differential equations

$$\begin{cases} -\dfrac{\partial u}{\partial t}(t, x) - \nu \Delta u(t, x) + H(x, m(t, x), \nabla u(t, x)) + \displaystyle\int_{\mathbb{T}^d} \dfrac{\partial H}{\partial m}(\xi, m(T, \xi), \nabla u(t, \xi)) m(T, \xi) d\xi = 0, & \text{in } [0, T) \times \mathbb{T}^d, \\[2mm] \dfrac{\partial m}{\partial t}(t, x) - \nu \Delta m(t, x) - \operatorname{div}\left(m(t, \cdot) H_p(\cdot, m(t, \cdot), \nabla u(t, \cdot))\right)(x) = 0, & \text{in } (0, T] \times \mathbb{T}^d, \\[2mm] u(T, x) = \phi(x, m(T, x)) + \displaystyle\int_{\mathbb{T}^d} \dfrac{\partial \phi}{\partial m}(\xi, m(T, \xi)) m(T, \xi) d\xi, \qquad m(0, x) = m_0(x), & \text{in } \mathbb{T}^d. \end{cases}$$

Here, $\frac{\partial}{\partial m}$ denotes a derivative with respect to the argument m, which is a real number because the dependence on the distribution is purely local in the setting considered here. When the Hamiltonian depends on the distribution m in a non-local way, one needs to use a suitable notion of derivative with respect to a probability density or a probability measure, see e.g. [4, 38, 40] for detailed explanations.

4.7.2 Numerical Simulations

Here we model a situation in which a crowd of pedestrians is driven to leave a given square hall (whose side is 50 m long) containing rectangular obstacles: one can imagine for example a situation of panic in a closed building, in which the population tries to reach the exit doors. The chosen geometry is represented on Fig. 4.7.

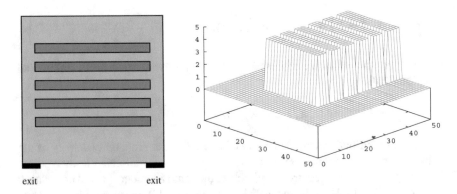

Fig. 4.7 Left: the geometry (obstacles are in red). Right: the density at $t = 0$

The aim is to compare the evolution of the density in two models:

1. Mean field games: we choose $v = 0.05$ and the Hamiltonian takes congestion effects into account and depends locally on m; more precisely:

$$H(x, m, p) = -\frac{8|p|^2}{(1+m)^{\frac{3}{4}}} + \frac{1}{3200}.$$

The MFG system of PDEs (4.3) becomes

$$
\begin{cases}
\dfrac{\partial u}{\partial t} + 0.05\,\Delta u - \dfrac{8}{(1+m)^{\frac{3}{4}}}\,|\nabla u|^2 = -\dfrac{1}{3200}, & \text{(4.59a)} \\[3mm]
\dfrac{\partial m}{\partial t} - 0.05\,\Delta m - 16\,\mathrm{div}\left(\dfrac{m\nabla u}{(1+m)^{\frac{3}{4}}}\right) = 0. & \text{(4.59b)}
\end{cases}
$$

The horizon T is $T = 50$ min. There is no terminal cost, i.e. $\phi \equiv 0$.

There are two exit doors, see Fig. 4.7. The part of the boundary corresponding to the doors is called Γ_D. The boundary conditions at the exit doors are chosen as follows: there is a Dirichlet condition for u on Γ_D, corresponding to an exit cost; in our simulations, we have chosen $u = 0$ on Γ_D. For m, we may assume that $m = 0$ outside the domain, so we also get the Dirichlet condition $m = 0$ on Γ_D.

The boundary Γ_N corresponds to the solid walls of the hall and of the obstacles. A natural boundary condition for u on Γ_N is a homogeneous Neumann boundary condition, i.e. $\frac{\partial u}{\partial n} = 0$ which says that the velocity of the pedestrians is tangential to the walls. The natural condition for the density m is that $v\frac{\partial m}{\partial n} + m\frac{\partial \tilde{H}}{\partial p}(\cdot, m, \nabla u) \cdot n = 0$, therefore $\frac{\partial m}{\partial n} = 0$ on Γ_N.

2. Mean field type control: this is the situation where pedestrians or robots use the same feedback law (we may imagine that they follow the strategy decided by a leader); we keep the same Hamiltonian, and the HJB equation becomes

$$\frac{\partial u}{\partial t} + 0.05\,\Delta u - \left(\frac{2}{(1+m)^{\frac{3}{4}}} + \frac{6}{(1+m)^{\frac{7}{4}}}\right)|\nabla u|^2 = -\frac{1}{3200}. \qquad (4.60)$$

while (4.59b) and the boundary conditions are unchanged.

The initial density m_0 is piecewise constant and takes two values 0 and 4 people/m², see Fig. 4.7. At $t = 0$, there are 3300 people in the hall.

We use the method described in Sect. 4.5.1, i.e., Newton iterations with the finite difference scheme originally proposed in [9], see [16] for some details on the implementation.

On Fig. 4.8, we plot the density m obtained by the simulations for the two models, at $t = 1, 2, 5$, and 15 min. With both models, we see that the pedestrians rush

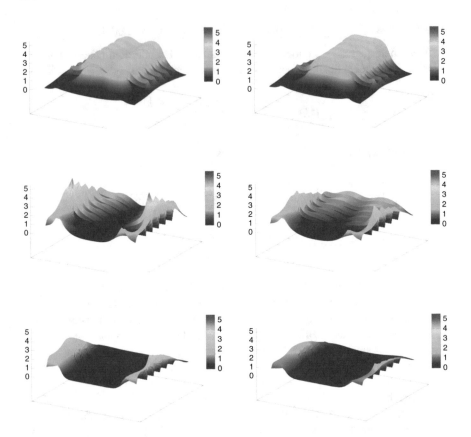

Fig. 4.8 The density computed with the two models at different dates, $t = 1$, 5 and 15 min. (from top to bottom). Left: Mean field game. Right: Mean field type control

towards the narrow corridors leading to the exits, at the left and right sides of the hall, and that the density reaches high values at the intersections of corridors; then congestion effects explain why the velocity is low (the gradient of u) in the regions where the density is high, see Fig. 4.9. We see on Fig. 4.8 that the mean field type control leads to lower peaks of density, and on Fig. 4.10 that it leads to a faster exit of the room. We can hence infer that the mean field type control performs better than the mean field game, leading to a positive *price of anarchy*.

4.8 MFGs in Macroeconomics

The material of this section is taken from a work of the first author with Jiequn Han, Jean-Michel Lasry, Pierre-Louis Lions, and Benjamin Moll, see [43], see also [44]. In economics, the ideas underlying MFGs have been investigated in the so-called

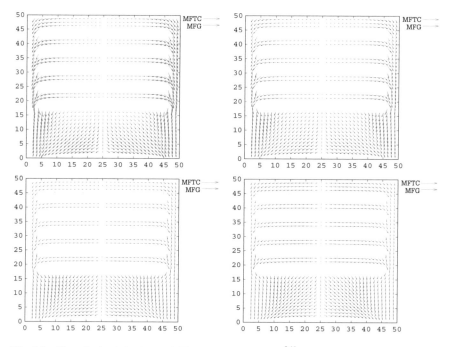

Fig. 4.9 The velocity $(v(t,x) = -16\nabla u(t,x)(1 + m(t,x))^{-3/4})$ computed with the two models at different dates, $t = 1, 2, 5$ and $t = 15$ min. (from top-left to bottom-right). Red: Mean field game. Blue: Mean field type control

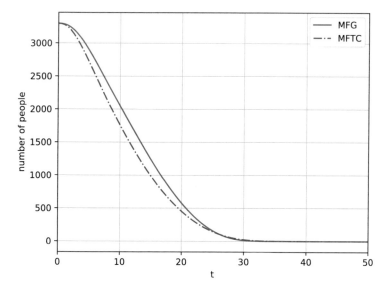

Fig. 4.10 Evolution of the total remaining number of people in the room for the mean field game (red line) and the mean field type control (dashed blue line)

heterogeneous agents models, see for example [45–48], for a long time before the mathematical formalization of Lasry and Lions. The latter models mostly involve discrete in time optimal control problems, because the economists who proposed them felt more comfortable with the related mathematics. The connection between heterogeneous agents models and MFGs is discussed in [44].

We will first describe the simple yet prototypical Huggett model, see [45, 46], in which the constraints on the state variable play a key role. We will then discuss the finite difference scheme for the related system of differential equations, putting the stress on the boundary conditions. Finally, we will present numerical simulations of the richer Aiyagari model, see [47].

Concerning HJB equations, state constraints and related numerical schemes have been much studied in the literature since the pioneering works of Soner and Capuzzo Dolcetta-Lions, see [49–51]. On the contrary, the boundary conditions for the related invariant measure had not been addressed before [43, 44]. The recent references [52, 53] contain a rigorous study of MFGs with state constraints in a different setting from the economic models considered below.

4.8.1 A Prototypical Model with Heterogeneous Agents: The Huggett Model

4.8.1.1 The Optimal Control Problem Solved by an Agent

We consider households which are heterogeneous in wealth x and income y. The dynamics of the wealth of a household is given by

$$dx_t = (rx_t - c_t + y_t)dt,$$

where r is an interest rate and c_t is the consumption (the control variable). The income y_t is a two-state Poisson process with intensities λ_1 and λ_2, i.e. $y_t \in \{y_1, y_2\}$ with $y_1 < y_2$, and

$$\mathbb{P}(y_{t+\Delta t} = y_1 | y_t = y_1) = 1 - \lambda_1 \Delta t + o(\Delta t), \quad \mathbb{P}(y_{t+\Delta t} = y_2 | y_t = y_1) = \lambda_1 \Delta t + o(\Delta t),$$

$$\mathbb{P}(y_{t+\Delta t} = y_2 | y_t = y_2) = 1 - \lambda_2 \Delta t + o(\Delta t), \quad \mathbb{P}(y_{t+\Delta t} = y_1 | y_t = y_2) = \lambda_2 \Delta t + o(\Delta t).$$

Recall that a negative wealth means debts; there is a *borrowing constraint*: the wealth of a given household cannot be less than a given *borrowing limit* $\underline{x}$. In the terminology of control theory, $x_t \geq \underline{x}$ is a constraint on the state variable.

A household solves the optimal control problem

$$\max_{\{c_t\}} \mathbb{E} \int_0^\infty e^{-\rho t} u(c_t)dt \quad \text{subject to} \quad \begin{cases} dx_t = (y_t + rx_t - c_t)dt, \\ x_t \geq \underline{x}, \end{cases}$$

where

- ρ is a positive discount factor
- u is a utility function, strictly increasing and strictly concave, e.g. the CRRA (constant relative risk aversion) utility:

$$u(c) = c^{1-\gamma}/(1-\gamma), \qquad \gamma > 0.$$

For $i = 1, 2$, let $v_i(x)$ be the value of the optimal control problem solved by an agent when her wealth is x and her income y_i. The value functions $(v_1(x), v_2(x))$ satisfy the system of differential equations:

$$- \rho v_1 + H(x, y_1, \partial_x v_1) + \lambda_1 v_2(x) - \lambda_1 v_1(x) = 0, \quad x > \underline{x}, \qquad (4.61)$$

$$- \rho v_2 + H(x, y_2, \partial_x v_2) + \lambda_2 v_1(x) - \lambda_2 v_2(x) = 0, \quad x > \underline{x}, \qquad (4.62)$$

where the Hamiltonian H is given by

$$H(x, y, p) = \max_{c \geq 0} \Big((y + rx - c)p + u(c) \Big).$$

The system of differential equations (4.61)–(4.62) must be supplemented with boundary conditions connected to the state constraints $x \geq \underline{x}$. Viscosity solutions of Hamilton–Jacobi equations with such boundary conditions have been studied in [49–51]. It is convenient to introduce the non-decreasing and non-increasing envelopes $H^\uparrow$ and $H^\downarrow$ of H:

$$H^\uparrow(x, y, p) = \max_{0 \leq c \leq y+rx} \Big((y + rx - c)p + u(c) \Big),$$

$$H^\downarrow(x, y, p) = \max_{\max(0, y+rx) \leq c} \Big((y + rx - c)p + u(c) \Big).$$

It can be seen that

$$H(x, y, p) = H^\uparrow(x, y, p) + H^\downarrow(x, y, p) - \min_{p \in \mathbb{R}} H(x, y, p).$$

The boundary conditions associated to the state constraint can be written

$$- \rho v_1(\underline{x}) + H^\uparrow(\underline{x}, y_1, \partial_x v_1) + \lambda_1 v_2(\underline{x}) - \lambda_1 v_1(\underline{x}) = 0, \qquad (4.63)$$

$$- \rho v_2(\underline{x}) + H^\uparrow(\underline{x}, y_2, \partial_x v_2) + \lambda_2 v_1(\underline{x}) - \lambda_2 v_2(\underline{x}) = 0. \qquad (4.64)$$

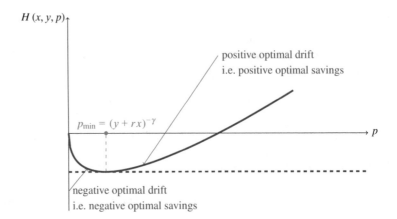

Fig. 4.11 The bold line is the graph of the function $p \mapsto H(x, y, p)$. The dashed and bold blue lines form the graph of $p \mapsto H^{\downarrow}(x, y, p)$. The dashed and bold red lines form the graph of $p \mapsto H^{\uparrow}(x, y, p)$

Consider for example the prototypical case when $u(c) = \frac{1}{1-\gamma}c^{1-\gamma}$ with $\gamma > 0$. Then

$$H(x, y, p) = \max_{0 \le c}\left((y + rx - c)p + \frac{1}{1-\gamma}c^{1-\gamma}\right) = (y + rx)p + \frac{\gamma}{1-\gamma}p^{1-\frac{1}{\gamma}}.$$

In Fig. 4.11, we plot the graphs of the functions $p \mapsto H(x, y, p)$, $p \mapsto H^{\uparrow}(x, y, p)$ and $p \mapsto H^{\downarrow}(x, y, p)$ for $\gamma = 2$.

4.8.1.2 The Ergodic Measure and the Coupling Condition

In Huggett model, the interest rate r is an unknown which is found from the fact that the considered economy neither creates nor destroys wealth: the aggregate wealth ! can be fixed to 0.

On the other hand, the latter quantity can be deduced from the ergodic measure of the process, i.e. m_1 and m_2, the wealth distributions of households with respective income y_1 and y_2, defined by

$$\lim_{t \to +\infty} \mathbb{E}(\phi(x_t)|y_t = y_i) = \langle m_i, \phi \rangle, \quad \forall \phi \in C_b([\underline{x}, +\infty)).$$

The measures m_1 and m_2 are defined on $[\underline{x}, +\infty)$ and obviously satisfy the following property:

$$\int_{x \ge \underline{x}} dm_1(x) + \int_{x \ge \underline{x}} dm_2(x) = 1. \tag{4.65}$$

Prescribing the aggregate wealth amounts to writing that

$$\int_{x\geq \underline{x}} x\,dm_1(x) + \int_{x\geq \underline{x}} x\,dm_2(x) = 0. \tag{4.66}$$

In fact (4.66) plays the role of the coupling condition in MFGs. We will discuss this point after having characterized the measures m_1 and m_2: it is well known that they satisfy the Fokker–Planck equations

$$-\partial_x \left(m_1 H_p(\cdot, y_1, \partial_x v_1)\right) + \lambda_1 m_1 - \lambda_2 m_2 = 0,$$
$$-\partial_x \left(m_2 H_p(\cdot, y_2, \partial_x v_2)\right) + \lambda_2 m_2 - \lambda_1 m_1 = 0,$$

in the sense of distributions in $(\underline{x}, +\infty)$.

Note that even if m_1 and m_2 are regular in the open interval $(\underline{x}, +\infty)$, it may happen that the optimal strategy of a representative agent (with income y_1 or y_2) consists of reaching the borrowing limit $x = \underline{x}$ in finite time and staying there; in such a case, the ergodic measure has a singular part supported on $\{x = \underline{x}\}$, and its absolutely continuous part (with respect to Lebesgue measure) may blow up near the boundary.

For this reason, we decompose m_i as the sum of a measure absolutely continuous with respect to Lebesgue measure on $(\underline{x}, +\infty)$ with density g_i and possibly of a Dirac mass at $x = \underline{x}$: for each Borel set $A \subset [\underline{x}, +\infty)$,

$$m_i(A) = \int_A g_i(x)\,dx + \mu_i \delta_{\underline{x}}(A), \quad i = 1, 2. \tag{4.67}$$

Here, g_i is a measurable and nonnegative function.

Assuming that $\partial_x v_i(x)$ have limits when $x \to \underline{x}_+$ (this can be justified rigorously for suitable choices of u), we see by definition of the ergodic measure, that for all test-functions $(\phi_1, \phi_2) \in \left(C_c^1([\underline{x}, +\infty))\right)^2$,

$$0 = \int_{x>\underline{x}} g_1(x)\left(\lambda_1\phi_2(x) - \lambda_1\phi_1(x) + H_p(x, y_1, \partial_x v_1(x))\partial_x\phi_1(x)\right)dx$$

$$+ \int_{x>\underline{x}} g_2(x)\left(\lambda_2\phi_1(x) - \lambda_2\phi_2(x) + H_p(x, y_2, \partial_x v_2(x))\partial_x\phi_2(x)\right)dx$$

$$+ \mu_1\left(\lambda_1\phi_2(\underline{x}) - \lambda_1\phi_1(\underline{x}) + H_p^{\uparrow}(\underline{x}, y_1, \partial_x v_1(\underline{x}_+))\partial_x\phi_1(\underline{x})\right)$$

$$+ \mu_2\left(\lambda_2\phi_1(\underline{x}) - \lambda_2\phi_2(\underline{x}) + H_p^{\uparrow}(\underline{x}, y_2, \partial_x v_2(\underline{x}_+))\partial_x\phi_2(\underline{x})\right).$$

It is also possible to use $\phi_1 = 1$ and $\phi_2 = 0$ as test-functions in the equation above. This yields

$$\lambda_1 \int_{x \geq \underline{x}} dm_1(x) = \lambda_2 \int_{x \geq \underline{x}} dm_2(x).$$

Hence

$$\int_{x \geq \underline{x}} dm_1(x) = \frac{\lambda_2}{\lambda_1 + \lambda_2}, \qquad \int_{x \geq \underline{x}} dm_2(x) = \frac{\lambda_1}{\lambda_1 + \lambda_2}. \tag{4.68}$$

4.8.1.3 Summary

To summarize, finding an equilibrium in the Huggett model given that the aggregate wealth is 0 consists in looking for (r, v_1, v_2, m_1, m_2) such that

- the functions v_1 and v_2 are viscosity solutions to (4.61), (4.62), (4.63), and (4.64)
- the measures m_1 and m_2 satisfy (4.68), $m_i = g_i L + \mu_i \delta_{x=\underline{x}}$, where L is the Lebesgue measure on $(\underline{x}, +\infty)$, g_i is a measurable nonnegative function $(\underline{x}, +\infty)$ and for all $i = 1, 2$, for all test function ϕ:

$$\left. \begin{array}{c} \displaystyle\int_{x > \underline{x}} \left(\lambda_j g_j(x) - \lambda_i g_i(x) \right) \phi(x) dx + \int_{x > \underline{x}} H_p(x, y_i, \partial_x v_i(x)) \partial_x \phi(x) dx \\[2mm] + \mu_i \left(-\lambda_i \phi(\underline{x}) + H_p^\uparrow(\underline{x}, y_i, \partial_x v_i(\underline{x}_+)) \partial_x \phi(\underline{x}) \right) + \mu_j \lambda_j \phi(\underline{x}) \end{array} \right\} = 0, \tag{4.69}$$

  where $j = 2$ if $i = 1$ and $j = 1$ if $i = 2$.
- the aggregate wealth is fixed, i.e. (4.66). By contrast with the previously considered MFG models, the coupling in the Bellman equation does not come from a coupling cost, but from the implicit relation (4.66) which can be seen as an equation for r. Since the latter coupling is only through the constant value r, it can be said to be weak.

4.8.1.4 Theoretical Results

The following theorem, proved in [43], gives quite accurate information on the equilibria for which $r < \rho$:

Theorem 5 Let $c_i^*(x)$ be the optimal consumption of an agent with income y_i and wealth x.

1. *If $r < \rho$, then the optimal drift (whose economical interpretation is the optimal saving policy) corresponding to income y_1, namely $rx + y_1 - c_1^*(x)$, is negative for all $x > \underline{x}$.*
2. *If $r < \rho$ and $-\frac{u''(y_1 + rx)}{u'(y_1 + rx)} < +\infty$, then there exists a positive number v_1 (that can be written explicitly) such that for x close to $\underline{x}$,*

$$rx + y_1 - c_1^*(x) \sim -\sqrt{2v_1}\,\sqrt{x - \underline{x}} < 0,$$

so the agents for which the income stays at the lower value y_1 hit the borrowing limit in finite time, and

$$\mu_1 > 0, \qquad g_1(x) \sim \frac{\gamma_1}{\sqrt{x - \underline{x}}},$$

for some positive value γ_1 (that can be written explicitly).
3. *If $r < \rho$ and $\sup_c \left(-\frac{cu''(c)}{u'(c)}\right) < +\infty$, then there exists $\underline{x} \le \overline{x} < +\infty$ such that*

$$rx + y_2 - c_2^*(x) < 0, \qquad \text{for all } x > \overline{x},$$
$$rx + y_2 - c_2^*(x) > 0, \qquad \text{for all } \underline{x} < x < \overline{x}.$$

Moreover $\mu_2 = 0$ and if $\overline{x} > \underline{x}$, then for some positive constant ζ_2, $rx + y_2 - c_2^*(x) \sim \zeta_2(\overline{x} - x)$ for x close to $\overline{x}$.

From Theorem 5, we see in particular that agents whose income remains at the lower value are drifted toward the borrowing limit and get stuck there, while agents whose income remains at the higher value get richer if their wealth is small enough. In other words, the only way for an agent in the low income state to avoid being drifted to the borrowing limit is to draw the high income state. Reference [43] also contains existence and uniqueness results.

4.8.2 A Finite Difference Method for the Huggett Model

4.8.2.1 The numerical scheme

We wish to simulate the Huggett model in the interval $[\underline{x}, \tilde{x}]$. Consider a uniform grid on $[\underline{x}, \tilde{x}]$ with step $h = \frac{\tilde{x} - \underline{x}}{N_h}$: we set $x_i = \underline{x} + ih$, $i = 0, \ldots, N_h$. The discrete approximation of $v_j(x_i)$ is named $v_{i,j}$. The discrete version of the Hamiltonian H involves the function $\mathcal{H} : \mathbb{R}^4 \to \mathbb{R}$,

$$\mathcal{H}(x, y, \xi_r, \xi_\ell) = H^\uparrow(x, y, \xi_r) + H^\downarrow(x, y, \xi_\ell) - \min_{p \in \mathbb{R}} H(x, y, p).$$

We do not discuss the boundary conditions at $x = \tilde{x}$ in order to focus on the difficulties coming from the state constraint $x_t \ge \underline{x}$. In fact, if $\tilde{x}$ is large enough, state constraint boundary conditions can also be chosen at $x = \tilde{x}$. The numerical monotone scheme for (4.61), (4.62), (4.63), and (4.64) is

$$- \rho v_{i,j} + \mathcal{H}\left(x_i, y_j, \frac{v_{i+1,j} - v_{i,j}}{h}, \frac{v_{i,j} - v_{i-1,j}}{h}\right) + \lambda_i \left(v_{i,k} - v_{i,j}\right) = 0, \qquad \text{for } 0 < i < N_h, \quad (4.70)$$

$$-\rho v_{0,j} + H^\uparrow\left(\underline{x}, y_j, \frac{v_{1,j} - v_{0,j}}{h}\right) + \lambda_i \left(v_{0,k} - v_{0,j}\right) = 0, \qquad \text{for } i = 0, \quad (4.71)$$

where $k = 2$ if $j = 1$ and $k = 1$ if $j = 2$. In order to find the discrete version of the Fokker–Planck equation, we start by writing (4.70)–(4.71) in the following compact form:

$$- \rho V + \mathcal{F}(V) = 0, \qquad (4.72)$$

with self-explanatory notations. The differential of $\mathcal{F}$ at V maps W to the grid functions whose (i, j)-th components is:

- if $0 < i < N_h$:

$$H_p^\uparrow\left(x_i, y_j, \frac{v_{i+1,j} - v_{i,j}}{h}\right) \frac{w_{i+1,j} - w_{i,j}}{h} + H_p^\downarrow\left(x_i, y_j, \frac{v_{i,j} - v_{i-1,j}}{h}\right) \frac{w_{i,j} - w_{i-1,j}}{h} + \lambda_i \left(w_{i,k} - w_{i,j}\right). \qquad (4.73)$$

- If $i = 0$:

$$H_p^\uparrow\left(\underline{x}, y_j, \frac{v_{1,j} - v_{0,j}}{h}\right) \frac{w_{1,j} - w_{0,j}}{h} + \lambda_i \left(w_{0,k} - w_{0,j}\right). \qquad (4.74)$$

We get the discrete Fokker–Planck equation by requiring that $\langle D\mathcal{F}(V)W, M \rangle = 0$ for all W, where $M = (m_{i,j})_{0 \le i \le N_h, j=1,2}$; more explicitly, we obtain:

$$0 = \begin{cases} \lambda_k m_{i,k} - \lambda_j m_{i,j} \\ + \dfrac{1}{h}\left(m_{i,j} H_p^\downarrow\left(x_i, y_j, \dfrac{v_{i,j} - v_{i-1,j}}{h}\right) - m_{i+1,j} H_p^\downarrow\left(x_{i+1}, y_j, \dfrac{v_{i+1,j} - v_{i,j}}{h}\right)\right) \\ - \dfrac{1}{h}\left(m_{i,j} H_p^\uparrow\left(x_i, y_j, \dfrac{v_{i+1,j} - v_{i,j}}{h}\right) - m_{i-1,j} H_p^\uparrow\left(x_{i-1}, y_j, \dfrac{v_{i,j} - v_{i-1,j}}{h}\right)\right), \end{cases} \qquad (4.75)$$

for $0 < i < N_h$, and

$$0 = \lambda_k m_{0,k} - \lambda_j m_{0,j} - \frac{1}{h}\left(m_{0,j} H_p^\uparrow\left(\underline{x}, y_j, \frac{v_{1,j} - v_{0,j}}{h}\right) + m_{1,j} H_p^\downarrow\left(x_1, y_j, \frac{v_{1,j} - v_{0,j}}{h}\right)\right). \qquad (4.76)$$

Assuming for simplicity that in (4.67), $(g_j)_{j=1,2}$ are absolutely continuous with respect to the Lebesgue measure, equations (4.75) and (4.76) provide a consistent discrete scheme for (4.69) if $h m_{0,j}$ is seen as the discrete version of μ_j and if $m_{i,j}$ is the discrete version of $g_j(x_i)$. If we do not suppose that g_j is absolutely continuous with respect to the Lebesgue measure, then $m_{i,j}$ may be seen as a discrete version of $\frac{1}{h} \int_{x_i-h/2}^{x_i+h/2} dg_j(x)$.

The consistency of (4.75) for $i > 0$ is obtained as usual. Let us focus on (4.76): assume that the following convergence holds:

$$\lim_{h \to 0} \max_{i,j} |v_{i,j} - v(x_i, y_j)| = 0 \quad \text{and} \quad \lim_{h \to 0} \max_j \left| \frac{v_{1,j} - v_{0,j}}{h} - \partial_x v_j(\underline{x}) \right| = 0.$$

Assume that $H_p(\underline{x}, \partial_x v_j(\underline{x})) > \varepsilon$ for a fixed positive number $\varepsilon > 0$. Then for h small enough,

$$H_p^\uparrow \left(x_i, y_j, \frac{v_{1,j} - v_{0,j}}{h} \right) > 0 \quad \text{and} \quad H_p^\downarrow \left(x_i, y_j, \frac{v_{1,j} - v_{0,j}}{h} \right) = 0,$$

for $i = 0, 1$. Plugging this information in (4.76) yields that $m_{0,j}$ is of the same order as $h m_{0,k}$, in agreement with the fact that since the optimal drift is positive, there is no Dirac mass at $x = \underline{x}$. In the opposite case, we see that (4.76) is consistent with the asymptotic expansions found in Theorem 5.

4.8.2.2 Numerical Simulations

On Fig. 4.12, we plot the densities g_1 and g_2 and the cumulative distributions $\int_{\underline{x}}^x dm_1$ and $\int_{\underline{x}}^x dm_2$, computed by two methods: the first one is the finite difference descrived above. The second one consists of coupling the finite difference scheme described above for the HJB equation and a closed formula for the Fokker–Planck equation. Two different grids are used with $N_h = 500$ and $N_h = 30$. We see that for $N_h = 500$, it is not possible to distinguish the graphs obtained by the two methods. The density g_1 blows up at $x = \underline{x}$ while g_2 remains bounded. From the graphs of the cumulative distributions, we also see that $\mu_1 > 0$ while $\mu_2 = 0$.

4.8.3 The Model of Aiyagari

There is a continuum of agents which are heterogeneous in wealth x and productivity y. The dynamics of the wealth of a household is given by

$$dx_t = (w y_t + r x_t - c_t) dt,$$

where r is the interest rate, w is the level of wages, and c_t is the consumption (the control variable). The dynamics of the productivity y_t is given by the stochastic

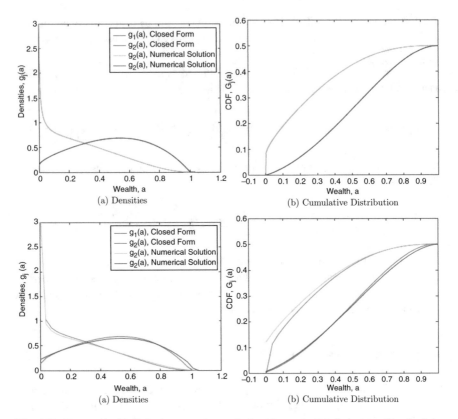

Fig. 4.12 Huggett model: Left: the measures g_i obtained by an explicit formula and by the finite difference method for two grid resolutions with 500 nodes (Top) and 30 nodes (Bottom). Right: the cumulative distributions $\int_{\underline{x}}^{x} dm_i$. Top: (**a**) Densities. (**b**) Cumulative Distribution. Bottom: (**a**) Densities. (**b**) Cumulative Distribution

differential equation in $\mathbb{R}_+$:

$$dy_t = \mu(y_t)dt + \sigma(y_t)dW_t,$$

where W_t is a standard one dimensional Brownian motion. Note that this models situations in which the noises in the productivity of the agents are all independent (idiosyncratic noise). Common noise will be briefly discussed below.

As for the Huggett model, there is a borrowing constraint $x_t \geq \underline{x}$, a household tries to maximize the utility $\mathbb{E}\int_0^\infty e^{-\rho t}u(c_t)dt$. To determine the interest rate r and the level of wages w, Aiyagari considers that the production of the economy is given by the following Cobb-Douglas law:

$$F(X, Y) = AX^\alpha Y^{1-\alpha},$$

for some $\alpha \in (0, 1)$, where, if $m(\cdot, \cdot)$ is the ergodic measure,

- A is a productivity factor
- $X = \int_{x \geq \underline{x}} \int_{y \in \mathbb{R}_+} x \, dm(x, y)$ is the aggregate capital
- $Y = \int_{x \geq \underline{x}} \int_{y \in \mathbb{R}_+} y \, dm(x, y)$ is the aggregate labor.

The level of wages w and interest rate r are obtained by the equilibrium relation

$$(X, Y) = \mathrm{argmax}\Big(F(X, Y) - (r + \delta)X - wY \Big),$$

where δ is the rate of depreciation of the capital. This implies that

$$r = \partial_X F(X, Y) - \delta, \qquad w = \partial_Y F(X, Y).$$

Remark 12 Tackling a common noise is possible in the non-stationary version of the model, by letting the productivity factor become a random process A_t (for example A_t may be a piecewise constant process with random shocks): this leads to a more complex setting which is the continuous-time version of the famous Krusell-Smith model, see [48]. Numerical simulations of Krusell-Smith models have been carried out by the authors of [43], but, in order to keep the survey reasonably short, we will not discuss this topic.

The Hamiltonian of the problem is $H(x, y, p) = \max_c \Big(u(c) + p(wy + rx - c) \Big)$ and the mean field equilibrium is found by solving the system of partial differential equations:

$$0 = \frac{\sigma^2(y)}{2} \partial_{yy} v + \mu(y) \partial_y v + H(x, y, \partial_x v) - \rho v, \qquad (4.77)$$

$$0 = -\frac{1}{2} \partial_{yy} \Big(\sigma^2(y) m \Big) + \partial_y \Big(\mu(y) m \Big) + \partial_x \Big(m H_p(x, y, \partial_x v) \Big), \qquad (4.78)$$

in $(\underline{x}, +\infty) \times \mathbb{R}_+$ with suitable boundary conditions on the line $\{x = \underline{x}\}$ linked to the state constraints, $\int_{x \geq \underline{x}} \int_{y \in \mathbb{R}_+} dm(x, y) = 1$, and the equilibrium condition

$$r = \partial_X F(X, Y) - \delta, \qquad w = \partial_Y F(X, Y),$$
$$X = \int_{x \geq \underline{x}} \int_{y \in \mathbb{R}_+} x \, dm(x, y) \quad Y = \int_{x \geq \underline{x}} \int_{y \in \mathbb{R}_+} y \, dm(x, y). \qquad (4.79)$$

The boundary condition for the value function can be written

$$0 = \frac{\sigma^2(y)}{2} \partial_{yy} v + \mu(y) \partial_y v + H^\uparrow(\underline{x}, y, \partial_x v) - \rho v, \qquad (4.80)$$

where $p \mapsto H^\uparrow(x, y, p)$ is the non-decreasing envelope of $p \mapsto H(x, y, p)$.

We are going to look for m as the sum of a measure which is absolutely continuous with respect to the two dimensional Lebesgue measure on $(\underline{x}, +\infty) \times \mathbb{R}_+$ with density g and of a measure η supported in the line $\{x = \underline{x}\}$:

$$dm(x, y) = g(x, y)dxdy + d\eta(y), \qquad (4.81)$$

and for all test function ϕ:

$$\left. \begin{aligned} \int_{x>\underline{x}} \int_{y\in\mathbb{R}_+} g(x, y)\left(\frac{\sigma^2(y)}{2}\partial_{yy}\phi(x, y) + H_p(x, y, \partial_x v(x, y))\partial_x\phi(x, y) + \mu(y)\partial_y\phi(x, y)\right) dxdy \\ + \int_{y\in\mathbb{R}_+} \left(\frac{\sigma^2(y)}{2}\partial_{yy}\phi(\underline{x}, y) + H_p^\uparrow(\underline{x}, y, \partial_x v(\underline{x}_+, y))\partial_x\phi(\underline{x}, y) + \mu(y)\partial_y\phi(\underline{x}, y)\right) d\eta(y) \end{aligned} \right\} = 0.$$

$$(4.82)$$

Note that it is not possible to find a partial differential equation for η on the line $x = \underline{x}$, nor a local boundary condition for g, because it is not possible to express the term $\int_{y\in\mathbb{R}_+} H_p^\uparrow(\underline{x}, y, \partial_x v(\underline{x}_+, y))\partial_x\phi(\underline{x}, y)d\eta(\underline{x}, y)$ as a distribution acting on $\phi(\underline{x}, \cdot)$.

The finite difference scheme for (4.77), (4.80), (4.81), (4.82), and (4.79) is found exactly in the same spirit as for Huggett model and we omit the details. In Fig. 4.13, we display the optimal saving policy $(x, y) \mapsto wy + rx - c^*(x, y)$ and the ergodic measure obtained by the above-mentioned finite difference scheme for Aiyagari model with $u(c) = -c^{-1}$. We see that the ergodic measure m (right part of Fig. 4.13) has a singularity on the line $x = \underline{x}$ for small values of the productivity y, and for the same values of y, the density of the absolutely continuous part of m with respect to the two-dimensional Lebesgue measure blows up when $x \to \underline{x}$. The economic interpretation is that the agents with low productivity are drifted to the borrowing limit. The singular part of the ergodic measure is of the same nature as the Dirac mass that was obtained for $y = y_1$ in the Huggett model. It corresponds to the zone where the optimal drift is negative near the borrowing limit (see the left part of Fig. 4.13).

4.9   Conclusion

In this survey, we have put stress on finite difference schemes for the systems of forward-backward PDEs that may stem from the theory of mean field games, and have discussed in particular some variational aspects and numerical algorithms that may be used for solving the related systems of nonlinear equations. We have also addressed in details two applications of MFGs to crowd motion and macroeconomics, and a comparison of MFGs with mean field control: we hope that these examples show well on the one hand, that the theory of MFGs is quite relevant for modeling the behavior of a large number of rational agents, and on the

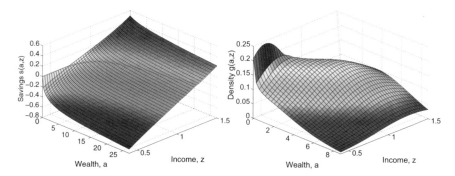

Fig. 4.13 Numerical simulations of Aiyagari model with $u(c) = -c^{-1}$. Left: the optimal drift (optimal saving policy) $wy + rx - c^*(x, y)$. Right: the part of the ergodic measure which is absolutely continuous with respect to Lebesgue measure

other hand, that several difficulties must be addressed in order to tackle realistic problems.

To keep the survey short, we have not addressed the following interesting aspects:

- Semi-Lagrangian schemes for the system of forward-backward PDEs. While semi-Lagrangian schemes for optimal control problems have been extensively studied, much less has been done regarding the Fokker–Planck equation. In the context of MFGs, semi-Lagrangian schemes have been investigated by F. Camilli and F. Silva, E. Carlini and F. Silva, see the references [54–57]. Arguably, the advantage of such methods is their direct connection to the underlying optimal control problems, and a possible drawback may be the difficulty to address realistic boundary conditions.
- An efficient algorithm for ergodic MFGs has been proposed by in [58, 59]. The approach relies on a finite difference scheme and a least squares formulation, which is then solved using Gauss-Newton iterations.
- D. Gomes and his coauthors have proposed gradient flow methods for solving deterministic mean field games in infinite horizon, see [60, 61], under a monotonicity assumption. Their main idea is that the solution to the system of PDEs can be recast as a zero of a monotone operator, an can thus be found by following the related gradient flow. They focus on the following example:

$$
\begin{cases}
\lambda + H_0(\cdot, \nabla u) - \log(m) = 0, & \text{in } \mathbb{T}, & (4.83a) \\
- \operatorname{div}\left(m \partial_p H_0(\cdot, m, \nabla u)\right) = 0, & \text{in } \mathbb{T}, & (4.83b) \\
\int_{\mathbb{T}} m = 1, \quad \int_{\mathbb{T}} u = 0, & m > 0 \text{ in } \mathbb{T}, & (4.83c)
\end{cases}
$$

where the ergodic constant λ is an unknown. They consider the following monotone map :

$$A \begin{pmatrix} u \\ m \end{pmatrix} = \begin{pmatrix} -\text{div} \left(m(\cdot) \partial_p H_0(\cdot, m(\cdot), \nabla u(\cdot)) \right) \\ -H_0(\cdot, \nabla u) + \log(m) \end{pmatrix}.$$

Thanks to the special choice of the coupling cost $\log(m)$, a gradient flow method applied to A, i.e.

$$\frac{d}{d\tau} \begin{pmatrix} u_\tau \\ m_\tau \end{pmatrix} = -A \begin{pmatrix} u_\tau \\ m_\tau \end{pmatrix} - \begin{pmatrix} 0 \\ \lambda_\tau \end{pmatrix},$$

preserves the positivity of m (this may not be true with other coupling costs, in which case additional projections may be needed). The real number λ_τ is used to enforce the constraint $\int_{\mathbb{T}} m_\tau = 1$.

This idea has been studied on the aforementioned example and some variants but needs to be tested in the stochastic case (i.e., second order MFGs) and with general boundary conditions. The generalization to finite horizon is not obvious.

- Mean field games related to impulse control and optimal exit time have been studied by C. Bertucci, both from a theoretical and a numerical viewpoint, see [62]. In particular for MFGs related to impulse control problems, there remains a lot of difficult open issues.

- High dimensional problems. Finite difference schemes can only be used if the dimension d of the state space is not too large, say $d \leq 4$. Very recently, there have been attempts to use machine learning methods in order to solve problems in higher dimension or with common noise, see e.g. [63–66]. The main difference with the methods discussed in the present survey is that these methods do not rely on a finite-difference scheme but instead use neural networks to approximate functions with a relatively small number of parameters. Further studies remain necessary before it is possible to really evaluate these methods.

- Numerical methods for the master equation when the state space is finite, see works in progress by the first author and co-workers.

To conclude, acknowledging the fact that the theory of mean field games have attracted a lot of interest in the past decade, the authors think that some of the most interesting open problems arise in the actual applications of this theory. Amongst the most fascinating aspects of mean field games are their interactions with social sciences and economics. A few examples of such interactions have been discussed in the present survey, and many more applications remain to be investigated.

Acknowledgments The research of the first author was partially supported by the ANR (Agence Nationale de la Recherche) through MFG project ANR-16-CE40-0015-01.

References

1. J.M. Lasry, P.-L. Lions, Jeux à champ moyen. I. Le cas stationnaire. C. R. Math. Acad. Sci. Paris **343**(9), 619–625 (2006)
2. J.-M. Lasry, P.-L. Lions, Jeux à champ moyen. II. Horizon fini et contrôle optimal. C. R. Math. Acad. Sci. Paris **343**(10), 679–684 (2006)
3. J.-M. Lasry, P.-L. Lions, Mean field games. Jpn. J. Math. **2**(1), 229–260 (2007)
4. P. Cardaliaguet, *Notes on Mean Field Games* (2011), preprint
5. P. Cardaliaguet, P. Jameson Graber, A. Porretta, D. Tonon, Second order mean field games with degenerate diffusion and local coupling. NoDEA Nonlinear Differential Equations Appl. **22**(5), 1287–1317 (2015)
6. A. Porretta, Weak Solutions to Fokker–Planck Equations and Mean Field Games. Arch. Ration. Mech. Anal. **216**(1), 1–62 (2015)
7. P.-L. Lions, Cours du Collège de France (2007–2011). https://www.college-de-france.fr/site/en-pierre-louis-lions/_course.htm
8. P. Cardaliaguet, F. Delarue, J.-M. Lasry, P.-L. Lions, *The Master Equation and the Convergence Problem in Mean Field Games*. Annals of Mathematics Studies, vol. 201 (Princeton University, Princeton, 2019)
9. Y. Achdou, I. Capuzzo-Dolcetta, Mean field games: numerical methods. SIAM J. Numer. Anal. **48**(3), 1136–1162 (2010)
10. Y. Achdou, J.-M. Lasry, Mean field games for modeling crowd motion, in *Contributions to Partial Differential Equations and Applications*, chapter 4 (Springer, Berlin, 2019), pp. 17–42
11. Y. Achdou, F. Camilli, I. Capuzzo-Dolcetta, Mean field games: numerical methods for the planning problem. SIAM J. Control Optim. **50**(1), 77–109 (2012)
12. Y. Achdou, F. Camilli, I. Capuzzo-Dolcetta, Mean field games: convergence of a finite difference method. SIAM J. Numer. Anal. **51**(5), 2585–2612 (2013)
13. Y. Achdou, A. Porretta, Convergence of a finite difference scheme to weak solutions of the system of partial differential equations arising in mean field games. SIAM J. Numer. Anal. **54**(1), 161–186 (2016)
14. P. Cardaliaguet, P. Jameson Graber, Mean field games systems of first order. ESAIM Control Optim. Calc. Var. **21**(3), 690–722 (2015)
15. L.M. Briceño Arias, D. Kalise, Z. Kobeissi, M. Laurière, Á.M. González, F.J. Silva, On the implementation of a primal-dual algorithm for second order time-dependent mean field games with local couplings. ESAIM: ProcS **65**, 330–348 (2019)
16. Y. Achdou, Finite difference methods for mean field games, in *Hamilton-Jacobi Equations: Approximations, Numerical Analysis and Applications*. Lecture Notes in Mathematical, vol. 2074 (Springer, Heidelberg, 2013), pp. 1–47
17. L.M. Briceño Arias, D. Kalise, F.J. Silva, Proximal methods for stationary mean field games with local couplings. SIAM J. Control Optim. **56**(2), 801–836 (2018)
18. R.T. Rockafellar, *Convex Analysis*. Princeton Landmarks in Mathematics (Princeton University, Princeton, 1997). Reprint of the 1970 original, Princeton Paperbacks
19. M. Fortin, R. Glowinski, *Augmented Lagrangian methods*. Studies in Mathematics and its Applications, vol. 15 (North-Holland, Amsterdam, 1983). Applications to the numerical solution of boundary value problems, Translated from the French by B. Hunt and D.C. Spicer
20. J.-D. Benamou, Y. Brenier, A computational fluid mechanics solution to the Monge-Kantorovich mass transfer problem. Numer. Math. **84**(3), 375–393 (2000)
21. J.-D. Benamou, G. Carlier, Augmented Lagrangian methods for transport optimization, mean field games and degenerate elliptic equations. J. Optim. Theory Appl. **167**(1), 1–26 (2015)
22. R. Andreev, Preconditioning the augmented Lagrangian method for instationary mean field games with diffusion. SIAM J. Sci. Comput. **39**(6), A2763–A2783 (2017)
23. Y. Achdou, M. Laurière, Mean field type control with congestion (II): an augmented Lagrangian method. Appl. Math. Optim. **74**(3), 535–578 (2016)

24. J. Eckstein, D.P. Bertsekas, On the Douglas-Rachford splitting method and the proximal point algorithm for maximal monotone operators. Math. Program. **55**(3, Ser. A), 293–318 (1992)
25. A. Chambolle, T. Pock, A first-order primal-dual algorithm for convex problems with applications to imaging. J. Math. Imaging Vision **40**(1), 120–145 (2011)
26. U. Trottenberg, C.W. Oosterlee, A. Schüller, *Multigrid* (Academic Press, San Diego, 2001). With contributions by A. Brandt, P. Oswald, K. Stüben
27. Y. Achdou, V. Perez, Iterative strategies for solving linearized discrete mean field games systems. Netw. Heterog. Media **7**(2), 197–217 (2012)
28. H.A. van der Vorst, Bi-CGSTAB: a fast and smoothly converging variant of Bi-CG for the solution of nonsymmetric linear systems. SIAM J. Sci. Statist. Comput. **13**(2), 631–644 (1992)
29. A. Porretta, E. Zuazua, Long time versus steady state optimal control. SIAM J. Control Optim. **51**(6), 4242–4273 (2013)
30. P. Cardaliaguet, J.-M. Lasry, P.-L. Lions, A. Porretta, Long time average of mean field games. Netw. Heterog. Media **7**(2), 279–301 (2012)
31. P. Cardaliaguet, J.-M. Lasry, P.-L. Lions, A. Porretta, Long time average of mean field games with a nonlocal coupling. SIAM J. Control Optim. **51**(5), 3558–3591 (2013)
32. J.-F. Chassagneux, D. Crisan, F. Delarue, Numerical method for FBSDEs of McKean-Vlasov type. Ann. Appl. Probab. **29**(3), 1640–1684 (2019)
33. A. Angiuli, C.V. Graves, H. Li, J.-F. Chassagneux, F. Delarue, R. Carmona, Cemracs 2017: numerical probabilistic approach to MFG. ESAIM: ProcS **65**, 84–113 (2019)
34. H.P. McKean, Jr. A class of Markov processes associated with nonlinear parabolic equations. Proc. Nat. Acad. Sci. USA **56**, 1907–1911 (1966)
35. A.-S. Sznitman, Topics in propagation of chaos, in *École d'Été de Probabilités de Saint-Flour XIX—1989*. Lecture Notes in Mathematical, vol. 1464 (Springer, Berlin, 1991), pp. 165–251
36. R. Carmona, F. Delarue, A. Lachapelle, Control of McKean-Vlasov dynamics versus mean field games. Math. Financ. Econ. **7**(2), 131–166 (2013)
37. R. Carmona, F. Delarue, Mean field forward-backward stochastic differential equations. Electron. Commun. Probab. **18**(68), 15 (2013)
38. R. Carmona, F. Delarue, *Probabilistic Theory of Mean Field Games with Applications. I.* Probability Theory and Stochastic Modelling, vol. 83 (Springer, Cham, 2018). Mean field FBSDEs, control, and games
39. A. Bensoussan, J. Frehse, Control and Nash games with mean field effect. Chin. Ann. Math. Ser. B **34**(2), 161–192 (2013)
40. A. Bensoussan, J. Frehse, S.C.P. Yam, *Mean Field Games and Mean Field Type Control Theory*. Springer Briefs in Mathematics (Springer, New York, 2013)
41. Y. Achdou, M. Laurière, Mean Field Type Control with Congestion. Appl. Math. Optim. **73**(3), 393–418 (2016)
42. Y. Achdou, M. Laurière, On the system of partial differential equations arising in mean field type control. Discrete Contin. Dyn. Syst. **35**(9), 3879–3900 (2015)
43. Y. Achdou, J. Han, J.-M. Lasry, P.-L. Lions, B. Moll, *Income and Wealth Distribution in Macroeconomics: A Continuous-time Approach*. Technical report, National Bureau of Economic Research (2017)
44. Y. Achdou, F.J. Buera, J.-M. Lasry, P.-L. Lions, B. Moll, Partial differential equation models in macroeconomics. Philos. Trans. R. Soc. Lond. Ser. A Math. Phys. Eng. Sci. **372**(2028), 20130397, 19 (2014)
45. T. Bewley, Stationary monetary equilibrium with a continuum of independently fluctuating consumers, in *Contributions to Mathematical Economics in Honor of Gerard Debreu*, ed. by W. Hildenbrand, A. Mas-Collel (North-Holland, Amsterdam, 1986)
46. M. Huggett, The risk-free rate in heterogeneous-agent incomplete-insurance economies. J. Econ. Dyn. Control. **17**(5–6), 953–969 (1993)
47. S.R. Aiyagari, Uninsured idiosyncratic risk and aggregate saving. Q. J. Econ. **109**(3), 659–84 (1994)
48. P. Krusell, A.A. Smith, Income and wealth heterogeneity in the macroeconomy. J. Polit. Econ. **106**(5), 867–896 (1998)

49. H.M. Soner, Optimal control with state-space constraint. I. SIAM J. Control Optim. **24**(3), 552–561 (1986)
50. H.M. Soner, Optimal control with state-space constraint. II. SIAM J. Control Optim. **24**(6), 1110–1122 (1986)
51. I. Capuzzo-Dolcetta, P.-L. Lions, Hamilton-Jacobi equations with state constraints. Trans. Am. Math. Soc. **318**(2), 643–683 (1990)
52. P. Cannarsa, R. Capuani, *Existence and Uniqueness for Mean Field Games with State Constraints* (2017)
53. P. Cannarsa, R. Capuani, P. Cardaliaguet, *Mean Field Games with State Constraints: From Mild to Pointwise Solutions of the PDE System* (2018)
54. F. Camilli, F. Silva, A semi-discrete approximation for a first order mean field game problem. Netw. Heterog. Media **7**(2), 263–277 (2012)
55. E. Carlini, F.J. Silva, A fully discrete semi-Lagrangian scheme for a first order mean field game problem. SIAM J. Numer. Anal. **52**(1), 45–67 (2014)
56. E. Carlini, F.J. Silva, A semi-Lagrangian scheme for a degenerate second order mean field game system. Discrete Contin. Dyn. Syst. **35**(9), 4269–4292 (2015)
57. E. Carlini, F.J. Silva, On the discretization of some nonlinear Fokker-Planck-Kolmogorov equations and applications. SIAM J. Numer. Anal. **56**(4), 2148–2177 (2018)
58. S. Cacace, F. Camilli, C. Marchi, A numerical method for mean field games on networks. ESAIM Math. Model. Numer. Anal. **51**(1), 63–88 (2017)
59. S. Cacace, F. Camilli, A. Cesaroni, C. Marchi, An ergodic problem for mean field games: qualitative properties and numerical simulations. Minimax Theory Appl. **3**(2), 211–226 (2018)
60. N. Almulla, R. Ferreira, D. Gomes, Two numerical approaches to stationary mean-field games. Dyn. Games Appl. **7**(4), 657–682 (2017)
61. D.A. Gomes, J. Saúde, Numerical methods for finite-state mean-field games satisfying a monotonicity condition. Appl. Math. Optim. **1**, 1–32 (2018)
62. C. Bertucci, A remark on uzawa's algorithm and an application to mean field games systems (2018). arXiv preprint: 1810.01181
63. R. Carmona, M. Laurière, Convergence analysis of machine learning algorithms for the numerical solution of mean field control and games: I–the ergodic case (2019). arXiv preprint:1907.05980
64. R. Carmona, M. Laurière, Convergence analysis of machine learning algorithms for the numerical solution of mean field control and games: Ii–the finite horizon case (2019). arXiv preprint:1908.01613
65. R. Carmona, M. Laurière, Z. Tan, Model-free mean-field reinforcement learning: mean-field MDP and mean-field Q-learning (2019). arXiv preprint:1910.12802
66. L. Ruthotto, S. Osher, W. Li, L. Nurbekyan, S.W. Fung, A machine learning framework for solving high-dimensional mean field game and mean field control problems (2019). arXiv preprint:1912.01825

LECTURE NOTES IN MATHEMATICS   🐎 Springer

Editors in Chief: J.-M. Morel, B. Teissier;

Editorial Policy

1. Lecture Notes aim to report new developments in all areas of mathematics and their applications – quickly, informally and at a high level. Mathematical texts analysing new developments in modelling and numerical simulation are welcome.

   Manuscripts should be reasonably self-contained and rounded off. Thus they may, and often will, present not only results of the author but also related work by other people. They may be based on specialised lecture courses. Furthermore, the manuscripts should provide sufficient motivation, examples and applications. This clearly distinguishes Lecture Notes from journal articles or technical reports which normally are very concise. Articles intended for a journal but too long to be accepted by most journals, usually do not have this "lecture notes" character. For similar reasons it is unusual for doctoral theses to be accepted for the Lecture Notes series, though habilitation theses may be appropriate.

2. Besides monographs, multi-author manuscripts resulting from SUMMER SCHOOLS or similar INTENSIVE COURSES are welcome, provided their objective was held to present an active mathematical topic to an audience at the beginning or intermediate graduate level (a list of participants should be provided).

   The resulting manuscript should not be just a collection of course notes, but should require advance planning and coordination among the main lecturers. The subject matter should dictate the structure of the book. This structure should be motivated and explained in a scientific introduction, and the notation, references, index and formulation of results should be, if possible, unified by the editors. Each contribution should have an abstract and an introduction referring to the other contributions. In other words, more preparatory work must go into a multi-authored volume than simply assembling a disparate collection of papers, communicated at the event.

3. Manuscripts should be submitted either online at www.editorialmanager.com/lnm to Springer's mathematics editorial in Heidelberg, or electronically to one of the series editors. Authors should be aware that incomplete or insufficiently close-to-final manuscripts almost always result in longer refereeing times and nevertheless unclear referees' recommendations, making further refereeing of a final draft necessary. The strict minimum amount of material that will be considered should include a detailed outline describing the planned contents of each chapter, a bibliography and several sample chapters. Parallel submission of a manuscript to another publisher while under consideration for LNM is not acceptable and can lead to rejection.

4. In general, **monographs** will be sent out to at least 2 external referees for evaluation.

   A final decision to publish can be made only on the basis of the complete manuscript, however a refereeing process leading to a preliminary decision can be based on a pre-final or incomplete manuscript.

   Volume Editors of **multi-author works** are expected to arrange for the refereeing, to the usual scientific standards, of the individual contributions. If the resulting reports can be forwarded to the LNM Editorial Board, this is very helpful. If no reports are forwarded or if other questions remain unclear in respect of homogeneity etc, the series editors may wish to consult external referees for an overall evaluation of the volume.

5. Manuscripts should in general be submitted in English. Final manuscripts should contain at least 100 pages of mathematical text and should always include

   – a table of contents;
   – an informative introduction, with adequate motivation and perhaps some historical remarks: it should be accessible to a reader not intimately familiar with the topic treated;
   – a subject index: as a rule this is genuinely helpful for the reader.
   – For evaluation purposes, manuscripts should be submitted as pdf files.

6. Careful preparation of the manuscripts will help keep production time short besides ensuring satisfactory appearance of the finished book in print and online. After acceptance of the manuscript authors will be asked to prepare the final LaTeX source files (see LaTeX templates online: https://www.springer.com/gb/authors-editors/book-authors-editors/manuscriptpreparation/5636) plus the corresponding pdf- or zipped ps-file. The LaTeX source files are essential for producing the full-text online version of the book, see http://link.springer.com/bookseries/304 for the existing online volumes of LNM). The technical production of a Lecture Notes volume takes approximately 12 weeks. Additional instructions, if necessary, are available on request from lnm@springer.com.

7. Authors receive a total of 30 free copies of their volume and free access to their book on SpringerLink, but no royalties. They are entitled to a discount of 33.3 % on the price of Springer books purchased for their personal use, if ordering directly from Springer.

8. Commitment to publish is made by a *Publishing Agreement*; contributing authors of multiauthor books are requested to sign a *Consent to Publish form*. Springer-Verlag registers the copyright for each volume. Authors are free to reuse material contained in their LNM volumes in later publications: a brief written (or e-mail) request for formal permission is sufficient.

Addresses:
Professor Jean-Michel Morel, CMLA, École Normale Supérieure de Cachan, France
E-mail: moreljeanmichel@gmail.com

Professor Bernard Teissier, Equipe Géométrie et Dynamique,
Institut de Mathématiques de Jussieu – Paris Rive Gauche, Paris, France
E-mail: bernard.teissier@imj-prg.fr

Springer: Ute McCrory, Mathematics, Heidelberg, Germany,
E-mail: lnm@springer.com

Printed in the United States
By Bookmasters